Skeletal Tissue Mechanics

Springer

New York
Berlin
Heidelberg
Barcelona
Budapest
Hong Kong
London
Milan
Paris
Singapore
Tokyo

R. Bruce Martin   David B. Burr   Neil A. Sharkey

Skeletal Tissue Mechanics

With 175 Illustrations

Springer

R. Bruce Martin
University of California Davis Medical
    Center
Orthopaedics Research Laboratories
Research Building I
4635 Second Avenue
Sacramento, CA 95817
USA

David B. Burr
Indiana University School of Medicine
Department of Anatomy, MS 259
635 Barnhill Drive
Indianapolis, IN 46202
USA

Neil A. Sharkey
The Pennsylvania State University
Center for Locomotion Studies
29 Recreation Building
University Park, PA 16802-5702
USA

Library of Congress Cataloging in Publication Data
Martin, R. Bruce, 1940–
      Skeletal tissue mechanics / R. Bruce Martin, David B. Burr, Neil
   A. Sharkey
         p.   cm.
      Includes bibliographical references and index.
      ISBN 0-387-98474-7 (hardcover : alk. paper). — ISBN 0-387-98474-7
   (hardcover : alk. paper)
         1. Bones—Mechanical properties. 2. Cartilage—Mechanical
   properties. I. Burr, David B. II. Sharkey, Neil A. III. Title.
         [DNLM: 1. Bone and Bones—physiology. 2. Bone and Bones—
   physiology examination questions. 3. Joints—physiology.
   4. Joints—physiology examination questions. 5. Cartilage—
   physiology. 6. Cartilage—physiology examination questions.
   7. Biomechanics. WE 200 M382s 1998]
   QP88.2.M184 1998
   612.7'5—dc21
   DNLM/DLC   98-2906

Printed on acid-free paper.

Acquiring Editor: Robin Smith
Production coordinated by Laura Carlson Co. and managed by Terry Kornak; manufacturing
supervised by Joe Quatela.
Typeset by Laura Carlson Co., Yellow Springs, OH, from the authors' electronic files.
Printed and bound by Maple-Vail, York, PA.
Printed in the United States of America.

9 8 7 6 5 4 3 2 1

ISBN 0-387-98474-7 Springer-Verlag New York Berlin Heidelberg   SPIN 10657906

Dedication

Many books are dedicated to the authors' children, but in this case it seems all the more appropriate because of the cornucopia of biology and mechanics present in a child's developing skeletal tissues. Rather than just listing our children's names, we have chosen to present them in the context of this book by using the accompanying graph. As you will learn, teams of cells called Basic Multicellular Units continually repair and remodel the interiors of our bones. The graph plots the activity of these multicellular units as a function of age. Bone remodeling is most rapid in toddlers like Corrine, and decreases logarithmically on the way to adulthood. Allowing for individual variations, everyone slides down this curve in the process of growing up.

We want to express here our heart-felt gratitude to our particular tribe of decidedly unbasic "multicellular units." Activated varying numbers of years ago, they have repaired our fatigue damage, adapted us to stress, and generally remodeled our lives for the better ever since. In an age of information technology, new and wondrous instruments of learning appear almost daily, but not one compares with the ability of children to educate the adults around them. We have learned much from our children, and dedicate this book to them.

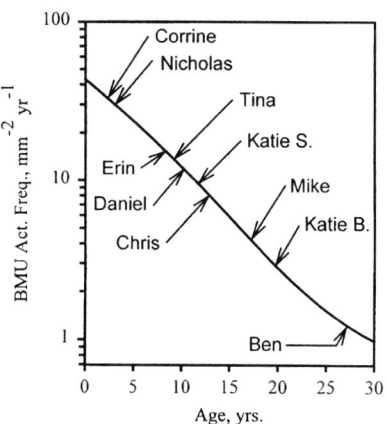

Preface

This book was written primarily as a textbook for graduate and advanced undergraduate students. It grew out of the lecture notes for a course called *Skeletal Tissue Mechanics,* which is part of the biomedical engineering curriculum at the University of California at Davis. Most of the many books available on skeletal biomechanics fall into one of two categories: textbooks dealing primarily with the biomechanics of locomotion and sport, or edited collections of reviews or tutorials on tissue mechanics. Although several of the latter are excellent resources for teaching at the graduate level, they generally lack a didactic presentation of the various topics and do not include exercises for the student. There are also textbooks aimed at teaching orthopaedic biomechanics to clinicians. However, very few of these attempt to analyze the biologic processes and mechanical behaviors of skeletal tissues in an integrated fashion. The integration of anatomy and physiology with structural and material behavior was one of this book's principal objectives.

Another objective was to develop an introduction to tissue mechanics that can be shared by students from several disciplines, including human and veterinary medicine, physical anthropology, mechanical engineering, and zoology, as well as bioengineering and biomechanics. In the interest of this diverse audience, time is taken to introduce basic mechanical and biologic concepts, and the approaches used for some of the engineering analyses are purposefully limited. We have tried to make the book an effective bridge between these classically dissociated disciplines. In the classroom we have observed spontaneous and enlightening discussions between students from dissimilar backgrounds. If the book sparks similar discussions in other classrooms, then we will have achieved at least a portion of our goal. In recent years, "interdisciplinary" has become a byword of productive research, and nowhere is this more evident than in the field of skeletal biomechanics. Traditional boundaries have dissolved, and today's graduate students, be they engineers or physiologists, clinicians or basic scientists, are required to synthesize knowledge across disciplines to become tomorrow's successful researchers. We hope that this text helps students to be more receptive to the possibilities for scientific enlightenment in disciplines beyond their chosen area of specialization.

The exercises at the end of each chapter endeavor to maintain this diversity of application. Also, it should be noted that few of the exercises are simply practice at solving problems demonstrated in the book. Instead, most of the exercises are designed to stimulate the student to synthesize concepts based on material discussed in the text and to develop analytical problem-solving skills that will serve them in the future. Thus, it is important for the student to work out a reasonable number of the exercises.

The third reason for writing this book was a selfish one: it has given us the benefit of stretching our brain cells again, in the way that only this kind of a writing project can. It is an exciting challenge to write a textbook in which the goal is to weave a fabric from the many disparate threads of a broad field, and to do so in a manner that is comprehensible to students as well as researchers. This statement does not imply that we suffer the illusion of having accomplished so much. We hope we have a proper sense of the limitations of this book, but at the same time we cannot help but believe that we have learned much in the writing of it. We look forward to learning still more as we invite comments from other instructors and students that may improve future editions.

We hope the book also serves as a useful resource for researchers. In a sense, it is a sequel to the monograph *The Structure, Function and Adaptation of Cortical Bone* (Martin and Burr, 1989). The present volume covers much of the same material, but has been updated and expanded to include trabecular bone, cartilage, synovial joints, and tendons and ligaments as well.

We thank the many colleagues who have contributed directly or indirectly to this effort; your insights and critiques were invaluable. James Paul Maganito was extremely helpful, and patient, in preparing the illustrations, as were the editors and production staff at Springer-Verlag in bringing the book to print. Finally, many thanks are owed to the students in the skeletal tissue mechanics course at UC Davis who helped refine earlier versions of this work.

R. Bruce Martin
Neil A. Sharkey
David B. Burr

Contents

1

Forces in Joints

...mechanical science is of all the noblest and most useful, seeing that by means of this all animate bodies which have movement perform all their actions...

Leonardo Da Vinci (1452–1519)

1.1 Introduction

The skeleton is first and foremost a mechanical organ. Its primary functions are to transmit forces from one part of the body to another and to protect certain other organs (e.g., the brain) from mechanical forces that could damage them. Therefore, the principal biologic role of skeletal tissues is to bear loads with limited amounts of deformation. To appreciate the mechanical attributes that these tissues must have to perform this role, it is necessary to learn something about the forces which whole bones normally carry. In most cases, these forces result from loads being passed from the part of the body in contact with a more or less rigid environmental surface (e.g., the heel on the ground when walking) through one or more bones to the applied or supported load (e.g., the torso). In addition to the forces transmitted in bone-to-bone contact, muscle and ligament forces act on the bones, and these forces (especially the muscle forces) are large and important.

Most muscle, ligament, and bone-to-bone forces act in or near the body's major diarthroidal joints. The purpose of this chapter is to explain how conventional engineering analysis may be used to estimate joint forces, and to provide some practical experience in solving such problems.

1.2 Static Analysis of Forces in Joints

Forces in the Elbow Joint

To introduce this subject, we consider the human elbow joint. We choose this joint because it works more or less as a simple hinge, and because it is familiar to us, conforming to the image that we acquired in grade school of the skeleton as a system of levers. Although we use this simplicity in solving for the forces, we should spend a few moments examining the anatomy to appreciate the simplifying assumptions that we are making.

A lateral view of the bones of the elbow is shown in Fig. 1.1A. The olecranon of the ulna wraps around the distal end of the humerus to form the major part of the elbow's hinge joint. The proximal joint surface of the radius articulates with the distal joint surface of the humerus and also with a cartilage-covered notch (the radial notch) on the lateral aspect of the ulna; this means there are three articulations in the elbow joint. These articulations allow for flexion and extension of this joint as well as pronation and supination[1] of the forearm. The radius and ulna are bound together by the interosseous membrane along the central part of their length and by several ligaments at their proximal and distal ends (not shown). We assume that these structures cause the two forearm bones to act as one structural unit. We also assume that the wrist joint is

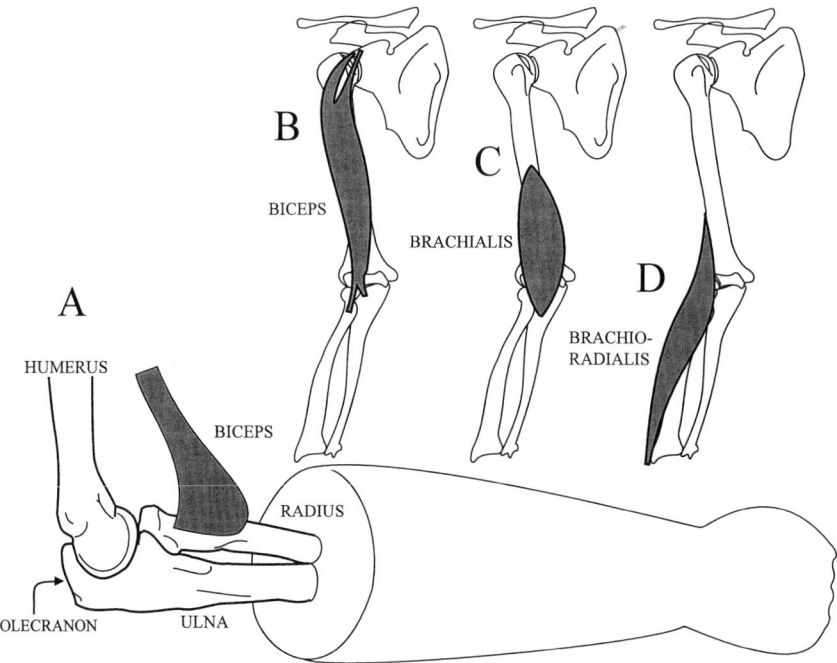

FIGURE 1.1A–D. Anatomical drawings of the elbow joint. **A** Overall view of the flexed joint and the biceps inserting primarily on the radius. **B, C, D** Approximate locations of the biceps, brachialis, and brachioradialis muscles.

---

[1]For the definition of these and other anatomical terms, see a human gross anatomy text, e.g., Grant's (Agur, 1991) or Gray's (Williams, 1995).

stabilized by its musculature so that the hand and forearm flex about the elbow as a simple hinge joint.

When the elbow is flexed with the palm upward, two muscles are primarily responsible: the biceps and the brachialis (Fig. 1.1B and 1.1C). The biceps has two heads (i.e., points of origin). The short head originates on the coracoid process of the scapula, while the long head runs through the shoulder joint to attach to the superior lip of the glenoid fossa (the scapular part of the shoulder joint). The biceps divides distally into a tendon that inserts on the proximal radius and an aponeurosis (tough band of connective tissue) which blends with other muscles in the proximal forearm (Fig. 1.1B). The brachialis originates along the anterior surface of the central humerus and inserts on the proximal ulna (Fig. 1.1C). When the forearm is flexed rapidly, or a large force must be applied during flexion, the brachioradialis also acts between the distal humerus and the distal radius (Fig. 1.1D). Other muscles also cross the elbow and participate in flexion. However, to simplify matters, and because the present problem is for a static situation, we assume that the biceps and brachialis muscles act as one in holding the elbow in a flexed position against the action of a weight held in the hand. We want to find the force required in the "biceps-brachialis muscle" to support the flexed forearm and the total force exerted on the distal end of the humerus by the radius and ulna. These forces are expressed as multiples of the weight in the hand.

To solve this problem and others like it, three steps are necessary:

1. Draw a free-body diagram.
2. Write the equations of static equilibrium.
3. Solve these equations simultaneously to obtain the unknown forces.

We simplify the analysis by assuming that it is two dimensional and that the forces all act in a sagittal plane containing the humerus and the forearm. The free-body diagram is constructed by isolating the structure being analyzed such that the internal forces to be determined are exposed and replacing all the forces acting on the structure by vectors. The free-body diagram for the elbow problem is shown in Fig. 1.2. The forearm is made

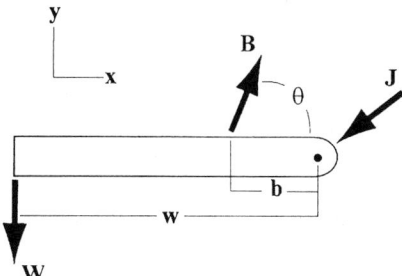

FIGURE 1.2. Free-body diagram for the elbow force calculation. **J** is the joint reaction force, **W** is the weight in the hand, and **B** is the biceps force.

a free body by "amputating" through the elbow joint, whose internal forces we wish to reveal. The forearm lies in the x–y plane. The weight in the hand, the muscle force, and the joint reaction force are represented by the vectors **W**, **B**, and **J**, respectively.[2] The joint force is assumed to pass through a fixed center of rotation for the joint, shown by a dot. The insertion point of the muscle is **b** meters (m) along the forearm from the joint center, and the center of gravity of the weight in the hand is **w** meters from this point. The equations for static equilibrium in two dimensions are

$$\Sigma M = 0$$
$$\Sigma F_X = 0 \qquad (1.1)$$
$$\Sigma F_Y = 0$$

where M stands for moments about an arbitrary point, and F_X and F_Y are force components in the x- and y-directions. Because, in two dimensions, there are three such equations, one may solve for three unknown force *components*. Alternatively, as usually happens, one may solve for one force *vector* and the *magnitude* of another force of known direction.

For the elbow problem, taking moments about the joint center, the equilibrium equations are

$$\Sigma M = wW - bB \sin \Theta = 0 \qquad (1.2)$$

$$\Sigma F_X = B \cos \Theta - J_X = 0 \qquad (1.3)$$

$$\Sigma F_Y = B \sin \Theta - W - J_Y = 0 \qquad (1.4)$$

(Here, any moment of the x-component of **B** about the joint center is ignored.) Solving these equations yields

$$B = wW/b \sin \Theta \qquad (1.5)$$

$$J_X = B \cos \Theta \qquad (1.6)$$

$$J_Y = B \sin \Theta - W \qquad (1.7)$$

If $\Theta = 75°$, $w = 0.35$ m, and $b = 0.04$ m, then $B = 9.1$ W, $J_X = 2.3$ W, and $J_Y = 7.8$ W. The magnitude of the joint reaction force is $J = (J_x^2 + J_y^2)^{1/2} = 8.1$ W, and its orientation is $\arctan(J_y/J_x) = 74°$ with respect to the x-axis. Thus, the muscle and joint reaction forces are eight to nine times greater than the weight held in the hand. This result is typical of virtually all the joints in the body in that the skeleton works at a mechanical disadvantage in terms of force, and as a consequence the forces acting on bones are high relative to the forces applied by (or to) the environment.

There is, of course, a good reason for this. Muscles can only contract about 30% of their length. In the case of the biceps, for example, the overall length is about 25 cm, so the maximum contraction is 7–8 cm. The lever

---

[2]In this chapter, vector quantities are represented in bold type and scalars (such as vector magnitudes) in ordinary type.

action of the forearm magnifies this distance by the ratio w/b in Fig. 1.2, enabling the hand to move much further (and also much faster). Of equal importance, the muscles insert proximally on the radius and ulna and do not create a "web" of flesh across the front of the elbow. Therefore, magnification of motion enables larger movements to be accomplished with a compact structure. The price that is paid for this compactness and magnification of motion is high forces in the muscles, across the joint surfaces, and within the bones.

Box 1.1 Historical Note
Giovanni Borelli on the Movement of Animals

Giovanni Alfonso Borelli (1608–1679) was the greatest of early biomechanicians. He held the chair in mathematics at Pisa in Italy, where he was a close friend of Malpighi, who was the professor of theoretical medicine. Together, they did much to pursuade seventeenth-century physicians of the importance of physics in understanding medicine and physiology. Borelli's treatise, *De Motu Animalium (On the Movement of Animals)*, has been translated into English and is a marvelous testament to his genius and ability to explain musculoskeletal mechanics clearly and simply. In the preface to his translation, Maquet points out that later biomechanicians unwittingly duplicated several of Borelli's important discoveries. While some of his results contain errors, it must be remembered that Newton's laws were not published until 1687. Thus, although Borelli understood very well the principle of balancing moments about a fulcrum, he could not have been expected to fully understand static equilibrium as we do, based on Newton's laws. One of Borelli's greatest achievements was the discovery that animals' joints work at a mechanical disadvantage in terms of force. He discusses this in the following excerpt.

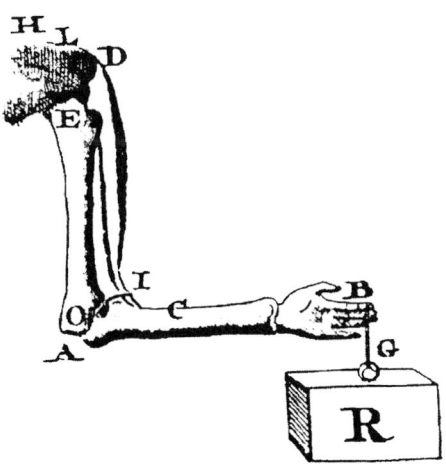

Diagram of elbow force problem from Borelli's *On the Movement of Animals* (translation by Maquet, 1989).

Proposition VIII

"It is commonly thought that Nature raises considerable weights by using the machines of the muscles with a weak moving force.

"The magnitude of the vital force of the muscles ... sustains, raises, and moves not only an arm or a leg, but the whole animal machine, enabling it even to dance. Besides the mass of the animal, heavy enough by itself, this force carries, pulls and pushes considerable weights.

"Aristotle...did not recognize the muscles but imagined spirits which pull and push the limbs. [He] remarked how difficult it would be for the huge mass of an elephant to be moved...by tenuous spirit or wind. He met the difficulty by saying that Nature moves the joints and limbs of the animal by using very small force...and said that the operation is carried out by way of a lever. Therefore, it is not surprising that huge weights can be moved...by a small force. Lucretius used the same example....Galen also says that a tendon is like a lever. He thinks that, consequently, a small force of the animal faculty can pull and move heavy weights.

"This general opinion seems to be so likely that, to my knowledge, not surprisingly, it has been questioned by nobody. Who indeed would be stupid enough to look for a machine [in the body] to move a very light weight with a great force, i.e., ...not to save forces but rather to spend forces? And if this is rightly considered as stupid, how is it possible that wise Nature, everywhere looking for economy, simplicity and facility, builds with great efficiency in animals machines to move, not heavy weights with a small force, but on the contrary, light weights with almost boundless force? ...I shall demonstrate that multiple and different machines actually are used in the motions of animals but that light weights are carried by large and strong force rather than heavy weights being supported by small force."

Forces in the Hip Joint

The hip joint is the articulation between the femur and the acetabulum of the pelvis. The hip is one of the most important joints in the body from a medical perspective. Especially in aging individuals, overall health is promoted by the exercise that goes with walking. The ability to walk depends on having a healthy, painfree hip. Two kinds of diseases, both very common among the elderly, preferentially affect the hip. *Osteoarthritis* is the primary reason that about 200,000 total[3] hip replacement procedures are performed each year in the United States, and *osteoporosis* in the femoral neck results in several hundred thousand hip fractures each year, virtually all of which require surgical treatment with a metal fixation device. To better understand the mechanics of the hip and the demands on the implants used to correct its problems, it is important to know the forces across the hip during walking and their determining factors.

---

[3]The word "total" refers to the replacement of both the femoral and acetabular sides of the joint. Sometimes only the femoral side is replaced.

When one is walking, the lower extremities and other parts of the body are moving (i.e., accelerating), so the conditions of static equilibrium are not satisfied. However, most people do not usually walk very vigorously, so the accelerations are small relative to the forces produced by muscle pulls and gravity. Therefore, the problem is usually solved as though the person were simply standing on one leg, assuming that this approximates the conditions during the "single leg stance phase" of gait, that is, when all the weight is being carried by one leg. In addition to this assumption, we assume that the problem is two dimensional, in the frontal plane, and that only one muscle is acting. If you try to stand on one leg, the force pulling your center of gravity downward tends to rotate your torso toward the medial side of the leg you are standing on. The muscles that resist this movement are the same ones which you use to abduct your thigh when lying down (i.e., move your lower limb away from the body's center line). Again, several muscles act to do this, but we can lump them all together and simply call them the "abductor muscles."

Again, the first step in calculating the forces in the joint is to draw a free-body diagram, "amputating" through the joint in question to "reveal" the force vectors acting there. In this case we delete the lower extremity and study the remaining portion of the body. Figure 1.3 shows this situation. It is

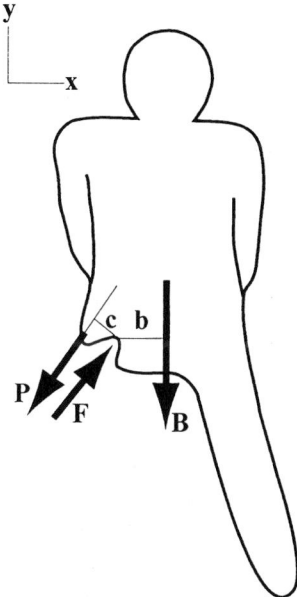

FIGURE 1.3. Free-body diagram for the calculation of the hip joint force while walking. The "amputated" leg is the one supporting the body. The leg in the free-body diagram is not in contact with the ground. **B** is the weight of the body (minus the amputated leg), **P** is the abductor muscle force, and **F** is the joint reaction force.

important to understand that the subject was standing on the leg that has been "amputated," and that the other leg, of which only a portion is shown, was not touching the ground. The abductor muscle force is represented by the force vector **P**. The joint reaction force acting on the middle of the acetabulum in the pelvis is **F**. The weight of the body, represented by the vector **B**, is actually the entire body weight, W, minus the weight of the leg supporting the body. Because each lower extremity weighs about $1/6$ W, we let $B = 5W/6$. This force acts downward slightly to the right of the centerline of the body. Taking moments about the center of the acetabula, we have

$$\Sigma M = cP - bB = 0 \tag{1.8}$$

$$P = (b/c)B = (b/c)(5/6)W \tag{1.9}$$

The lengths of the moment arms b and c have been estimated from anteroposterior pelvic radiographs. It was found that the b/c ratio ranges between 2 and 3.5. Following the lead of Frankel and Burstein (1970), we choose a conservative value of 2.4 and obtain the convenient result that $P = 2W$. That is, the force required in the abductor muscles to balance the body on the head of the weight-bearing femur during the stance phase of gait is twofold body weight.

To solve the rest of the problem, we write the equations for force equilibrium, assuming that the x-direction is horizontal and the y-direction is vertical:

$$\Sigma F_x = F_x - P_x = F_x - 2W \sin \Theta = 0 \tag{1.10}$$

$$\Sigma F_y = F_y - P_y - B = F_y - 2W \cos \Theta - 5W/6 = 0 \tag{1.11}$$

where $\Theta \approx 30°$ is the angle that the abductor muscle line of action makes with the y-axis. Because $\sin 30° = 0.5$, the components of the joint reaction force F are

$$F_x = W \quad \text{and} \quad F_y = (2 \cos 30° + 5/6)W = 2.57W \tag{1.12}$$

The *horizontal* force of the femur on the pelvis is equal to body weight, and the vertical force is 2.5 times as much. The total joint reaction force is $F = (F_x^2 + F_y^2)^{1/2} = 2.75W$. This force acts at an angle to the horizontal $\Theta = \arctan(F_y/F_x) = 68.7°$.

We have seen that the magnitude of the forces in the hip joint depends critically on the ratio of the body weight moment arm to the abductor muscle moment arm. Thus, anything that increases the former or decreases the latter increases the abductor muscle force required for gait and consequently the force on the head of the femur as well. People with short femoral necks have higher hip forces, other things being equal. More significantly, people with a wide pelvis also have larger hip forces. This tendency means that women have larger hip forces than men because their pelves must accommodate a birth canal (Burr et al., 1977). This fact may be one reason that women have more hip fractures and hip replacements because of arthritis than men do. It is also conceivable that this places women at a biomechanical disadvantage with respect to some athletic activities, although studies do not always show gender differences in the biomechanics of running, particularly endurance running (Atwater, 1990).

Box 1.2 Technical Note
Measurement of Hip Joint Forces In Vivo

To verify the estimates of hip joint forces made using free-body calculations, four groups of investigators have implanted devices that allowed measurement of hip joint forces into hip replacement patients (Rydell, 1965, 1966; English and Kilvington, 1979; Davy et al., 1988; Bergmann et al., 1988, 1993). When humans are used as experimental subjects, ethical concerns demand that the subject's health not be endangered in any way. This ethical mandate has extensive technical consequences for measurement of hip joint forces. To begin with, a healthy human subject cannot be used; experimentation must be restricted to a subject with a deseased or injured hip who will be undergoing replacement of the proximal femur with a prosthesis by way of treatment for this condition. This condition provides the opportunity for using an instrumented prosthesis that can measure joint forces instead of a standard prosthesis. However, the instrumented prosthesis must be as strong and as durable as a standard one, which imposes important constraints on the instrumentation design. Moreover, wires cannot be run from the prosthesis to the surface of the skin, where they could pose a risk of infection or other problems. Consequently, the instrument must be fully contained within the prosthesis, with a working space of a few cubic centimeters, and it must be capable of transmitting data to a receiver outside the body while the patient walks.

The solution to this technical problem has been to use strain gauges (3 to 12 in number) to measure strains on interior surfaces of the prosthesis, and microelectronic circuitry to process the strain signal, which is broadcast as an FM signal to a receiver held against the skin. The device must be fully calibrated in the laboratory before insertion into the subject, so that the measured strains can be converted to force components on the ball or head of the prosthesis. The power source in the prosthesis used by Davy et al. (1988) was a battery located in the stem of the prosthesis and activated by a magnetic switch turned on by a magnet held outside the patient's leg (left figure). Because batteries carry a slight risk of releasing toxic components, and because they have a finite life, other investigators (Bergman et al., 1993) used an inductive power source. A magnetic coil wrapped around the subject's leg induced current in a receiver coil within the prosthesis.

One might question whether a patient who has a degenerated joint, has recently had surgery on it, and is in a laboratory environment, is capable of walking "normally"—but these are the best data available. The findings of all four studies are consistent with one another and with the theoretical estimates. During the single leg stance phase of gait, hip joint forces between 2.5 and 3.3 times body weight were measured (right figure). The highest force recorded was

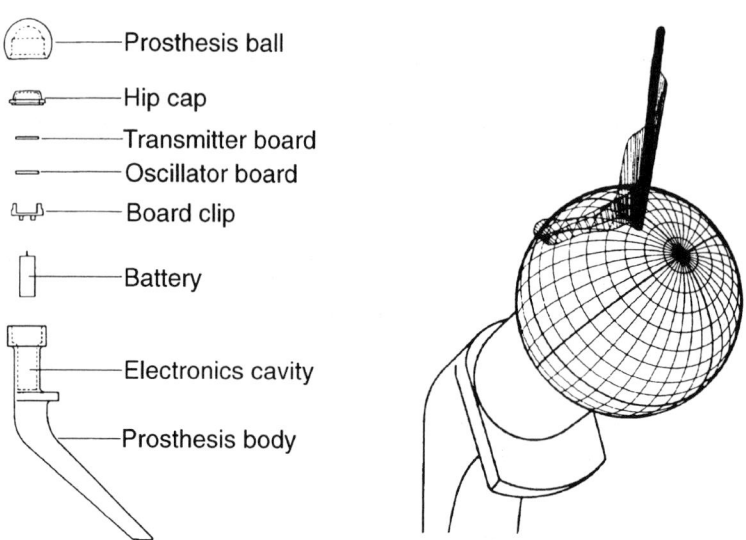

—Prosthesis ball

—Hip cap

—Transmitter board
—Oscillator board

—Board clip

—Battery

—Electronics cavity

—Prosthesis body

Left figure: Exploded view of telemetry electronics inside an instrumented hip prosthesis. (Reproduced with permission from Davy et al., 1988.) *Right figure:* Three-dimensional plot of hip joint force vector during a gait cycle (with crutches). Orientations and lengths of lines emanating from the surface of the prosthesis head indicate force direction and magnitude, respectively. (Reproduced with permission from Davy et al., 1988.)

in an individual who was attempting static single-leg balancing; during recovery from a momentary loss of balance, a force of 5.5 body weight was measured in the hip (Davy et al., 1988).

Clinical Significance of High Joint Forces

Because diarthroidal joints work at a mechanical force disadvantage so that limbs can move far and rapidly with short muscle contractions, high stresses are produced in the tissues of the bones and joints. Normally, these tissues carry their loads without causing pain, but various diseases and injuries can damage the tissues so that the deformations associated with loading are painful. To some degree, the pain is proportional to the amount of force carried by the tissues; in other words,

$$\text{Pain} = \text{Force} \times \text{Disease}$$

There are no nerves in cartilage, and the source of joint pain is poorly understood, but experience shows that reducing joint forces often alleviates pain. Often the physician is not able to do much about the disease, so the first consideration in controlling pain may be to reduce the forces in the joint. For example, the patient can lose weight, or walk with a cane, and thereby reduce the forces transmitted across the joint.

1.3 Hip Forces in Human Ancestors

Physical anthropologists have analyzed hip joint forces in skeletons of various hominids. Of particular interest is *Australopithecus* because of collateral evidence (footprints) that these individuals may have walked very much like modern humans. Figure 1.4 is a depiction of the famous *Australopithecus* skeleton known as Lucy. The differences between the anatomy of the femur and pelvis of *Australopithecus* and *Homo sapiens* include factors that seem to be very pertinent to hip joint force. For example, the neck of the *Australopithecus* femur (stippled in Fig. 1.5, overlaying the outline of a modern femur) was proportionately longer than ours, but the bone was smaller overall. The pelvis was also smaller overall, but the ilium (shown stippled in Fig. 1.6, on one side of a modern human pelvis of similar size) was more outwardly flared, moving the line of action of the abductor muscles away from the hip joint. These factors affect the moment arms of both the abductors and the body weight vector. The analysis of Lovejoy and co-workers (1973) indicates that the hip joint force in *Australopithecus* was about 2.5 fold body weight, a value somewhat less

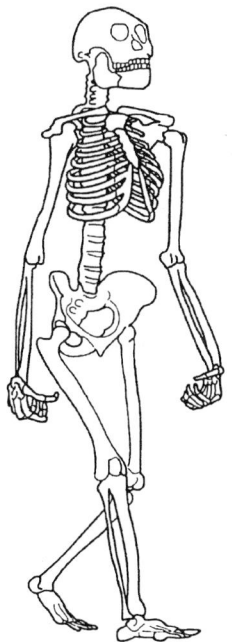

FIGURE 1.4. Artist's conception of the skeleton of *Australopithecus afarensis* ("Lucy") walking. (Reproduced from McHenry, 1991.)

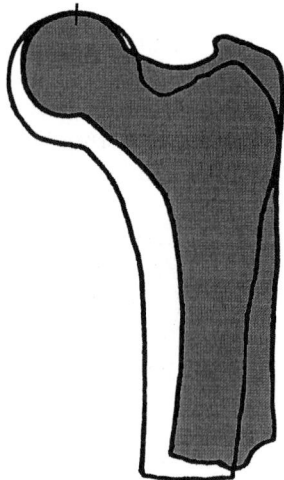

FIGURE 1.5. Lucy's proximal femur *(stippled)* compared to that of *Homo sapiens.* (Redrawn with permission from Lovejoy et al., 1973.)

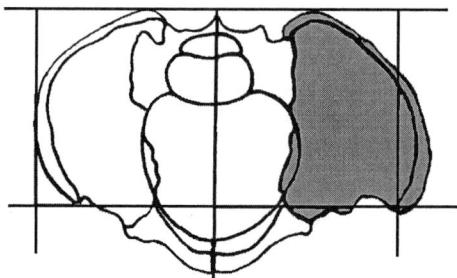

FIGURE 1.6. Lucy's ilium (*stippled,* on *right*) compared to that of *Homo sapiens* (on *left*). The left and right *vertical lines* would contain the modern pelvis; notice that Lucy's ilium projects well beyond the line on the right. (Redrawn with permission from Lovejoy et al., 1973.)

than that of modern humans. However, when they performed a similar analysis on the skeletons of Native Americans excavated at archeological sites (and thus more comparable in their condition to fragmented *Australopithecus* skeletons), they obtained a value of 2.5 for these as well. Therefore, Lovejoy et al. concluded that *Australopithecus* and modern humans experienced similar hip forces. They then considered the *pressure* on the head of the femur in these two species. Considering the degree to which the head of the *Australopithecus* femur was smaller than ours, and the estimated difference in body weights, it was concluded that the pressure on the *Australopithecus* cartilage was about half that on our femoral heads. However, these estimates are quite tenuous because of the scarcity and fragmented condition of the *Australopithecus* skeletons. Later in their paper, Lovejoy et al. conceded that other observations suggest that the *Australopithecus* hip force may have been substantially less than that of *Homo sapiens.* Two of the exercises at the end of this chapter allow you to explore this application of skeletal biomechanics.

Box 1.3 Technical Note
Reproducing Joint Forces in the Laboratory

Cadaver specimens are frequently used to model normal, pathologic, or reconstructed joints. For these laboratory simulations, researchers often rely on static calculations to approximate the loading conditions experienced by the joint in life. For example, a typical experiment examining the stability of a new hip prosthesis might use cadaveric hip joints implanted with the new component and then mounted into a testing machine for loading. But in what direction should the load be applied? What should the load magnitude be? Static analysis can be used to answer these questions.

   Once the investigator has determined the direction and magnitude of the joint reaction force for a given condition, it is simply a matter of aligning the specimen such that the vector of the joint force coincides with the axis of the load delivery system (left figure). Many important biomechanical studies have been executed in just this fashion. However, like static analysis itself, this type of simulation incorporates several assumptions and the results should be interpreted with caution.

   Let us revisit the hip example. A more physiologically correct model might use the whole pelvis and incorporate abductor muscle forces to reproduce the in vivo loading environment (middle and right figures). In this case the proper joint reaction vector is produced by simulating abductor contraction using a cable system attached at the muscle origin and insertion sites. Shortening the cable produces a moment about the hip that is opposed at the sacrum. In theory, when the vertically directed force at the sacrum achieves upper-body weight, the joint reaction force at the hip will be the same as for the extracted hip joint model and the same as that produced during single-leg stance in life.

   In a comparison of these two laboratory methods (Bay et al., 1997a), whole intact pelves were first loaded under simulated muscle action and the contact pressure distribution on the articular surfaces was recorded. The hips were then extracted from the pelvis and reloaded. The direction and magnitude of the joint reaction force was identical for both cases, but the distribution of contact pressure within the joint was quite different. Why? Remember, static analysis assumes rigid bodies but musculoskeletal tissues do not always behave as rigid bodies. In the intact model the bones of the pelvis were free to flex and deform

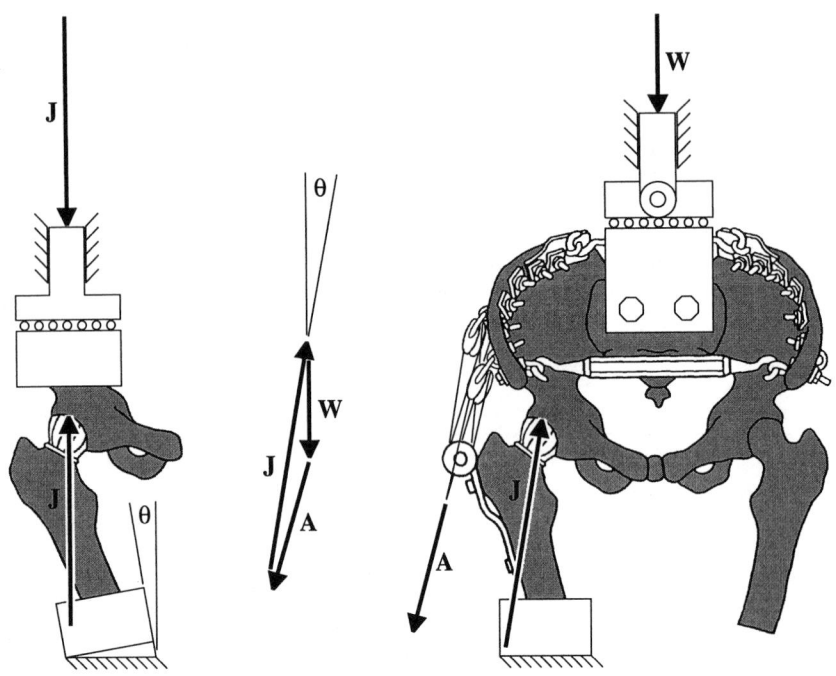

The *left* and *right* diagrams show two different experimental setups for simulating hip joint loading using cadaver bones. The vector diagram in the *middle* is for the case at *right*. (Reproduced with permission from Bay et al., 1997a.)

under load, while in the simpler joint model the pelvis became more rigid because of its altered geometry and the constraints placed upon it.

1.4 The Three-Force Rule

A simple rule is useful in solving certain problems by inspection, or graphically. To illustrate this principle, we consider the forces acting in the knee joint when a person is walking up stairs. Figure 1.7 shows a free-body diagram of the leg during this situation. It is again assumed that the forces are confined to a single plane, and that static equilibrium is approximated because the movement is relatively slow. The "ground reaction force" (**G**) is now the reaction of the step on the foot, assumed to be directed vertically upward and equal to body weight. This force acts to rotate the leg counterclockwise, flexing the knee joint. This moment is resisted by the quadriceps muscles of the thigh, which pull on the patellar ligament and resist knee flexion. It is assumed that the magnitude of this force **L** is unknown, but that its direction can be obtained from drawing a line

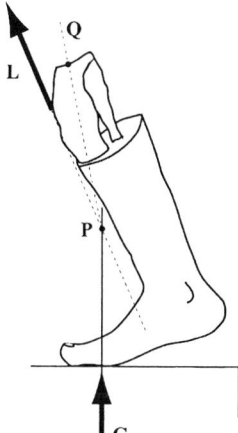

FIGURE 1.7. Partial free-body diagram of leg and foot to illustrate the three-force rule. **G** is ground reaction force, **L** is patellar tendon force, *P* is the point where their lines of action meet, and *Q* is the point where the joint force acts on the tibia.

between its attachment points on a radiograph. This direction is shown by the vector **L** in Fig. 1.7. The question that the three-force rule can answer is, "What is the direction of the force, **R**, exerted by the femoral condyles on the top of the tibia?"

To see what the direction of this force must be, it is only necessary to realize that the first two forces, **L** and **G**, produce no moments about the point *P*, where their lines of action meet. Because **R** is the only other force acting in the problem, its vector, when extended, must also pass through the point *P*. Otherwise, **R** would produce a net moment about *P* and ΣM would not be zero. We know that the vector **R** must act on a point *Q* within the knee joint. To find **R**'s line of action it is only necessary to construct the line *PQ* so that it passes through both *P* and the place where **R** acts on the tibia. Once this is done, the problem can be solved graphically in a second diagram (Fig. 1.8). First, a vertical vector

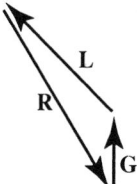

FIGURE 1.8. Vector diagram for three-force rule example. **G** and **L** are defined in Figure 1.7, **R** is the joint reaction force acting at *Q* in that figure.

is constructed having a length proportional to body weight (i.e., the vector **G**). Then, a vector of the appropriate length and direction to represent **L** is drawn from the head of **G**. Finally, vector **R** is drawn from the head of **L** to the tail of **G** so that the three vectors form a closed triangle; this determines the length (magnitude) of **R**. Note that this graphical solution to a two-dimensional joint mechanics problem has again only allowed us to find three unknowns: the magnitude of **L** and the magnitude and direction of **R**. If we had not known the direction of **L**, we would not have been able to solve the problem.

1.5 Indeterminate Joint Problems

In the problems we have considered so far, we have limited ourselves to two-dimensional analyses in which it was possible to solve for three unknowns. To reduce the number of unknowns to the number of equations, we had to combine some muscles with others and to ignore others. We could have obtained more equations by extending our analysis to three dimensions, but then we would have had to consider additional muscles as well. All real joints present static equilibrium problems that are mathematically indeterminate because there are always more unknowns than equilibrium equations. Thus, to solve for joint forces more realistically, it is necessary to find more equations to use in the solution.

To see an example of how this might be done, we return to the elbow problem. Previously, we noted that the primary flexors of the forearm are the brachialis and the biceps, and we combined their actions into a single force vector. Now, we separate these muscle forces. In addition, we include the third muscle, the brachioradialis (see Fig. 1.1). Table 1.1 shows the moment arms of the three muscles with respect to the center of rotation of the elbow when it is flexed at 90° (Winter, 1990). It also shows the

TABLE 1.1. Elbow muscle data

Muscle	Moment arm, cm	PCA, cm^2
Biceps	a = 4.6	α = 4.6
Brachialis	b = 3.4	β = 7.0
Brachioradialis	c = 7.5	γ = 1.5

PCA, physiologic cross-sectional area.
From Winter, 1990.

physiologic cross-sectional area (PCA) of each muscle. It is approximately true that the maximal force that a muscle can exert is proportional to its physiologic cross-sectional area. Another way to say this is that muscles have an upper limit to the stress that they can generate within themselves. Measured maximum stresses vary from 20 to 100 N/cm^2, depending on conditions (isometric or dynamic, parallel or pennate fibers, etc.). We will use 40 N/cm^2 as a rule of thumb for the maximal stress (Morris, 1948; Maughan et al., 1983).

With all three muscles in the problem, there are five unknowns (three muscle force magnitudes and two components of joint force) but still only three equilibrium equations. We need two other relationships between the variables to obtain a solution. One way to get such additional equations would be to assume that all three muscles work together to equalize their individual stresses. That is, each muscle force is proportional to the muscle's physiologic cross-sectional area. Then the three muscle forces are

$$A = k\alpha \ \text{(biceps)}$$
$$B = k\beta \ \text{(brachialis)} \qquad (1.13)$$
$$C = k\gamma \ \text{(brachioradialis)}$$

where α, β, and γ are the respective cross-sectional areas and $k = 40 \ N/cm^2$ is the constant muscle stress. This may be rewritten as

$$A/\alpha = B/\beta = C/\gamma \qquad (1.14)$$

which constitutes the two additional equations that were needed. Using these to substitute for B and C in the moment equilibrium equation,

$$\Sigma M = aA + bB + cC - wW = 0 \qquad (1.15)$$

one has

$$A = [\alpha w/(a\alpha + b\beta + c\gamma)]W = 2.86 \ W$$
$$B = (\beta/\alpha)A = 4.36 \ W \qquad (1.16)$$
$$C = (\gamma/\alpha)A = 0.93 \ W$$

This solution indicates that the biceps, the muscle with the intermediate cross-sectional area and moment arm, also exerts the intermediate force. The brachioradialis, which has the largest moment arm and the smallest cross-sectional area, exerts the least force, and vice versa for the brachialis. Whether this has anything to do with the actual distribution of forces in the muscles is debatable! Many other equally plausible, but not necessarily so mathematically trivial, "additional conditions" may be imagined. For example, the force distribution may shift around to give different muscle fibers a rest, or to keep the weight from tipping sideways in the hand, or both. A number of investigators have speculated that the body resolves the indeterminacy of such problems by acting to optimize or minimize some biologically important variable, such as the total energy required by all the muscles acting. Collins (1995) provides additional discussion of this subject.

Box 1.4 Technical Note
Surgical Reconstruction of the Hip Joint

Total hip arthroplasty, which involves replacement of the ball and socket of the hip joint with manufactured components, is one of the most successful surgical operations developed in this century. The procedure has relieved joint pain and increased the quality of life for millions of people. Before the development of successful hip replacement surgery by Sir John Charnley (1973), patients with debilitating degenerative arthritis of the hip were forced to either suffer with the condition or undergo hip joint fusion (called *arthrodesis*). As described in Section 1.2, the loads imposed on bones and joints are dramatic as a result of the mechanical disadvantage under which they function. Forces at the hip can easily exceed fivefold body weight. In addition, the hip joints of an average individual experience about a million load cycles (steps) each year. Thus total hip components should be both strong and durable.

A total hip replacement consists of two parts: a femoral component, the ball, and an acetabular component, the socket (see figure). Femoral components are made from high-strength alloys such as cobalt-chrome or titanium, and consist of a highly polished head atop an intramedullary stem. The stem is inserted into the canal of the femur and is usually fixed with an acrylic plastic (polymethyl-methacrylate, PMMA) called *bone cement*. This material does not actually cement the implant to the bone, however; it simply fills space so that the fit does not have to be exact, "grouting" the implant in place. More recent designs utilize a porous stem surface designed for fixation by bone ingrowth. The acetabular component is fabricated from ultrahigh molecular weight polyethylene, with or without a metal backing. A design with a metal backing is illustrated. Other combinations of materials for the bearing surface are sometimes used, including metal-on-metal, ceramic-on-ceramic, and ceramic-on-polyethylene. The PMMA and metal-on-polyethylene bearing surfaces innovations were important factors enabling Charnley to develop a system that featured good material and structural strength, good resistance to fatigue damage, and low friction at the bearing surface. Careful surgical technique remains important, but with current methods most patients, being elderly and relatively sedentary, can expect satisfactory performance for 15 years or more.

Despite the operation's overall success, failures do occur, more frequently in young, active patients. The high forces imposed on the joint make maintenance of implant fixation difficult; component loosening and migration are not uncommon. High joint reaction forces also produce relatively high frictional forces at the metal–plastic interface. Over time, frictional wear debris, both metallic and polyethylene, elicits biologic responses that resorb bone and may further compromise

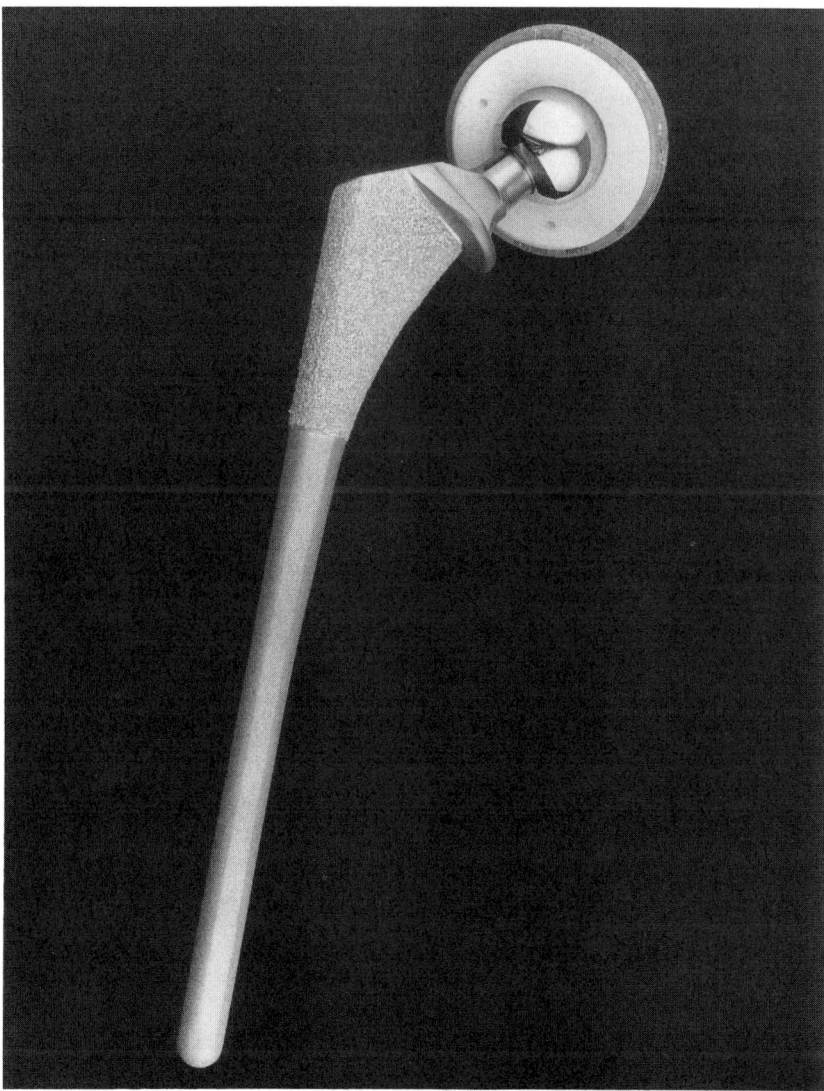

Photograph of modern hip joint replacement prosthesis. Spherical, polished, cobalt-chromium alloy ball at top articulates with ultrahigh molecular weight polyethylene socket having a backing of similar metal. The long metal stem is placed inside the medullary canal of the femur. In this example, the proximal stem and socket backing have porous surfaces designed for bone ingrowth in lieu of using polymethyl-methacrylate (PMMA) for fixation to the bone.

implant stability. Improving the fixation and wear characteristics of total joint components is a major focus of orthopaedic research.

1.6 Equine Fetlock Forces

Horses, and in particular racehorses, place extraordinarily high loads on their limbs. This is especially true of the forelimbs, which are thought to carry about 60% of the animal's weight. Many racehorses are seriously injured each year by mechanical failures of the structures in the distal forelimb. Because it is so difficult to repair these failures in such a way that the horse can stand and walk during healing, and because horses must be erect and mobile to survive, many animals die as a result of injuries that would not be life threatening in humans or other smaller animals.

Figure 1.9 shows a sketch of the anatomy of the distal forelimb of the horse. Remember, the horse does not have feet the way humans do; it walks on the tips of its "fingers." The other bones of the foot (or hand) serve to lengthen the leg. P1 and P2 mark the first and second phalangeal bones in the figure; the third phalangeal bone is inside the hoof. The equine third metacarpal bone (MC3) has become extremely stout, and carries loads from the phalangeal complex up to the carpal bones and the radius, seen at the top of the diagram. (The second and fourth metacarpal bones are vestigial struts stuck to the sides of MC3; the other metacarpal bones no longer exist.) The joint between MC3 and P1 is called the *fetlock*; its biomechanical importance derives largely from the fact that the bones above it rise straight up to the torso, but the phalangeal bones below it

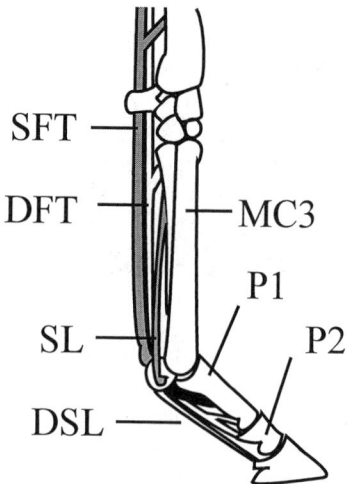

FIGURE 1.9. Equine distal forelimb anatomy, *sagittal view*. *SFT*, superficial digital flexor tendon; *DFT*, deep digital flexor tendon; *SL*, suspensory ligament; *DSL*, distal sesmoidean ligaments; *MC3*, third metacarpal bone; *P1*, proximal phalanx; *P2*, middle phalanx.

angle sharply forward in a springy cantilever arrangement. The joint between P1 and P2 is called the *pastern joint.*[4]

Behind the third metacarpus, which is also called the *cannon bone*, two tendons come down from the superficial and deep digital flexor muscles. The superficial flexor tendon (SFT) inserts on P2, and the deep flexor tendon (DFT) inserts on P3 inside the hoof. On their way to these attachment sites, these tendons pass over two small bones located just behind the distal end of the cannon bone. These bones are called the *proximal sesamoid bones*; their cartilaginous dorsal (anterior) surfaces articulate with the palmar (posterior) distal condyles of the cannon bone. The fetlock's sesamoid bones function very much like the patella in your knee, moving the line of action of the tendons away from the joint center so that they exert a larger moment about it. In addition to these tendons, there is a set of four ligaments that help stabilize the fetlock. The *suspensory ligament* (SL) runs from the proximal posterior surface of MC3 down to the sesamoid bones. From there, three *distal sesamoidian ligaments* connect to P1 and P2. These four suspensory ligaments, along with the sesamoid bones, are often called the *suspensory apparatus.* The cannon bone brings the weight of the horse down to the fetlock, which sits in the suspensory apparatus in much the same way as a person lies in a hammock. The flexor muscles and tendons help carry this load, with the sesamoid bones providing a bearing surface to protect the distal cannon bone, as well as increasing the moment arms of the tendons and ligaments, as previously mentioned.

To analyze the forces in the fetlock joint, we construct the free-body diagram shown in Fig. 1.10. The analysis is two dimensional in the sagittal plane. The "amputation" has been done in such a way that the free body includes the sesamoid bones and everything distal to the cannon bone. These structures are all assumed to be rigidly connected; the pastern joint is assumed to be fixed. The forces resulting from the flexor tendons and the suspensory ligament are all combined into a vector, **T**, which has components in the x- and y-directions. Considering the large weight of the horse, the weight of the free body is ignored. The remaining forces acting on the free body are the two components of the ground reaction force, **G**, and the two components of the joint reaction force, **F**. **F** is assumed to act at the joint's center of rotation. Taking moments about this center, we have

$$\Sigma M = g_1 G_y - g_2 G_x - tT_y = 0 \qquad (1.17)$$

where g_1 and g_2 are as defined in the lower portion of Fig. 1.10 and t is the moment arm of the vertical component of **T**. It is assumed that the x-component of **T** has a negligible moment arm with respect to the joint center.

---

[4]P1 is also called the pastern bone. The name apparently comes from the fact that to pasture a horse without benefit of fences, ranchers would hobble it with a short rope connecting the P1 bones.

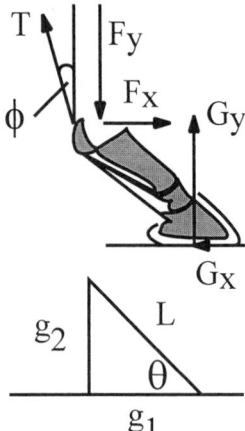

FIGURE 1.10. *Upper:* Free-body diagram for forces in the equine fetlock joint. G_x and G_y, components of the ground reaction force; **T**, tendon force; F_x and F_y, components of the joint force. *Lower:* Diagram for the moment arms, g_1 and g_2, of the ground reaction force. L, distance from the ground contact to the joint center; Θ, angle of the phalanges with respect to the ground.

The force equilibrium equations are

$$\Sigma F_x = -T_x - G_x + F_x = 0 \tag{1.18}$$

$$\Sigma F_y = T_y + G_y - F_y = 0 \tag{1.19}$$

These equations may be solved for various ground reaction forces and anatomical angles (Φ and Θ). First, let us assume that **G** and **T** are entirely vertical (e.g., when the horse is standing). Then $\Phi = 0$, $T = (g_1/t)G = GL \cos \Theta/t$ and $F = T + G$. If the horse weighs 500 kg (roughly 5000 N), and 60% of that is divided equally between the two forelimbs, then $G = G_y = 1500\ N$. For $L = 0.2$ m, $\Theta = 40°$, and $t = 0.05$ m:

$$T = (1500)(0.2)(0.766)/0.05 = 4596\ N$$

The force carried by the tendons and ligament behind the cannon bone is nearly equal to the entire weight of the animal! In fact, if the angle Θ sags to 33°, T becomes 5000 N. The joint reaction force is, of course, also vertical, and is $F = T + G = 6096\ N$, or 1.2 times body weight.

Now suppose that the ground reaction force is still entirely vertical, but the angle Φ is 20°. Then, from the moment equation,

$$T_y = (g_1/t)G = GL \cos \Theta/t \tag{1.20}$$

so that

$$T = T_y/\cos \Phi = (GL/t)(\cos \Theta/\cos \Phi) = 4891\ N \tag{1.21}$$

and from the force equations

$$F_y = T_y + G = [1 + (L/t) \cos \Theta] \, G = 6096 \ N \qquad (1.22)$$

$$F_x = T_x = T \sin \Phi = 1673 \ N \qquad (1.23)$$

The resultant joint force is 6321 N, and it acts at an angle of 15° to the vertical.

We return to this problem in the exercises that follow, and consider the effects of horizontal components of the ground reaction force on T and F. If this topic is of particular interest to you, you may want to read papers by Bartel et al. (1978) and Riemersma et al. (1988). A related work on the evolution of equine locomotion by Thomason (1991) is also of interest.

1.7 Summary and Further Reading

This chapter has considered the analysis of forces between one bone and another across joints. We have seen that such forces are always several times the external force being supported. This occurs because skeletal levers are arranged to magnify movement rather than force. Thus, muscle forces must be substantially greater than the forces being applied to the external environment. A person turning over in bed produces substantial hip joint forces, without supporting body weight, through the required muscle actions. Similarly, marine mammals such as dolphins and whales, whose buoyancy spares them the need to resist gravity, nevertheless have massive skeletons to transmit the muscle forces produced in swimming. The forces acting on bones can be estimated using the principles of static equilibrium. This method can provide very useful information, but significant approximations must be made to solve such problems. In general, the joints of vertebrate animals are statically indeterminate; that is, the number of unknowns (muscles acting to move or stabilize the joint) is almost always greater than the number of equations to be solved.

Suggestions for further reading in this field begin with the classic textbook *Biomechanics of Human Motion* (Williams and Lissner, 1977). Another well known textbook, aimed at the medical community, is *Basic Biomechanics of the Musculoskeletal System* by Nordin and Frankel (1989). Steindler's (1955) *Kinesiology of the Human Body Under Normal and Pathological Conditions* is excellent for its concentration on anatomy and pathomechanics. Early studies of the biomechanics of the human hip joint by Inman (1947) and McLeish and Charnley (1970), and Rydell's (1966) paper on measurement of hip joint forces using an instrumented prosthesis, will give you a greater appreciation for the complexities of experimental work in this area. *Animal Mechanics* (Alexander, 1968) provides an approach to this topic well suited to zoologists. If you are historically inclined, by all means seek out Maquet's (1989) translation of Giovanni Borelli's wonderful seventeenth-century work, *De Motu Animalium (On the Movement of Animals)*, sampled in Box 1.1.

1.8 Exercises

1.1. Solve the second part of the hip force problem (the force equilibrium equations) by considering a free-body diagram of the weight-bearing lower extremity and using the result $P = 2W$. Remember to include the weight of the extremity itself. What is the disadvantage of using this free-body diagram to solve the moment equation?

1.2. Suppose that the distance b in women's pelves is 10% greater than men's because of their greater pelvic width. How does this change their abductor muscle force magnitude and the magnitude and direction of their hip joint reaction force, assuming the moment arm of the abductor muscles is independent of gender?

1.3. Consider the force required in the erector spinae muscles to stabilize the head of a student leaning over his book. Also of interest is the force between the fifth and sixth cervical vertebrae. Figure 1.11 is a free-body diagram of this situation: **E** is the erector spinae force, **H** is the weight of the head and neck acting downward at their combined center of gravity, and **R** is the vertebral reaction force acting at a center of pressure on the inferior end plate of C5. Determine the magnitude of **E** and the magnitude and direction of **R**. Assume that **E** is parallel to the spine and $\Theta = 70°$. Let the moment arms of **E** and **H** with respect to the point of action of **R** be 0.02 and 0.10 m, respectively. Express your results as multiples of H.

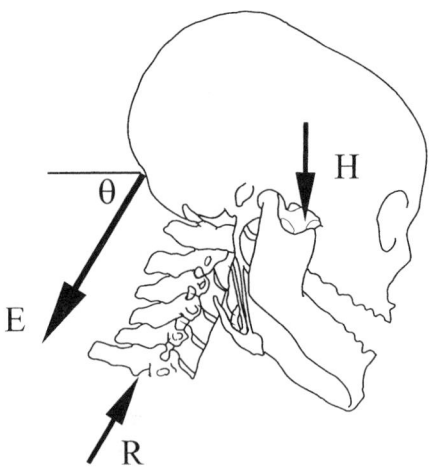

FIGURE 1.11. Diagram for Exercise 1.3, calculating the forces in the neck while studying. **H** is the weight of the head and neck acting at their center of gravity; **E** is the force of the erector spinae muscles and Θ is its angle; and **R** is the vertebral reaction force.

1.4. Consider the force produced in the temporomandibular joint when chewing. Figure 1.12 shows the location of the two major muscles active in a sagittal plane when chewing. The temporalis muscle arises from a broad area on the side of the skull, passes down through the space enclosed by the zygomatic arch, and inserts on the coronoid process of the mandible. In the picture of the skull, this point is hidden behind the other big chewing muscle, the masseter, which runs from the anteroinferior edge of the zygomatic arch down to the "angle" of the jaw. Ignoring other masticatory muscles, draw a free-body diagram of the mandible, assuming that the problem is two dimensional and the same chewing forces act on each side of the jaw. Let **C** be a vertical chewing force acting on a molar with a 0.06-m moment arm about the condylar process in the temporomandibular joint (TMJ). Assume that the masseter force, **M**, acts at an angle of 115° with respect to the x-direction, and that the temporalis muscle force, **T**, acts at an angle of 80°. Their moment arms with respect to the TMJ center are $b = 0.04$ m and $d = 0.02$ m, respectively. After each of these forces is represented on the free-body diagram, write the equations for static equilibrium. Under what conditions could these equations be solved? How do the direction and magnitude of the joint reaction force, **R**, depend on the relative magnitudes of **M** and **T**? Try to obtain solutions to the equilibrium equations using at least two different kinds of additional conditions. You can read more about the mechanics of the human jaw in papers by Barbenel (1972) and Osborn (1996).

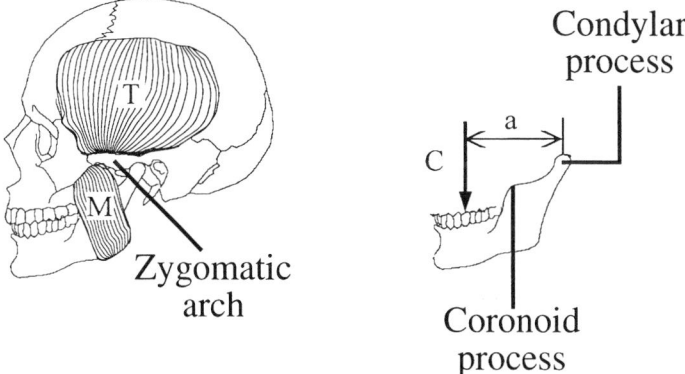

FIGURE 1.12. Diagram for Exercise 1.4. *Left:* Anatomical diagram showing locations of the temporalis (*T*) and masseter (*M*) muscles. The former passes through the zygomatic arch to pull the coronoid process up and to the right. *Right:* Detail of the mandible showing the assumed biting force, **C**, and its moment arm, *a*, with respect to the joint.

1.5. A male gymnast is performing the "iron cross" exercise on the rings, in which the arms are held straight out with a ring grasped in each hand, and the body hangs motionless and vertical between the rings (Fig. 1.13). Consult an anatomy book to determine the principal adductor muscles for the arm in this position, and estimate their angle of pull and insertion point. Draw a free-body diagram and calculate the approximate magnitude of the adductor muscle force and the magnitude and direction of the reaction force in the glenohumeral (shoulder) joint during the maneuver. State the principal assumptions necessary for your calculation.

1.6. Suppose, in Exercise 1.5, that the ropes holding the rings do not rise vertically from the gymnast's hands, but pull outward or inward at an angle θ. How does this angle affect the forces in the shoulder joint?

1.7. In a classic paper, McLeish and Charnley (1970) described in detail the considerations necessary to estimate the forces in the hip during the stance phase of gait. (As noted in Box 1.4, Sir John Charnley also developed the total hip replacement procedure in its modern form.) Read this remarkable paper. (To locate it, see the bibliography at the end of this book.)

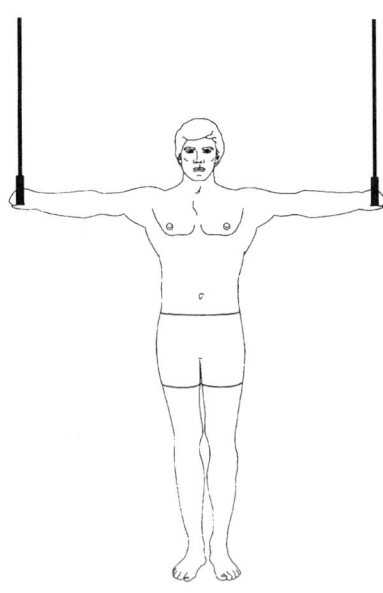

FIGURE 1.13. Sketch for Exercise 1.5 shows a gymnast performing the iron cross maneuver.

1.8. For the equine fetlock joint, work out the magnitude and direction of the joint force when a horizontal component of the ground reaction force (i.e., $\mathbf{G}_x$ in Fig. 1.10) is present. Let $\mathbf{G}_x = \pm 0.20\, \mathbf{G}_y$, and assume, as before, that $L = 0.2$ m, $\Phi = 0$, $\Theta = 40°$, and $t = 0.05$ m.

1.9. To compare *Homo sapiens* and *Australopithecus* hip joint mechanics using similar skeletal materials, Lovejoy and co-workers (1973) studied the skeletons of Native Americans recovered in the field rather than those from laboratory cadavers representing modern Americans. They estimated interfemoral head distance and abductor moment arm distance from the archeological remains of 20 *Homo sapiens* and 1 *Australopithecus*; these results are shown in Table 1.2. Assume that the body weight moment arm (b in Figs. 1.3 and 1.14) is slightly greater than half the interfemoral head distance (D_{IF}) shown in the second column of Table 1.2, i.e.,

$$b = k\, D_{IF} \tag{1.24}$$

where k is in the neighborhood of 0.5. Assume also that the abductor moment arm (c in Figs. 1.3 and 1.14) and angle are as shown in the third and fourth columns. Plot a graph of the force on the head of the femur as a function of k for the three kinds of skeletons shown. What can you deduce about the hip forces in Native American females and the *Australopithecus* individual compared to Native American males?

1.10. Another interesting factor in the development of human bipedalism is its relationship to cranial capacity. The diameter of the birth canal is primarily determined by the size of the infant's head at birth. As human evolution proceeded and cranial capacity increased, the birth canal had to become larger. Presumably, this led to widening of the pelvis and increased the moment arm of the body weight (b in Figs. 1.3 and 1.14) with respect to the hip joints

TABLE 1.2. Archeological skeletal data

Skeletons	Interfemoral head distance, mm	Abductor moment arm, mm	Abductor angle, degrees (°)
Australopithecus, STS 14	132	36	15
Homo sapiens, Amerindians, male ($n = 8$)	158.0 ± 10.3	50.6 ± 1.7	12.9 ± 2.2
Homo sapiens, Amerindians, female ($n = 12$)	165.8 ± 9.0	45.4 ± 3.8	13.3 ± 2.8

From Lovejoy et al., 1973.

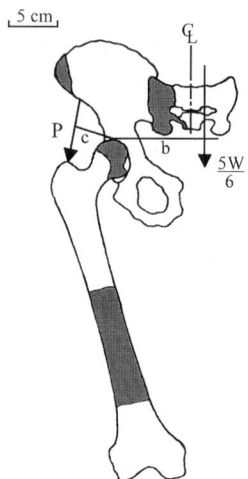

FIGURE 1.14. Diagram for *Australopithecus* hip force problem (Exercise 1.9). **P** is the abductor muscle force and *c* is its moment arm with respect to the joint center; *b* is the moment arm of the body weight minus the support leg (i.e., *5W/6*). *Shaded regions* indicate reconstructed portions of the bones. (Redrawn with permission from Lovejoy et al., 1973.

of females. Assume the other parameters of the hip joint were

abductor muscle angle = 15°
abductor muscle moment arm = c = 36 mm
interfemoral head distance = D_{IF} = 132 mm
body weight moment arm = b = $k\,D_{IF}$
k = 0.60

and remained constant (refer to Exercise 1.9). Determine and graph the relationship between the hip abductor force **P** and the newborn cranial volume V, assuming that the head is a sphere and its newborn volume increases from 500 to 1000 cm³. Does encephalization (increasing brain size) appear to be an important factor in increasing hip joint forces during bipedal walking? For additional discussion of this issue, see Leutenegger (1972) and Ruff (1995).

1.11. Go to the library and read the paper by Zihlman and Hunter (1972). It is a more complex, three-dimensional analysis of hip mechanics in *Australopithecus*. Write a paragraph discussing its strengths and weaknesses.

2

Skeletal Biology

It is true, if we come to torture a bone with the Fire, it seems to confess
that it consists of all the five Chymical Principles...

Clopton Havers (1691)

2.1 Introduction

Chapter 1 has shown us that the skeleton must withstand very high
forces; because our muscles can only contract a small percentage of
their length, we must amplify movements using levers that "spend
rather than save forces," as Borelli put it. In this chapter we begin to
develop an understanding of how the biology of bone and cartilage
provides tissues that can support these large forces day in and day out
for a lifetime.

As more is learned about the mechanics of the skeleton, it becomes
increasingly clear that mechanical demands heavily influence what happens biologically. Most of what we know about the basic cell- and
tissue-level mechanisms that enable bone to adjust to mechanical forces
has been learned since the 1960s. Before then, skeletal research was
focused on bone chemistry, for two reasons. First, mineral metabolism is
of great significance in physiology and medicine, and there was intense
interest in this subject in the middle of the twentieth century. Second,
bone is a difficult tissue to study histologically because it is calcified.
Although significant histologic studies of bone have been conducted
sporadically throughout this century, most histology laboratories still
only work with decalcified bone. Quantitative analyses of histologic sections of mineralized bone became common only in the 1970s.
Biochemists, on the other hand, could study bone mineral with existing
methods. They soon realized that bone served as a mineral reservoir for
the rest of the body, particularly with regard to calcium and phosphorus.
For quite a long time, the research of physiologists and biochemists on
this aspect of the skeleton strongly influenced medical interest in bone.
Bone biology was seen largely in terms of its role in transferring mineral to and from the skeleton. In this view, calcium was removed from the
reservoir through bone resorption by cells called *osteoclasts*. Calcium
was put into the reservoir by cells called *osteoblasts*, which laid down

new bone. These two kinds of cells were thought to work independently of one another and to be separately controlled by hormones. To greatly simplify the picture, if the serum calcium concentration fell, *parathyroid hormone* was released and stimulated osteoclasts to resorb some bone and put its calcium into the bloodstream. If serum calcium became too high, another hormone called *calcitonin* was released to produce the opposite effect. If the balance between resorption and formation of bone was not sufficiently controlled by the hormonal system, bone loss and *osteoporosis* (literally, "porous bones") resulted.

Coincidentally with the advent of more quantitative studies of bone's remodeling dynamics, in the second half of the century much more became known about the mechanical properties of bone. The "old" bone biology, dominated by interest in mineral metabolism, began to be replaced by a "new" bone biology based on broader perspectives. The lamp of this new bone biology has been fueled by contributions from many individuals, but the flame was kindled by an orthopaedic surgeon named Harold M. Frost. Using simple laboratory techniques, he began in the 1960s to make histologic sections of bone and look at them with an independent mind. He deduced that osteoclasts and osteoblasts usually do not work independently, but are coupled together in teams, with osteoblasts automatically following osteoclasts and replacing the bone that they removed. Frost called these teams of bone cells *basic multicellular units*, or *BMUs*. He went on to make many other observations and deductions about bone biology. Initially, his ideas were poorly understood and often rejected, but through the years many of his concepts have been accepted in one form or another. Central to these ideas is the concept that the skeleton is primarily a mechanical organ, rather than a calcium reservoir. Although bone plays an important role in calcium homeostasis, the main job of osteoclasts and osteoblasts is now seen as maintenance of the mechanical integrity of the skeleton. How this is done by teams of cells communicating by chemical messengers and other means, while still fulfilling the metabolic roles of the skeleton, is the principal focus of current orthopaedic research. It is also an important cornerstone of this book.

Our purpose here is to briefly summarize bone and cartilage biology as a foundation for our study of skeletal tissue mechanics. What does the skeleton look like inside? What is it made of? How are bones initially formed? How do they grow? How does a fractured bone heal? These are the questions that are addressed in this chapter.

2.2 The Shapes of Bones

While the shapes of our bones are quite variable, comparisons with those of other animals show that the architecture of a vertebrate skeleton is fairly conservative and stereotypical. We usually can recognize a femur or a tibia, and we recognize that they have a characteristic relationship, regardless of what animal they came from. Yet within these broad architectural constraints, the skeletal morphology is quite variable, and we can imagine a rubber archetypical skeleton being stretched and molded to create such diverse skeletons as those of a flamingo, a bat, an elephant, a whale, or a human.

Not only is skeletal architecture conservative, but so too is the topology of individual bones within a species. Thus, a physical anthropologist can recognize an isolated bone, or even a fragment of a bone, as belonging to a human rather than one of the other primates. These constraints on architecture and topology give the impression that bones are determinate, static structures. On a scale of centimeters, this is approximately true. However, if we increase the resolution of our investigation, and focus on scales of microns to millimeters, we see that bone is a highly dynamic tissue, continually adjusting to its physiologic and mechanical environment by changes in its composition and microscopic architecture. An important principle of skeletal physiology is that bones are able to sense the mechanical loads which they bear and modify their structures to suit changes in these loads. This principle is called *Wolff's law*, and we study it in detail in Chapter 6.

2.3 Types of Bone Tissue

Figure 2.1 is a sketch designed to show the general features found in most long bones, such as the humerus or metatarsal. This figure illustrates the relationships between many of the structural features of bones described in this chapter.

The nonmineralized spaces within a bone contain *marrow*, a tissue composed of blood vessels, nerves, and various types of cells. The chief function of the marrow is to generate the principal cells present in blood. The internal presence of marrow is a nearly universal feature of bones (the ossicles of the inner ear are an exception). The relationship of marrow and bone is both biologic and physical: they share common stem cells, and marrow never exists outside of bone. Bone can be made to form outside the normal skeleton (e.g., in a muscle) by implanting osteogenic materials, but when this happens, a space containing marrow automatically forms in the nodule of extraskeletal bone. Moreover, marrow itself is a highly osteogenic material and can stimulate bone formation if placed in an extraskeletal location (Tavassoli and Yoffey, 1983).

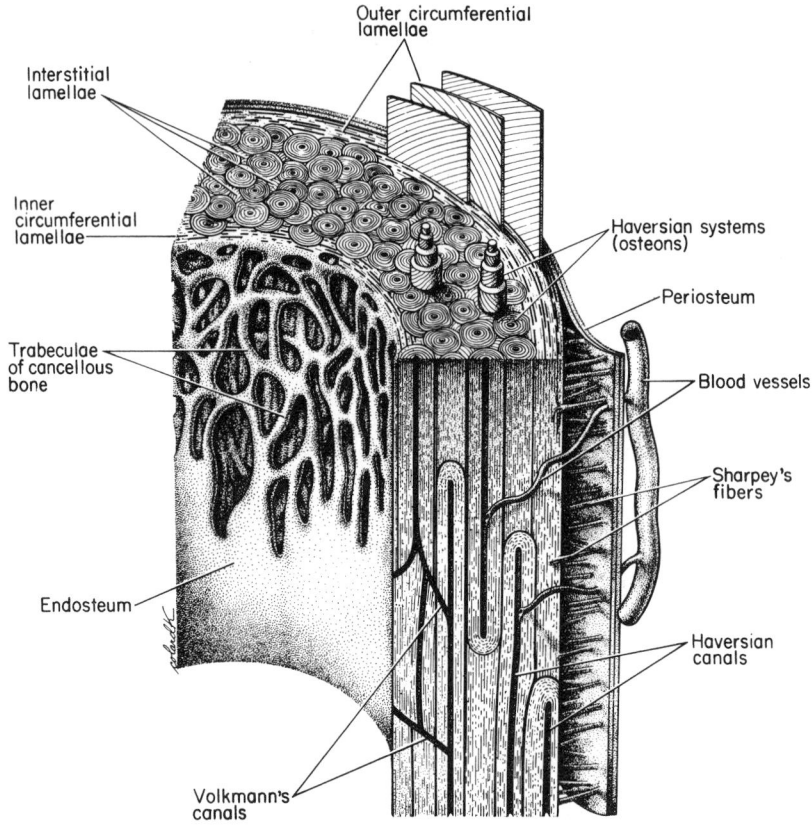

Outer circumferential
lamellae

Interstitial
lamellae

Inner
circumferential
lamellae

Haversian systems
(osteons)

Periosteum

Trabeculae
of cancellous
bone

Blood vessels

Sharpey's
fibers

Endosteum

Haversian
canals

Volkmann's
canals

FIGURE 2.1. Sketch of some important features of a typical long bone. (After Benninghoff, 1949.)

Trabecular Vs. Compact Bone

If you were to slice open a dozen bones of various sorts and wash away the marrow with a high-pressure stream of water, it would be clear that the remaining bone is of two distinct kinds, as determined by porosity (the volume fraction of soft tissues). While in principle the porosity of bone can vary continuously from zero to 100%, in fact most bone tissues are of either very low or very high porosity, with little bone of intermediate porosity. These two types of bone tissue are referred to as *compact bone* and *trabecular bone*, respectively.

Trabecular bone (also called *cancellous* or *spongy* bone) is porous bone found in the cuboidal bones (e.g., vertebrae), the flat bones, and the ends of long bones (Fig. 2.2); its porosity is 75%–95%. The pores are interconnected and filled with marrow. The bone matrix is in the form of plates or struts called *trabeculae*, each about 200 micrometers (μm) thick. The

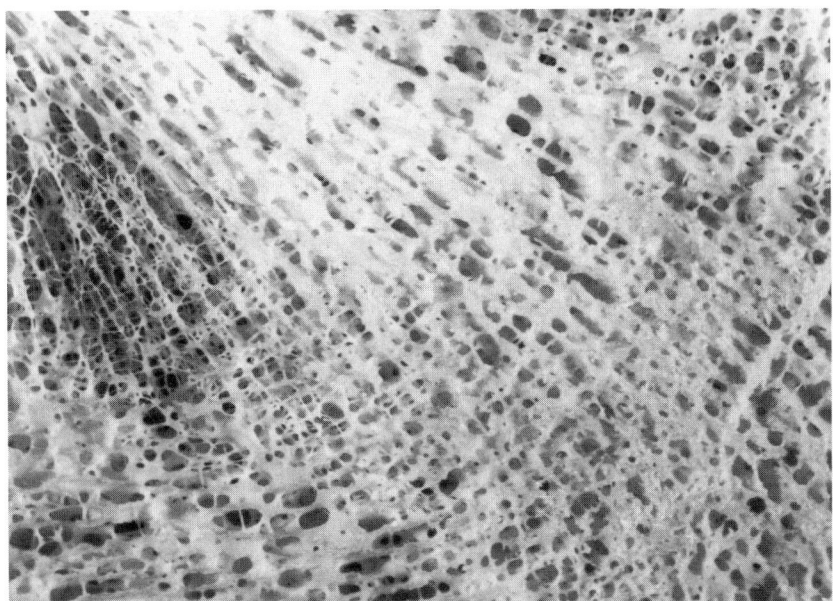

FIGURE 2.2. Example of trabecular bone structure in the distal end of a human femur. Field width, ~1 cm.

arrangement of the trabeculae is variable. Sometimes they appear to be organized into orthogonal arrays; often, they are more randomly arranged.

Compact bone is the dense bone found in shafts of long bones and forming a *cortex* or shell around vertebral bodies and other spongy bones (Fig. 2.3). Hence, it is also called *cortical bone*. Its porosity is 5%–10%, and its pores consist of spaces categorized as follows

Haversian canals are approximately aligned to the long axis of the bone, contain capillaries and nerves, and are about 50 μm in diameter (about the diameter of a human hair). Haversian canals are named after an English physician, Clopton Havers (1691).

Volkmann's canals are short, transverse canals connecting Haversian canals to each other and to the outside surfaces of the bone. These canals also contain blood vessels and probably nerves. They are named after Richard von Volkmann (1830–1889), a surgeon and early advocate of Lister's antiseptic surgical methods.

Resorption cavities are the temporary spaces created by osteoclasts in the initial stage of remodeling (described later). Resorption cavities are about 200 μm in diameter.

It is important to remember that bone is a **dynamic** porous structure; its porosity may change as the result of a pathologic condition or in a

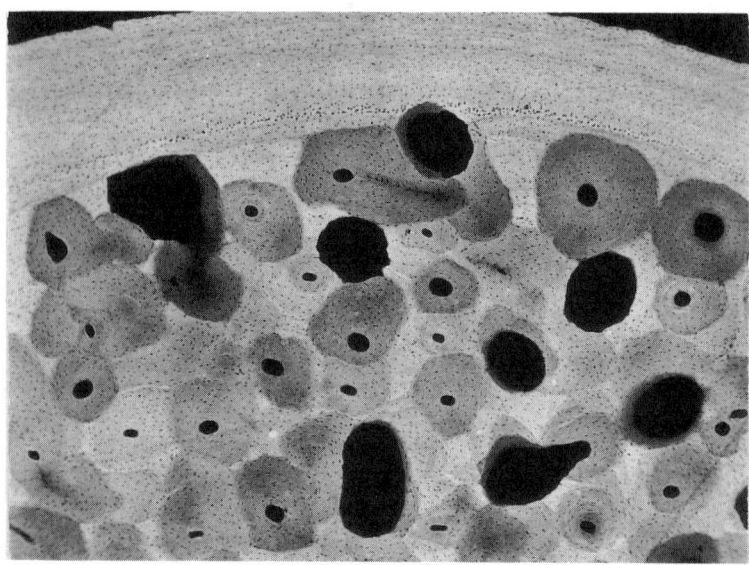

FIGURE 2.3. Microradiograph (X-ray image of a thin cross section) showing compact bone. Haversian canals and resorption spaces are *black*; note variable mineralization of osteons. Outermost region (at *top*) contains well-mineralized primary bone. Field width, ~2 mm. (Courtesy of Dr. Jenifer Jowsey.)

normal adaptive response to a mechanical or physiologic stimulus. Trabecular bone may become more compact, or compact bone may become more porous. Such changes strongly affect bone's mechanical properties.

Lamellar Vs. Woven Bone

Examining compact and trabecular bone at a still finer scale of resolution, it is evident that each may contain two major types of bone tissue.

Lamellar bone is slowly formed, highly organized bone consisting of parallel layers or lamellae comprising of an anisotropic matrix of mineral crystals and collagen fibers. Two fundamentally different kinds of "plywood" architecture coexist within the general complexity of bone's lamellar structure (Giraud-Guille, 1988). The first of these schemes corresponds to the classical view of lamellar structure: the collagen fibers are parallel in each lamella and change direction by 90° at the lamellar interface. That is, as a lamella is built up, layers of collagen are put down in one orientation. Then, when the next lamella is started, the orientation of collagen fibers suddenly changes so that they are laid down at right angles to the previous direction. In other regions of the bone, this orthogonal plywood-like structure is replaced by helicoidal plies

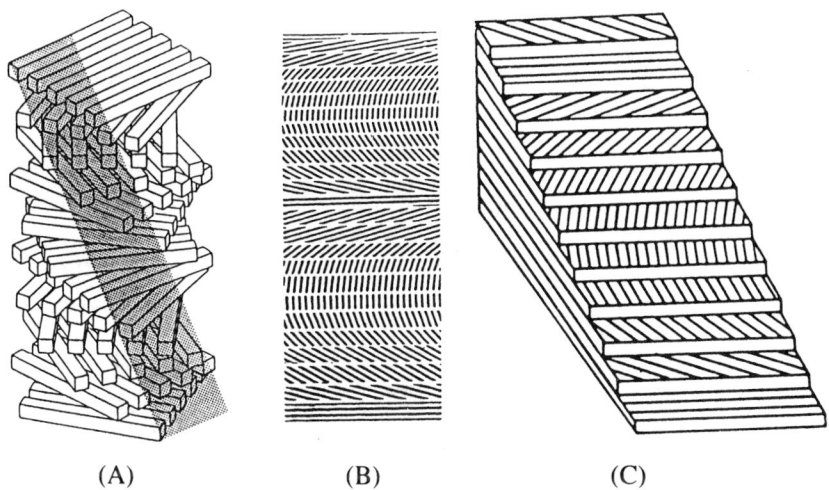

(A)   (B)   (C)

FIGURE 2.4A–C. Diagram of helicoidal plywood structure. **A** Three-dimensional structure. **B** Visual effect of arches seen when an oblique section (as indicated by the *shaded plane* in **A**) is cut. **C** The arches seen in an oblique section are formed by the helicoidal lamellae. (Reproduced with permission from Neville, 1984.)

(Fig. 2.4). In this scheme the collagen fibers **continuously** change their direction, so that in a sense there are no individual lamellae. However, the bone still shows a lamellar structure because as the orientation of the collagen fibers rotates through 180° cycles, the fiber orientation repeats itself, and layers appear when a histologic section is examined microscopically. In her landmark paper, Giraud-Guille (1988) showed that both these architectures are present in human compact bone, but their relative distribution and interspecies variation have not been determined.

Both these forms of lamellar structure give rise to *birefringence*, the capacity of some fibrous structures to interact with polarized light. When a histologic section of bone is transilluminated with polarized light and viewed through a polarizing filter oriented perpendicular to the vibration plane of this incident light, the section appears dark (i.e., the observer sees a "dark field") except where collagen fibers are parallel to the plane of the section. These collagen fibers rotate the light's plane of polarization so it is no longer perpendicular to the viewing polarizing filter. Therefore, the light is not blocked and reaches the viewer's eyes. Thus, in a bone section observed in a polarizing microscope, transversely oriented fibers are bright and longitudinally oriented fibers are dark. In the 1960s, Ascenzi and his co-workers used this phenomenon to categorize osteons as having bright, alternating, or dark lamellae (corresponding to their collagen fibers exhibiting mostly circumferential, alternating, or longitudinal orientation). These classifications are shown as parts a, b, and c, respectively, in Fig. 2.5. A dark "iron cross" pattern superimposed on the brighter osteons shows where the lamellae lie parallel to the polarization directions of the upper or lower polarizing filters.

Woven bone is a quickly formed, poorly organized tissue in which collagen fibers and mineral crystals are more or less randomly arranged. Woven bone may become more highly mineralized than lamellar bone, which, mechanically speaking, may help compensate for its lack of organization. Rats and other small animals also have *fine-fibered bone* in the cor-

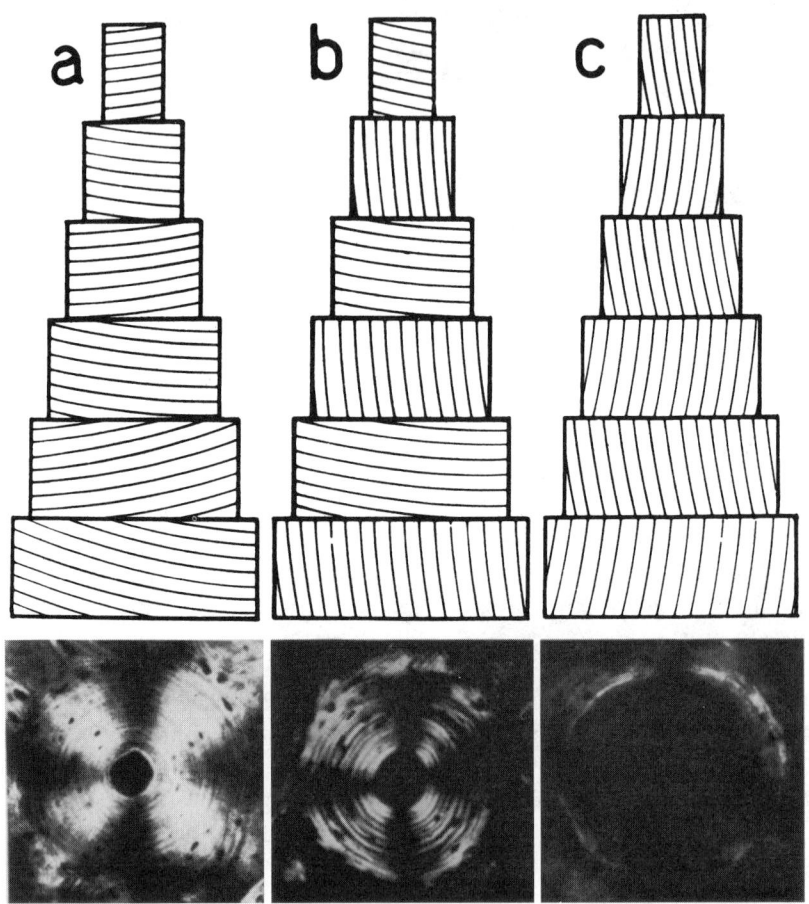

FIGURE 2.5. Three osteon types as defined by Ascenzi and Bonucci. Photomicrographs at *bottom* show appearance in plane-polarized light; diagrams *above* show hypothesized fiber arrangements in successive lamellae. a, Type T or transverse (i.e., circumferentially wrapped) fiber orientatio; b, type A or alternating fiber orientations; c, type L or longitudinal fiber orientation. (Reproduced with permission from Ascenzi and Bonucci, 1967.)

tices of their bones. In this bone the rather randomly arranged collagen fibers are smaller and closer together than in woven bone, so the tissue appears to be more organized. It may be that the lamellar organization of bone can vary continuously, depending on factors such as how fast it is made. The important generalization is that woven bone can be made more quickly than lamellar bone, but is weaker.

Primary Vs. Secondary Bone

Compact bone may be further characterized as **primary** or **secondary** bone.

Primary bone is tissue laid down de novo on an existing bone surface, such as the periosteal surface, during growth. It may be of two general types:

Circumferential lamellar bone, in which the lamellae are parallel to the bone surface. Figure 2.6 is a schematic diagram of circumferential lamellar bone beneath the periosteal surface. Blood vessels are incorporated into the lamellar structure such that each is surrounded by several circular lamellae, forming a *primary osteon* with a *primary Haversian canal* at its center.

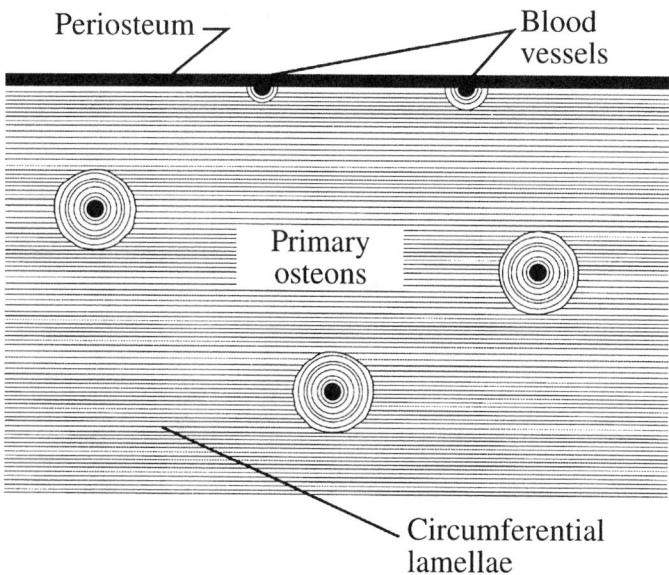

FIGURE 2.6. Sketch of primary circumferential lamellar bone structure. Primary osteons form when blood vessels on the bone surface become incorporated into the new periosteal bone. They usually have several concentric lamellae, but their cement line is not scalloped.

Plexiform bone, in which the rate of formation is greatly increased by continually constructing a trabecular network on the surface and filling in the gaps. The result is a mixture of woven bone (the trabeculae) and lamellar bone (the filled-in spaces). Plexiform bone contains rectilinear residual vascular spaces, which often produce a "brick wall" appearance. Figure 2.7 contains a low-magnification view of this structure. It typically occurs in large, fast-growing animals like cows. Racehorses, which put enormous stress on some of their bones, have a similar kind of bone that may have exceptional fatigue resistance (Stover et al., 1992).

Secondary bone results from the resorption of existing bone and its more or less immediate replacement by new, lamellar bone. This process is known as remodeling; it is discussed in detail later. In compact bone, secondary tissue consists of cylindrical structures known as *secondary osteons* or *Haversian systems* (Fig. 2.8). These are about 200 μm in diameter, and consist of about 16 cylindrical lamellae surrounding a central *Haversian canal.* The boundary between the osteon and the surrounding bone is known as the *cement line.* In adult humans, most compact bone is entirely composed of secondary bone, which may include whole osteons and the remnants of older osteons that have been partially resorbed (*interstitial bone*). Trabecular bone in adults is also mostly secondary bone, and this is true from an even earlier age, because it is remodeled, or "turned over," more rapidly than compact bone. However, the remodeling of trabecular bone rarely produces osteons because they usually do not fit inside individual trabeculae.

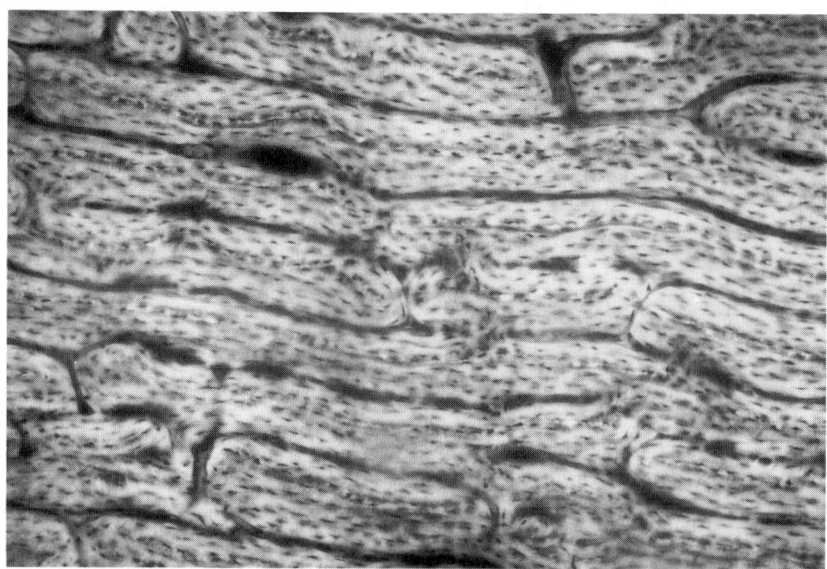

FIGURE 2.7. Photomicrograph of plexiform bone. Field width, ~500 μm.

FIGURE 2.8. Schematic diagram of secondary osteons on a field of primary bone. One osteon is still forming; it has overlapped the Haversian canal of an existing osteon.

2.4 Composition of Bone

Bone is composed of collagen, water, hydroxyapatite mineral, and small amounts of proteoglycans and noncollagenous proteins.

Collagen is a structural protein, widely distributed throughout the animal kingdom, that can spontaneously organize itself into strong fibers. More than a dozen types of collagen have been identified. The predominant collagen in bone is type I, which is also found in tendons, ligaments, and skin. Collagen gives bone flexibility and tensile strength. It also provides loci for the nucleation of bone mineral crystals, which give bone rigidity and compressive strength.

Mineral in bone consists almost entirely of *hydroxyapatite* crystals, $Ca_{10}(PO_4)_6(OH)_2$. The individual crystals are rods or plates with hexagonal symmetry, measuring about $50 \times 50 \times 400$ angstroms (1 micrometer [μm] = 10,000 angstroms [Å]) . Bone mineral is impure, containing many structural substitutions (e.g., carbonate, fluoride, citrate). These impurities are governed by the composition of body fluids and in turn affect the solubility of the bone mineral.

Ground substance of bone consists of *proteoglycans* (formerly called *mucopolysaccharides*). In particular, *decorin* and *biglycan* are small species of proteoglycans found in bone. Although the specific role of the proteoglycans in bone is unclear, decorin is known to modulate collagen fibril assembly. Proteoglycans may also function to control the location or rate of mineralization in bone through their calcium-binding properties.

Noncollagenous proteins include a number of molecules whose functions are also unclear. The most abundant noncollagenous protein is *osteocalcin*, which is secreted by osteoblasts and appears to be important in the mineralization of new bone. It also is a chemoattractant for bone cells, and

TABLE 2.1. Volumetric composition of whole bone in dogs

Component	Site or specific molecule	Volume, %
Water, 25%	Bonded to collagen	60
	Other	40
Organic matrix, 32%	Collagen	89
	Proteoglycan	1
	Other organic molecules: e.g., osteocalcin, <1%; osteonectin, <1%	10
Apatite mineral, 43%	In gaps between collagen ends	28
	Intrafibrillar	58
	interfibrillar	14

From Robinson and Elliott, 1957.

assays of its serum concentration are an excellent method of noninvasively determining rates of bone turnover. Other noncollagenous proteins in bone include *osteopontin* and *osteonectin*.

Some of the *water* in the calcified bone matrix is free, and some is bound to other molecules. The mineralization of *osteoid* (the organic portion of extracellular bone) displaces part of its water. Therefore, the water content of new bone tissue changes as it mineralizes.

Table 2.1 gives the approximate composition of bone tissue by volume in dogs, measured in primary or secondary lamellar bone several months after formation.

Quantitative Representation of Bone Composition

We have seen that bone is composed of three primary substances: water, mineral, and an organic matrix that is largely collagen. Clearly, these three substances have distinctive physical properties. Therefore, factors such as the strength of bone depend on the relative amounts of each substance. (Obviously, mechanical properties also depend on the architectural organization of each substance, but we discuss these in later chapters.) It is useful to consider here some variables that quantify the composition of bone.

Let V_T refer to the total volume of some region of bone. This volume can be divided into two parts: the hard, bony matrix (V_m) and voids filled with soft tissue (V_v). Thus,

$$V_T = V_m + V_v \qquad (2.1)$$

The volume fractions of these two portions are known, respectively, as the bone volume fraction

$$B_V = V_m/V_T \qquad (2.2)$$

(often abbreviated BVF in the bone literature), and the porosity,

$$p_v = V_v/V_T \qquad (2.3)$$

Clearly, $B_V + p_v = 1$. In cancellous bone, $p_v > 0.5$ and is largely marrow; in cortical bone, $p_v < 0.5$ and is largely composed of Haversian and Volkmann's canals. The smallest voids in bone, consisting of octeocyte lacunae and canaliculi, are usually included in V_m rather than V_v. The material inside V_m is often called the bone **tissue** to distinguish it from the term "bone," which usually refers to a region large enough to contain Haversian canals or other voids.

Another commonly used variable is bone's **apparent density**, ρ, which is defined as the mass of a volume of bone divided by its volume. Here, we are speaking of the mass of both the hard and soft tissues inside V_T. Typically, ρ would be measured simply by machining a cube of bone, weighing it, and dividing the mass by the cube's calculated volume. If ρ_m is the density of the bone tissue and ρ_v is the density of the soft tissues in the void spaces, then

$$\rho = (\rho_m V_m + \rho_v V_v)/V_T \tag{2.4}$$

$$\rho = \rho_m B_V + \rho_v p_v \tag{2.5}$$

As $B_V = 1 - p_v$

$$\rho = \rho_m - (\rho_m - \rho_v)p_v \tag{2.6}$$

Note that apparent density depends on both the porosity of the bone (p_v) and the density of the bone tissue (ρ_m). Based on measurements of cortical bone, the value $\rho_m = 2.0$ g/ml is suggested. Because the soft tissues in the voids are almost entirely water, $\rho_v \cong 1$ g/ml. Later, we rearrange Eq. 2.6 to estimate porosity from apparent density measurements.

If we now turn our attention to the composition of the bone tissue (i.e., the material in V_m), we may write

$$V_m = V_o + V_a + V_w \tag{2.7}$$

where the subscripts o, a, and w refer to the organic matrix (largely collagen), the mineral, and water, respectively. Representing the densities of the organic, ash, and water fractions of the bone tissue with the appropriate subscripts, we can also write

$$\rho_m = (\rho_o V_o + \rho_a V_a + \rho_w V_w)/V_m \tag{2.8}$$

Currey (1990) has suggested the values $\rho_o = 1.1$ g/ml and $\rho_a = 3.2$ g/ml. Of course, $\rho_w = 1.0$ g/ml.

If we dry a specimen of bone as it comes from the body in an oven at 100°C and then weigh it, its **dry mass** is $m_d \cong \rho_o V_o + \rho_a V_a$ because we evaporate not only the free water in the bone tissue but also that in the soft tissues. If we then ash the specimen by placing it in a furnace at about 800°C for 24 h, evaporating the organic matrix, the **ash mass** will be $m_a = \rho_a V_a$. The ratio of these two masses is called the **ash fraction**:

$$\text{Ash fraction} = m_a/m_d = \rho_a V_a/(\rho_o V_o + \rho_a V_a) \tag{2.9}$$

The ash fraction is a measure of the degree of mineralization of the bone tissue. Unlike apparent density, it is independent of the porosity of the bone. Typically, ash fractions are about 0.65 ± 0.03.

Basic Stereology

Stereology has been defined as "...methods for the exploration of three-dimensional space, when only two-dimensional sections through solid bodies or their projections on a surface are available." (H. Elias, quoted in Underwood, 1970). Stereology is extremely useful in measuring variables related to the structure and composition of bone. In the foregoing paragraphs, we referred to weighing specimens to determine apparent density and calculating porosity from this variable. A more direct way to measure porosity is to cut the specimen into histologic sections and measure the void volume fraction from the two-dimensional (2-D) image (Fig. 2.9). There are three ways to do this.

The first method is to measure the areas of all the voids in the image, or a set of images, and divide the total void area (A_V) by the entire area of the imaged bone (A_T). It can be shown that the 2-D or areal porosity (A_V/A_T) approaches the 3-D or volumetric porosity ($p_v = V_V/V_T$) as more and more measurements are accumulated (Fig. 2.9A),

$$V_V/V_T = A_V/A_T \tag{2.10}$$

provided A_V/A_T is averaged over a large enough sample. The second method is to randomly position "test lines" (Fig. 2.9B) on the image and measure the fractional length of the test lines that falls on void spaces. This length fraction, L_V/L_T, is a one-dimensional measure of void volume fraction, the mean value of which approaches the 2-D and 3-D values as the accumulated length of test lines increases.

Finally, one can use a "zero-dimensional" measurement technique in which a grid of points is randomly placed over the image and the fraction of the points that falls on voids, P_V/P_T, is recorded (Fig. 2.9C). In the limit of many measurements, one expects that

$$\text{Porosity} = p_v = V_V/V_T = A_V/A_T = L_V/L_T = P_V/P_T \tag{2.11}$$

In practice, P_V/P_T is commonly used if the measurements are being made "by eye" or if a computer is being used for image analysis. In the latter case, each test point is a pixel (picture element) in the computer image. The problem is programming the computer to recognize void space pixels, which is usually done by making the bone tissue light or dark in relation to the void space and using a threshold gray scale value to partition the pixels into tissue and void groups. This is not a trivial problem, however, and point counting by a human observer is often the most reliable method, especially if the void spaces need to be categorized (e.g., as Haversian or Volkmann's canals).

The equivalence of volume fraction measurements in zero, one, two, and three dimensions does not apply to another important variable in bone histomorphometry: internal surface area. Bone resorption and formation can only occur on bone surfaces, which can be external surfaces (i.e., periosteal or endosteal surfaces) or internal surfaces (i.e., Haversian or Volkmann's canal surfaces, or trabecular surfaces in cancellous bone). Measurements of internal surfaces are usually expressed per unit volume of bone; this variable, called *spe-*

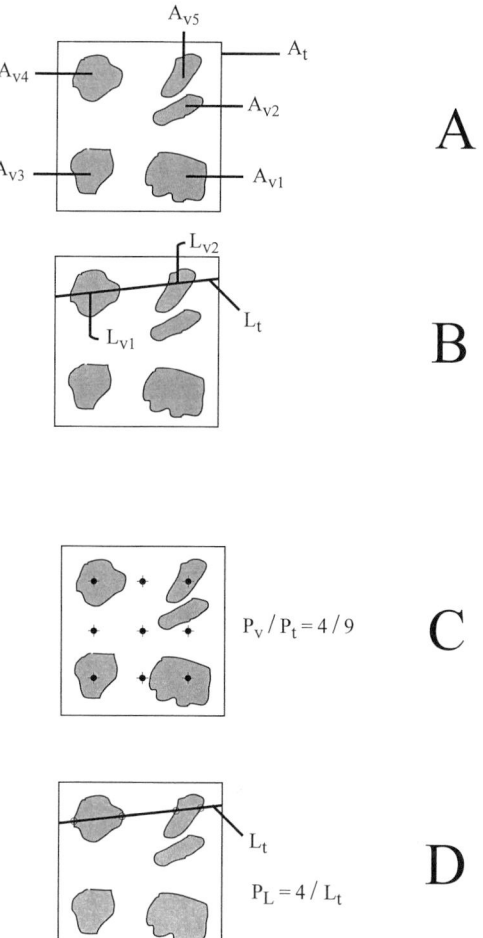

FIGURE 2.9A–D. Basic stereologic methods. *Shaded regions* represent soft tissue spaces in two-dimensional sections of bone. If i = 1, 2, 3, etc., A_{vi} = area of the *i*th void, L_{vi} = length of test line within *i*th void, P_v = number of test points falling in voids; P_t = total number of test points, and L_t = total length of test line. Images **A**, **B**, **C**, and **D** are further described in the text.

cific surface (S_V), has units of mm²/mm³ or mm⁻¹. The two-dimensional equivalent of S_V is the perimeter of voids in the image divided by the total area of the image, L_A. The one-dimensional version is the number of test line intercepts with void surfaces divided by the total length of test lines (P_L, Fig. 2.9D). These three measurements of specific surface area are related as follows:

$$S_V = (4/\pi)\ L_A = 2\ P_L \qquad (2.12)$$

where it is again assumed that sufficient measurements have been made of L_A or P_L for the mean to provide a good estimate of S_V. It is also important to realize that porosity measurements are independent of the shapes and orientations of the voids, but specific surface measurements are not. Consequently, test lines that uniformly represent all orientations are used to measure *average* specific surface, while test lines oriented in particular directions can be used to quantify the anisotropy of bone's internal structure.

2.5 Bone Cells

So far we have discussed the extracellular matrix, that part of bone which is outside of living cells and can be mineralized. Formation and maintenance of the matrix is carried out by the cells that make up a small but critical percentage of the bone volume. There are four types of bone cells. These cells fall into two categories: those that resorb bone, and those which form or have formed bone. The "resorbers" are closely related to *macrophages*, cells that migrate throughout all tissues of the body to remove debris and pathologic material. The "formers" are closely related to cells like *fibroblasts*, which produce structural molecules in other tissues.

Osteoclasts are the cells that resorb bone (Fig. 2.10). They are multinuclear because they are formed by fusion of cells called *monocytes* originating in

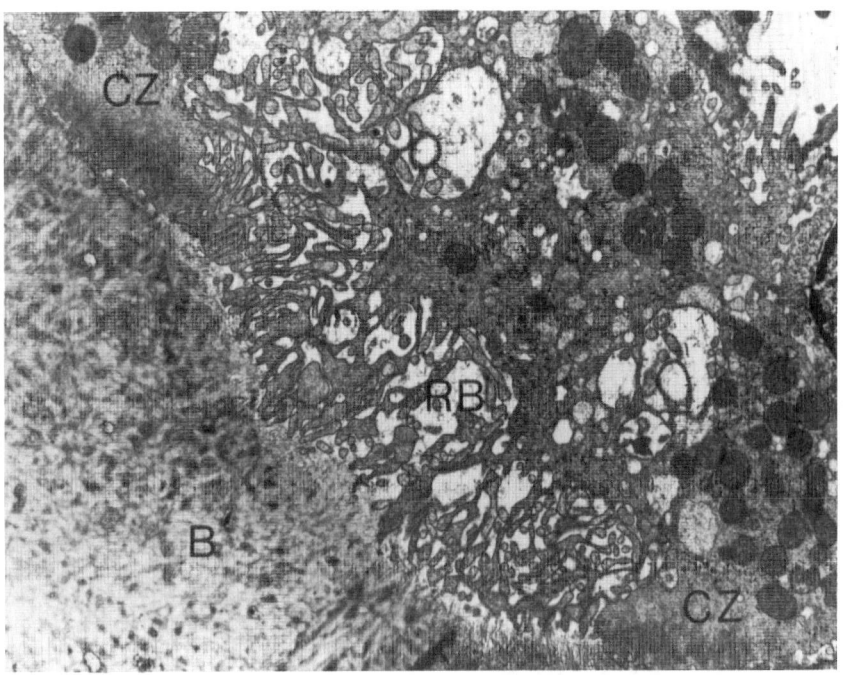

FIGURE 2.10. Electron photomicrograph of a portion of an osteoclast *(upper right)* resorbing bone *(lower left)*. *CZ* labels clear zones where the cell is sealed to the bone surface; *RB* labels the ruffled border where the cell releases chemicals that break down the calcified bone matrix *(B)*. (Reproduced with permission from Borysenko and Beringer, 1984.)

the hemopoietic portion of the bone marrow. Resorption occurs along a highly invaginated *brush* or *ruffled border* of the cell, which is sealed to the bone surface by a peripheral *clear zone*. Osteoclasts erode their way through bone at a rate of tens of micrometers per day by first demineralizing the adjacent bone with acids and then dissolving its collagen with enzymes. These destructive chemicals are manufactured by the cell and transported to the ruffled border in portable intracellular chambers called *secretory vesicles*.

Osteoblasts are mononuclear, cuboidal cells that produce osteoid, the organic portion of the bone matrix. Figure 2.11 is an electron photomicrograph showing the interface between an osteoblast (at top) and the bone surface

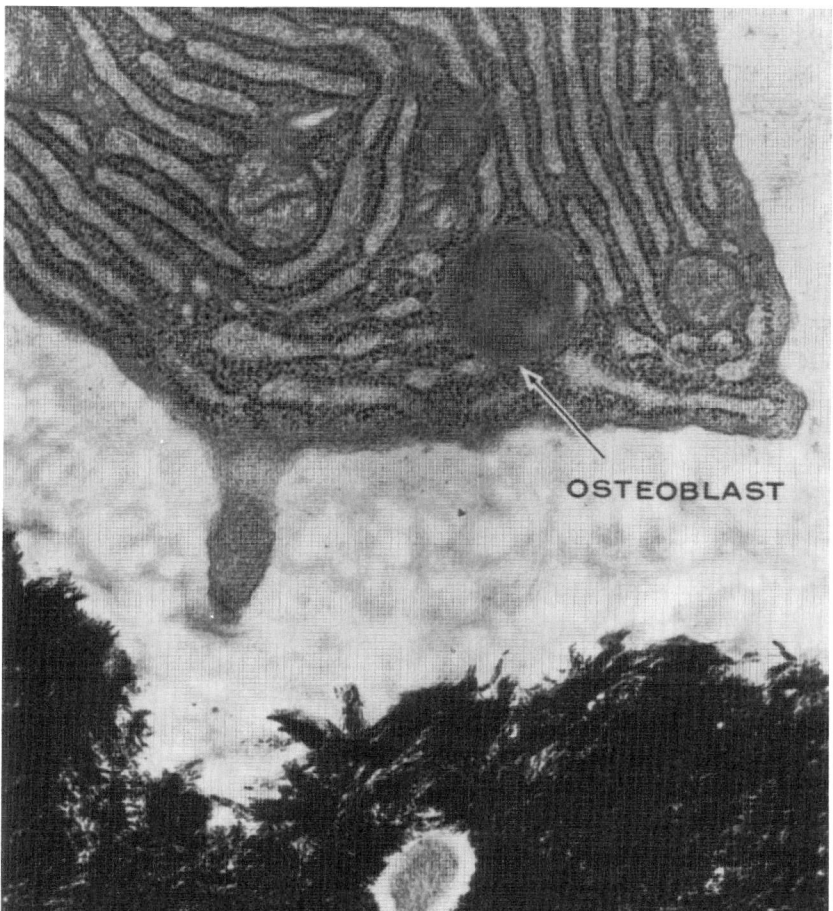

FIGURE 2.11. Electron photomicrograph of osteoblast forming bone. Dark material at bottom is mineralized bone. The lighter, interposed material is osteoid, production of which necessitates the extensive rough endoplasmic reticulum in the cell. A portion of a process protrudes from the cell; another is seen in cross section in the bone matrix. (From Reddi and Anderson, 1976, by copyright permission of The Rockefeller University Press.)

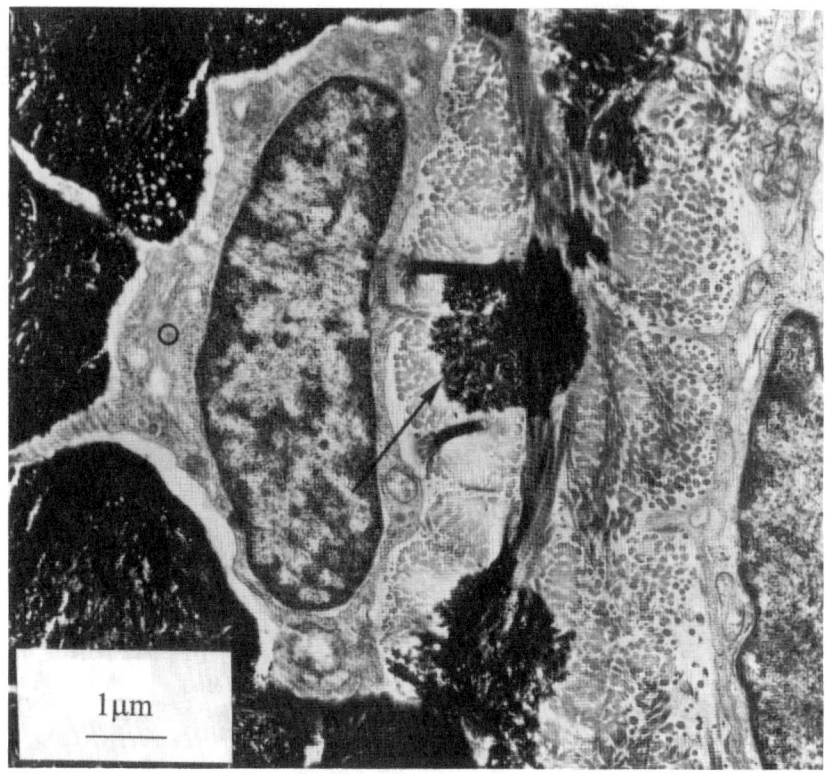

FIGURE 2.12. Electron photomicrograph of osteoblast forming bone (*far right*). It is burying another osteoblast (*O, left center*), which will become an osteocyte. Note cell processes at *far left* and collagen fibers both parallel and normal to the page. *Black material* is mineral; *arrow* indicates patch of new mineral in osteoid between the cells. (Reproduced with permission from Cooper et al., 1966.)

(below). Osteoid contains collagen and noncollagenous proteins, proteoglycans, and water. Much of the water is replaced by mineral as the osteoid calcifies. The boundary between osteoid and calcified bone is called the *mineralization* or *calcification front*. Collagen precursors and the other organic matrix molecules are produced in the osteoblast's rough endoplasmic reticulum. Also shown is a *cell process* extending from the osteoblast into the bone that it has helped produce. This process connects to an *osteocyte* within the calcified matrix, as described next.

Osteoblasts are differentiated from cells known as *mesenchymal cells*. Depending on where the bone formation occurs, these mesenchymal cells come from the deep or *cambium* layer of the *periosteal membrane* or from the stromal tissue of bone marrow. Differentiation of mesenchymal cells into osteoblasts is a multistage process requiring 2–3 days and

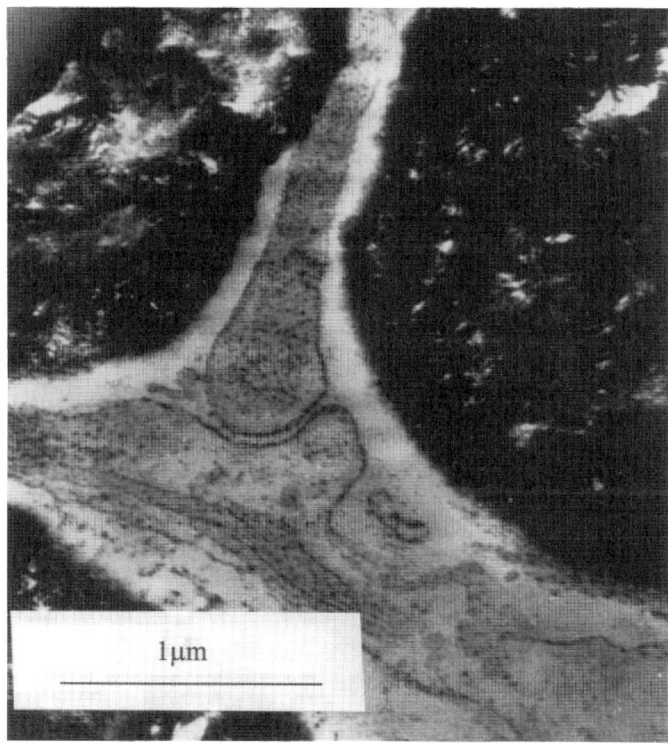

1 μm

FIGURE 2.13. Electron photomicrograph shows junction between two osteocyte processes. (Reproduced with permission from Cooper et al., 1966.)

appears to be triggered by mechanical stress to the tissue. Osteoblasts lay down osteoid at a rate of about 1 μm/day. This is called the bone *apposition rate*.

Osteocytes are former osteoblasts that have become buried in the bone which they and their neighbors have made. Osteocytes sit in cavities called *lacunae* and communicate with each other and with osteoblasts via processes passing through tunnels called *canaliculi* ("little canals"). Figure 2.12 shows a recently buried osteocyte and one of the osteoblasts that helped bury it. The black regions are calcified bone. The "paths" through these regions indicate the location of canaliculi containing intercellular processes. As shown in Fig. 2.13, processes from adjoining cells are connected by *gap junctions*, implying that these cells communicate and exchange substances to a significant degree. Figure 2.14 shows numerous osteocytes and their network of processes at a much lower magnification. There are about 15,000 lacunae per cubic millimeter (mm^3) of bone (Mullender et al., 1996), but because of their small size, the lacunae and canaliculi occupy only about 1% of the bone volume. Their surface area, on the other hand, is huge. Johnson (1966) estimat-

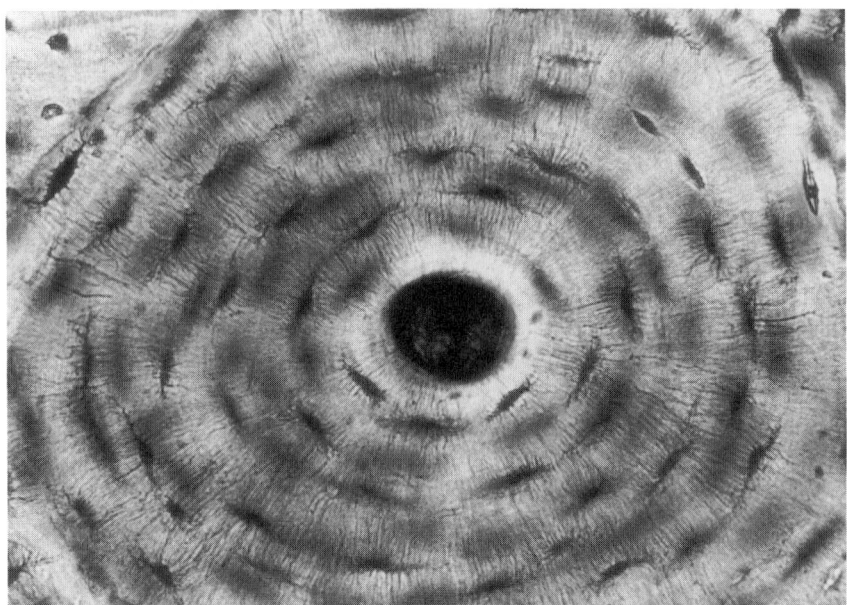

FIGURE 2.14. Photomicrograph shows the network of osteocyte lacunae (*dark ellipses*) and canaliculi (*fine dark lines* radiating from lacunae) within an osteon in basic fuchsin-stained, undemineralized cross section of equine third metacarpus. Field width, ~200 μm.

ed the total skeletal surface area of canaliculi in an adult male skeleton to be 1200 m^2, compared to 3.2 m^2 for Haversian and Volkmann's canals and 9 m^2 for cancellous bone surfaces. This intimate contact with virtually every nook and cranny of the skeleton is one reason that osteocytes are thought to be important in transporting mineral into and out of bone, and perhaps in sensing mechanical stress. However, their physiologic functions and significance are as yet poorly understood.

Bone lining cells are, like osteocytes, "retired" (quiescent) osteoblasts. These are the osteoblasts that escaped being buried in newly formed bone and remained on the surface when bone formation ceased. As production of bone matrix stops, bone lining cells become quiescent and flattened against the bone surface, but they do not form a continuous, gapfree barrier over the bone. They maintain communication with osteocytes and each other via gap-junctioned processes, and also appear to maintain their receptors for parathyroid hormone, estrogen, and other chemical messengers. Like osteocytes, they are thought to be responsible for transfers of mineral into and out of bone (Parfitt, 1987) and for sensing mechanical strain. They are also believed to initiate bone remodeling in response to various chemical and mechanical stimuli (Miller and Jee, 1992).

Box 2.1 Technical Note
Bone Histology

Microscopic examination of biologic tissues requires that they be infiltrated and embedded in a solid medium to retain proper shape and cellular anatomy during subsequent sectioning. Paraffin, a waxy mixture of hydrocarbons commonly used to manufacture candles, has been the traditional medium of choice because of its low melting point and its miscibility with other organic compounds used to dehydrate and fix the tissue. Manufacturers now add plastic polymers and dimethyl sulfoxide to the paraffin to improve its infiltration and cutting characteristics. In the histology lab, small pieces of tissue, usually about a cubic centimeter in size, are immersed in a mold containing melted paraffin that is allowed to solidify. The embedded "tissue block" is then mounted in a microtome and thin sections (approximately 5 μm thick) are cut, mounted on glass slides, and examined under light microscopy. Skilled histologists can cut and mount several ribbons of serial sections an hour; what they cut is essentially a block of soft wax with cellular and extracellular material of similar softness suspended within.

Because of their hardness, bones and teeth present special problems. Obviously, processing these tissues in the standard fashion would result in broken microtome blades, shabby sections, and haggard histologists. To circumvent the problem, diagnostic histology laboratories first demineralize the sample to make it soft enough for processing using standard embedding and sectioning techniques. Methods of decalcification (the traditional name despite the fact that phosphorous is being liberated along with the calcium) have been in use since the turn of the century, and usually employ solutions of acid or chelating agents to solubilize and extract the mineral. Acetic, chromic, formic, hydrochloric, and nitric acid have all been used successfully; ethylenediaminetetraacetic acid (EDTA) is the current chelating agent of choice. Although such methods make paraffin embedding possible and are adequate for routine examinations of the cellular elements in tumors and marrow, they do not render the complete microscopic picture, particularly in terms of bone as a dynamically changing structural tissue. To learn about bone as a tissue, researchers in the field usually use intact mineralized sections that require specialized equipment and procedures to process.

Before the advent of plastics in the 1930s, the only way to prepare acceptable mineralized sections of bone was by grinding down nonembedded rough-cut sections, most often by hand. Wet sandpaper was often used to prepare sections, especially thick sections of 100 μm or more; Frost's seminal discoveries were made using this method. For thinner sections, the specimen was sometimes sandwiched and ground with abrasive powder between two glass plates. Surprisingly beautiful histologic specimens only 5 μm thick can be produced in this way, however the technique is time consuming, tedious, and runs the risk of destroying trabecular architecture. Things changed for the better when ethyl and methyl methacrylates were introduced as embedding media around 1940. Much like paraffin embedding, samples of bone could be immersed and infiltrated with solutions of liquid media that were then polymerized. These hard plastics were a much closer match to the material properties of bone and thin sections (<10 μm) could be cut directly from the tissue blocks using specialized heavy-duty microtomes, or thick sections (100 μm) could be prepared without grinding or damage using a diamond wafering saw. Several new polymers and resins are available today for embedding and processing hard mineralized tissue.

In the later half of the twentieth century, examinations of undecalcified (and frequently unembedded) histologic sections of bone have elucidated the cellular machinery responsible for normal bone growth and repair. Histologic manifestations of metabolic bone diseases have been well defined. Dr. Frost and his colleagues developed staining and analysis techniques for mineralized sections of bone that enabled clear identification of pertinent cells, endosteum and periosteum, osteoid, cement lines, and canaliculi, as well as patterns of mineralization. These methods, combined with fluorochrome labeling (see Box 3.2), have given us insight into the complex cellular activity responsible for maintaining the mechanical integrity of our skeletons throughout life.

2.6 Cartilage

Like bone, cartilage comes in different forms, each suited to a particular application. For the sake of simplicity, we define three types of cartilage.

Hyaline cartilage is the most prevalent type of cartilage in the adult. It is found in the ventral ends of ribs, in the tracheal rings, and covering the joint surfaces of bones, where it is known as *articular cartilage*. In addition, the growth plates are composed of this cartilage (the growth plate is described in Section 2.7).

Elastic cartilage is the variety of cartilage found in the external ear, eustachian tubes, and epiglottis. It has greater opacity, flexibility, and elasticity than hyaline cartilage, and is yellowish rather than white. Its extracellular matrix is permeated with dense, branching elastic fibers unique to this type of cartilage.

Fibrocartilage is the type of cartilage occurring in intervertebral disks, the pubic symphysis, and in the bony attachments of certain tendons. It also may form when hyaline cartilage is damaged.

Composition of Cartilage

Articular cartilage is most often hyaline, although some joints contain fibro-cartilaginous disks (e.g., the menisci of the human knee). The extracellular matrix of articular cartilage is composed mostly of type II collagen and proteoglycans manufactured by cartilage cells (called *chondrocytes*), and then assembled outside the cell into a mesh of collagen interwoven with aggregated proteoglycan molecules called *aggrecan*. Like bone cells, chondrocytes live in chambers called lacunae. Sometimes single or multiple chondrocytes are found in "bags" within the matrix called *chondrons*.

Unlike bone cells, chondrocytes are encapsulated not in mineralized bone but in a viscoelastic extracellular matrix of collagen fibers and proteoglycan molecules. Although they are active metabolically, producing collagen, proteoglycans, and proteolytic enzymes, there is no direct cell–cell contact, nor do these cells have processes to help them communicate with each other. The mechanism for intercellular signal transduction, and how chondrocytes coordinate their activities, if they do, are unknown.

Cartilage matrix is ordinarily not mineralized and is about 70% (by mass) water. Approximately 40%–70% of the dry weight of the matrix is collagen and 15%–40% is proteoglycans. Chondrocytes make up less than 5% of the volume of articular cartilage.

Type II collagen is the predominant species (about 80%) of collagen found in articular cartilage, but several other types are also present (Table 2.2). Type XI collagen forms a core for the type II collagen fibril, and probably controls fibril growth. Type IX collagen, which is really a form of proteoglycan molecule, is found periodically along the type II collagen fibril and is covalently cross-linked to it. It is probably important to the structural integrity of the collagen fibril assembly, and it has been suggested that disruption of the type IX–type II bond may be one factor leading to the degeneration of articular cartilage and a condition called *osteoarthrosis*.[1]

Type VI collagen is found in small amounts in the pericellular region of the chondrocytes. Although its function is not precisely understood, it is thought to be an adhesion molecule that brings stability to the cell–extracellular matrix system. Type X collagen is found in small amounts within type II collagen fibers. However, it is primarily found forming a network around hypertrophic chondrocytes in the growth plate, where it is probably involved in controlling mineralization of the cartilage matrix. Types XII and XIV have also been recently discovered in cartilage, and may provide cohesion between the extrafibrillar matrix and collagen fibers.

Proteoglycans in aggregate form are the other primary polymer in the matrix of cartilage. These are giant molecules composed of a central hyaluronic acid chain to which are attached (by *link proteins*) side branches called *core proteins*. These core proteins are in turn the attachment points for *glycosaminoglycans* (GAGs) (Fig. 2.15). The GAG chains consist

TABLE 2.2. Types of collagen in articular cartilage and their functions

Collagen type	Function
II	Structural and mechanical; primary constituent of articular cartilage
VI	Pericellular adhesion molecule
IX	Fibril association; stabilizes type II
X	Hypertrophic zone of growth plate; role in calcification postulated
XI	Core of type II; controls fibril growth

---

[1]Osteoarthrosis is a term applied to joint degeneration that has a mechanical rather than an inflammatory cause. This condition is more commonly referred to as osteoarthritis, even though the "itis" in that term implies that inflammation is the initiating cause.

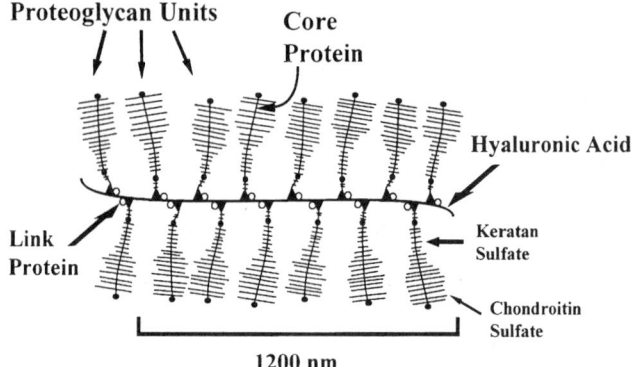

FIGURE 2.15. Diagram of proteoglycan aggregate structure.

of repeating disaccharide units in which a sugar is linked to a hexosamine. Chondroitin 4-sulfate, chondroitin 6-sulfate, and keratin sulfate, all of which carry a negative charge, are the three predominant GAGs found in cartilage. These molecules are covalently bound to the core protein to form the proteoglycan monomer which, when linked to the hyaluronic acid molecule, become the proteoglycan aggregate known as *aggrecan*.

Several smaller species of proteoglycan are also found in articular cartilage. *Decorin* "decorates" the collagen fibrils and apparently controls fibril assembly. It probably functions to reduce fibril diameter, particularly close to the articular surface, where it is found in greater abundance. *Fibromodulin* is found in increasing amounts in the cartilage matrix away from the chondrocyte, and seems to regulate fibril assembly. Other proteoglycans (lumican, epiphycan, versican, neurocan, and perlecan) are found in varying amounts in different kinds of cartilage.

Mechanical Significance of Cartilage

Articular cartilage has unique chemical properties allowing it to serve as a bearing surface. It is able to transfer loads from one bone to another while simultaneously allowing the load-bearing surfaces to articulate (i.e., roll and slide over one another) with very low friction. This behavior is studied in detail in Chapter 7. Here, however, we may ask, "How can a material that is 70% water and not mineralized support loads that are several times body weight?" The answer to this question lies in the proteoglycan portion of the cartilage matrix. Specifically, the proteoglycan aggregate *aggrecan* serves this function. The proteoglycan monomers attached to the hyaluronic acid backbone are all highly negatively charged because of the chemical structure of the sulfated GAGs that compose them. This charge causes the aggrecan molecules to repel one another and branch out into something with the appearance of a bottlebrush. It also causes the aggrecan to be very hydrophilic. This attraction for water molecules resists the tendency for water to flow out of the cartilage when it is compressed. As the "bristles" of the brush restrain the water, they are placed in tension, and when the compressive load on the cartilage is removed the aggrecan molecules reattract water to replace whatever has managed to escape. The large size and negative charge of the aggrecan molecule give cartilage its hydrophilic properties. Thus, compressive loads are borne by a volume of water that cannot escape, rather like a rubber hot water bottle filled with water. In this case, however, the water is constrained not by an outside membrane but by an internal structure that electrically confines the water molecules.

Organization of Articular Cartilage

Collagen and proteoglycans are dispersed differently through the thickness of the articular cartilage, with more collagen toward the joint surface and more proteoglycan in the deeper layers of cartilage matrix. This arrangement works well mechanically. Large tensile stresses within the articular surfaces and at the edges of joint contact areas are resisted primarily by tangentially oriented collagen fibers. The compressive and hydrostatic stresses found in the deeper layers of cartilage are resisted by the incompressibility of water, which is held in place by the hydrophilic aggrecan molecules.

The matrix of collagen fibers in cartilage is not organized randomly. Instead, the structure varies from place to place in response to functional demands, some of which are mechanical (Fig. 2.16). The surface layer, the *lamina splendens*, consists of closely packed, small diameter (~30-nm) fibrils, tangentially arranged. The lamina splendens acts as a barrier membrane to enzymes and large molecules such as hyaluronic acid. Thus, small molecules like glucose and hemoglobin in the synovial fluid can penetrate the lamina to nourish the chondrocytes deeper in the cartilage, but hyaluronic acid, the large, lubricating molecule, remains on the surface of the cartilage. Mechanically, the horizontal fiber arrangement allows the lamina splendens to resist the high tensile and

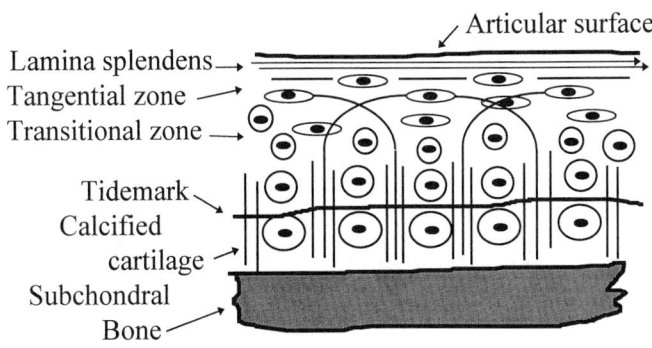

FIGURE 2.16. Sketch of articular cartilage structure.

shear forces in and around the joint contact area. However, with age, gaps appear in the lamina, allowing synovial fluid to carry in enzymes (e.g., collagenase or hyaluronidase) that can break down the cartilage matrix.

The lamina splendens is only a few micrometers thick and merges gradually into the *tangential zone* in which the collagen fibrils are slightly larger, although still arranged parallel to the articular surface. This layer is more cellular than the lamina, containing small, flattened chondrocytes. The tangential zone also contributes to the tensile stiffness of the cartilage matrix.

The *transitional zone* (or radial zone) contains collagen fibers that are larger than those in the tangential zone and also are more widely spaced and obliquely or pseudorandomly arranged. The chondrocytes tend to line up in radial columns (i.e., perpendicular to the surface), particularly in the deeper layers of this zone. These chondrocyte columns are parallel to the surrounding collagen fibers, and the chondrocytes within them actively synthesize structural molecules that replenish the cartilage matrix.

This zone ends at the *tidemark*, beyond which the cartilage has calcified with hydroxyapatite—the same mineral as found in bone. The calcified cartilage layer may serve to make the change in elastic modulus between articular cartilage and bone more gradual. In addition, collagen fibers from the transitional zone seem to cross the tidemark and reinforce against shear between stiff (calcified) and compliant (uncalcified, articular) cartilage layers.

For many years, investigators have used a simple method to visualize the orientation of collagen fibers in articular cartilage: simply sticking a pin into the cartilage surface at multiple points over the surface of a joint reveals the pattern because the material splits parallel to the fiber direction. Figure 2.17 illustrates the articular surface at the distal end of the human knee after this has been done.

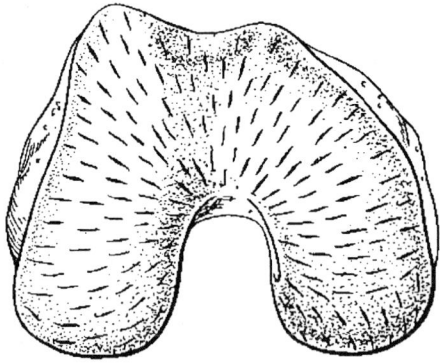

FIGURE 2.17. "Split line map" of the articular surface on the distal end of the human femur.

Noncalcified articular cartilage has no nerves and no blood vessels (cal-cified cartilage has both). The chondrocytes are nourished by diffusion through the extracellular matrix, which is largely water. This passive trans-port of nutrients may be aided by flows induced by mechanical loading, but the dependence on passive transport rather than a blood supply lim-its the capacity of cartilage to repair large defects. Thus, although chon-drocytes are constantly resupplying the extracellular matrix with collagen and proteoglycans as it degrades, the capacity to repair large amounts of damage is apparently absent.

The Role of Cartilage in Growth

Suppose that you were given the job of designing the mechanism by which a child's bones are to grow longer. The bones must do this while the child is active and loading them. The simplest solution might be to put osteoblasts on the ends of each bone, and let them form a layer of new bone there each day. The obvious problem with that solution is that the forces and motions between the bones would crush and grind the osteoblasts to bits. An alter-native would be to put the osteoblasts inside the bone and let them "push out" new bone matrix, expanding the volume from within. Some early bone scientists thought this was how longitudinal growth worked, but eventually it became clear that the mineralized bone matrix is too rigid for this to hap-pen. Bone can only be formed by cells on a bone surface because cells can only extrude new matrix into an unmineralized, "soft tissue" space. But how can such a soft tissue support loads during growth?

The solution to this dilemma is cartilage. Because it is largely water, cells can "push out" new matrix, expanding its volume, yet it is a tissue that can support very high loads **if they are primarily compressive**. As is described in Chapters 6 and 7, there is evidence that nature has arranged a sort of pact between cartilage and bone such that the former handles stresses that are mainly compressive (hydrostatic) and the latter shear (deviatoric) stresses.

This remarkable capacity for growth of a load-bearing material makes car-tilage especially suitable as a skeletal substance during development. Indeed, most of the skeleton is first formed in cartilage models, which later are replaced by bone. The occurrence of cartilage is more restricted in postfetal life, but it continues to play an indispensable role in the longitudinal growth of children's bones and in the maintenance of articular surfaces in adults.

2.7 Longitudinal Growth of Bones

Longitudinal growth of bones occurs in a nonmineralized region of growth near, but not at, each end of the bone. This region is called the *growth plate* or *physis*. The physis separates the bony *epiphysis* from the bony metaphysis. New cartilage is constantly formed by chondrocytes within the growth plate. It turns out that cartilage is viscous enough that cells within it can extrude more cartilage matrix, increasing its volume. This would increase its thickness except that the cartilage on the metaphyseal side of the growth plate mineralizes and becomes part of the metaphyseal bone, so that the length of the metaphyseal and diaphyseal parts of the bone is increased, while the thickness of the growth plate remains constant.

The growth plate may be divided into zones, each representing a stage in the life cycle of its chondrocytes (Fig. 2.18). (The number of these zones can vary, depending on the reference consulted.) In the following description, you can imagine that you are traveling down through these zones and into the metaphysis at a given point in time, leaving the upper surface of the growth plate behind. Alternatively, you can imagine that you are sitting at a point which is fixed in relation to the bone as a whole, and watching your surroundings change as time passes, so that the top of the growth plate leaves you behind.[2] Forming a clear mental image of this relativity between time and position will help you understand the process by which bones grow longitudinally. The growth plate's zones are, from the top (i.e., the epiphyseal side), as follow.

1. **Reserve or resting zone:** Cells of moderate size are scattered irregularly throughout this zone, which is anchored to the bone of the epiphysis and receives nourishment from epiphyseal blood vessels. These cells are not chondrocytes and are not really "resting," but are dividing slowly to provide chondrocytes for the remainder of the growth plate. Such cells are often termed *stem cells*. When they divide, one of the resulting pair remains a stem cell, and the other differentiates into a cell with a different function, in this case a chondrocyte.
2. **Proliferative zone:** Here, the chondrocytes divide repeatedly and arrange themselves in columns. They also become disklike, so that they look somewhat like a stack of coins. They produce proteoglycan and type II collagen and other molecules needed for the surrounding extracellular matrix. The collagen fibers are aligned parallel to the cell columns.
3. **Hypertrophic zone:** As they stop proliferating, the chondrocytes are said to "mature." They accumulate glycogen in their cytoplasm and secrete copious amounts of matrix, increasing the volume of the surrounding substance. They also *hypertrophy*; they increase their intracellular volume. Because the growth plate is radially constrained by bone around its perimeter, its volume changes are expressed primarily in the longitudinal direction. It is thought that the chondrocytes in this zone are in the early stages of *apoptosis* (programmed cell death). They stop producing cartilage matrix and begin producing molecules that prepare the adjacent cartilage for calcification. They also increase in size.

---

[2]Recall Einstein's famous question, "Does Zurich stop at this train?"

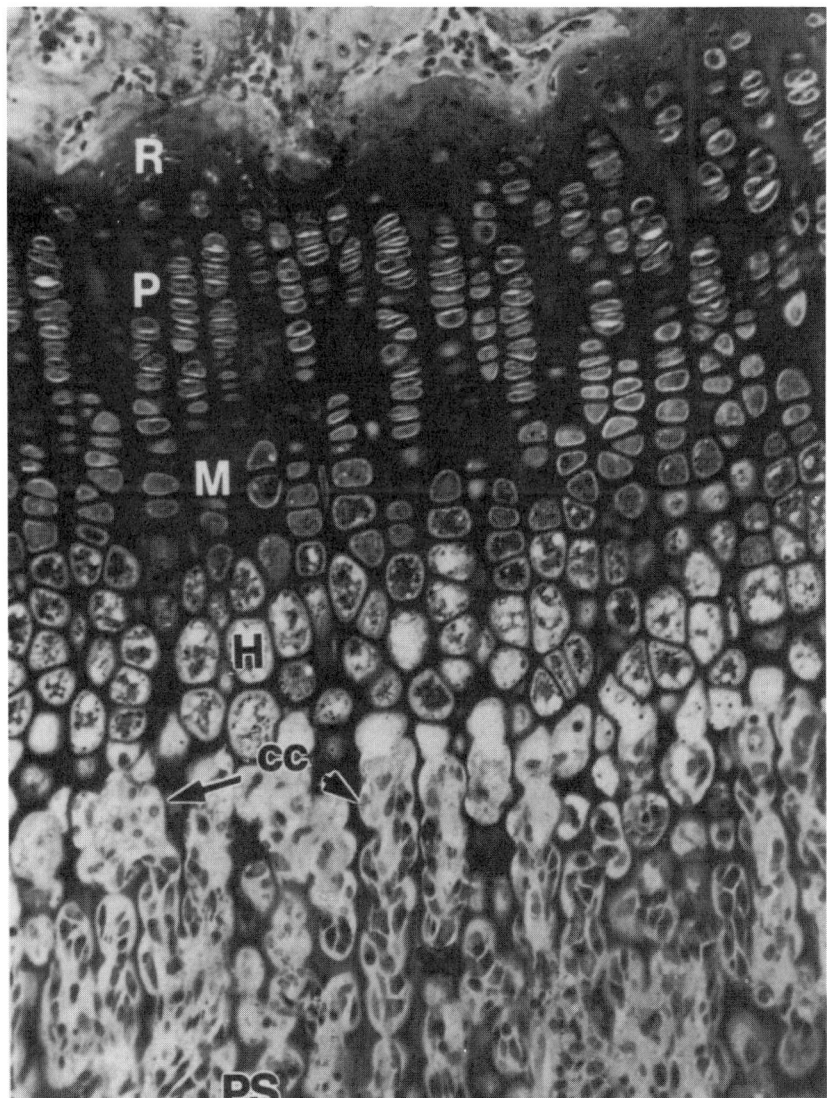

FIGURE 2.18. Zones of the growth plate: *R*, resting; *P*, proliferating; *M*, maturing; *H*, hypertrophying chondrocytes; *PS*, primary spongiosa of the metaphysis. Calcified cartilage cores in the newly formed trabeculae are labeled *cc*. (from Jee, 1988.)

4. **Zone of provisional calcification:** In this zone the degenerating chondrocytes continue to hypertrophy. As they reach the limit of this zone, where the growth plate ends and the metaphysis begins, the chondrocytes die. Simultaneously, the surrounding cartilage matrix is calcified.

The calcification mechanism is imperfectly understood, but a working hypothesis for this process is as follows. The hypertrophying chrondro-cytes synthesize type X collagen and a protein called *chondrocalcin*, which may help initiate calcification. Proteoglycans may also be disassembled by enzymes released by the degenerating cells. Mineral crystal formation commences within *matrix vesicles*, which are intact spherical membranes extruded from the hypertrophic chondrocytes. These vesicles contain the enzyme ATPase, which provides energy to transport calcium ions into the vesicle against a concentration gradient. In addition, other enzymes in the vesicles cleave phosphate and calcium from compounds to which they are bound, allowing them to form crystals. The initiation of apatite crystal growth is the most important function of the matrix vesicle. As the levels of calcium and phosphate increase, amorphous calcium–phosphate aggregates form, eventually spreading beyond the vesicle wall and accumulating more mineral by *epitaxy* (crystal growth that imitates the form of the substrate). At this point, apatite crystals begin to grow in such a fashion that they are integrated with the collagen molecules in the matrix, and water is displaced as the cartilage matrix mineralizes.

Development of Metaphyseal Trabeculae

The columns of enlarged lacunae left by the dead chondrocytes form the basis for tunnels, allowing blood vessels and cells from the metaphysis to gain access to the calcified cartilage at the bottom of the growth plate. As each chondrocyte hypertrophies, it dissolves some of the cartilage around it, so that only narrow walls, or *septa*, are left between the enlarged lacunae. Cells called *septoclasts*, another monocyte descendant, but perhaps not identical to osteoclasts, resorb the *septa* at the bottom of each chondrocyte column (Lee et al., 1995). Blood vessels from the metaphysis then invade these tunnel-like spaces, which are separated by columns or "trabeculae" of calcified cartilage. Other osteoclast-like cells (called *chondroclasts* by some authors) resorb some of the calcified cartilage on the walls of the tunnels. Osteoblasts then lay down small amounts of bone on these calcified cartilage trabeculae. This remodeling activity begins to convert calcified cartilage to bone and marks the transition from growth plate to metaphyseal "bone." At this early stage in the formation of spongy bone the tissue is called *primary spongiosa*; it still contains considerable calcified cartilage within the trabecular cores. As time goes by, and remodeling continues to replace the cartilaginous portions of the trabeculae with the metaphyseal bone, the trabecular bone becomes the *secondary spongiosa*. Eventually, the trabeculae become entirely bone. Those trabeculae in the center of the metaphysis are eventually entirely resorbed to form the medullary canal, with an arch of trabecular struts remaining to transmit loads from the central portion of the growth plate out to the cortices of the diaphysis.

Growth of the Physis

In the newborn, the physis is typically a flat, more or less circular plate. As the child grows and the bones become larger, the physis must increase in diameter. This increase is achieved by cell division at its circumference in a region called the *zone of Ranvier*. In addition, the physis loses its flat shape and becomes curved, often with a complex system of ridges, valleys, and lappets. The physis, being composed of cartilage, is weak in shear and tension, and physeal injuries in children are relatively common; the epiphysis of the proximal femur sliding off the shaft at the physis (*slipped capital femoral epiphysis*) is a well-known example. One consequence of the development of an irregularly shaped contour to the physis in the growing child is that the complex interlocking geometry of the epiphysis and metaphysis confers some protection against failure of the physis in response to shearing forces.

Most long bones have two physes, one at each end (although some bones, such as the metacarpals and phalanges, have only one functional physis). Usually one of these contributes more than the other to longitudinal bone growth. For example, in the femur the distal physis contributes more to growth, whereas in the tibia the proximal physis is more active. Historically, the relative growth rates of the two physes of a long bone have been measured by comparing bones of different age with respect to some naturally occurring marker in the diaphysis, usually the entrance point of the nutrient artery.

Closure of the Physes

As a child matures, bones reach their adult length and the physis is no longer required. At this point, the physes "close" by ossification, connecting the epiphysis to the metaphysis with bone. The blood circulatory systems of the epiphysis and metaphysis, formerly independent, also unite. This region of bone remodels over many years, but the physeal "scar" or "ghost" (a plate of bone at the site of the former physis) may persist into old age.

Although we generally consider the role of the physis to be longitudinal growth of bones, its ability to *stop* longitudinal growth by closure is also important. In animals that lack an ossified epiphysis (many reptiles, e.g., crocodiles), there is no physeal closure and the bones continue to increase in length indefinitely, so that great longevity becomes associated with great size.

Closure of the various physes occurs in a specific sequence, and by examining a radiograph (usually of the left hand and wrist) one can determine the *skeletal age* of an individual by observing which physes have closed. Skeletal age is not the same as chronological age; the growth plates close earlier in some individuals than others. In girls, physes close several years earlier than in boys; this contributes to the shorter average stature of women compared to men.

2.8 Modeling and Remodeling of Bone

We have just described the way in which most bones grow in length. This kind of bone formation, in which cartilage is formed first, calcified, and then replaced by bone, is known as *endochondral ossification*. Another mode of bone formation, in which osteoblasts make bone directly, is called *intramembranous ossification*. So-called "flat bones" (skull, scapulae, pelvis) are formed in this way. The word "intramembraneous" refers to the fact that in the embryo, this kind of ossification occurs adjacent to membrane-like layers of mesenchymal cells that differentiate into osteoblasts. The same sort of direct formation of bone by osteoblasts occurs in periosteal modeling, where it is still called "intramembraneous bone formation" because the osteoblasts and their precursor cells are found in the deep layer of the periosteal membrane. If the osteoblasts are not part of a membrane (i.e., if they are on an endosteal, trabecular, or Haversian canal surface), the bone formation is said to be *appositional* rather than intramembraneous.

Modeling Vs. Remodeling

As the long bones develop, their shafts must grow in diameter as well as length. We have seen how growth in length throughout childhood is accomplished by continuation of endochondral ossification in the growth plate. Growth in diameter is accomplished by periosteal intramembranous ossification. However, as growth occurs, it is not enough just to add material to make the bone organs longer and larger. The bones must also be shaped in various ways. That means that bone must be **removed in some places while it is added in others**. This sculpting, involving osteoclastic activity (bone resorption) in some places and osteoblastic activity (bone formation) in others, has become known as *modeling*. In addition, fatigue damage must be repaired inside bones and their internal architecture must be adjusted to varying load conditions, both while the skeleton is developing and throughout the remainder of one's life. This repair involves removal and replacement of bone in particular places by the coupled actions of osteoclasts and osteoblasts working at the same site. This replacement process has become known as *remodeling*.[3]

Thus, modeling and remodeling refer to the actions of osteoblasts and osteoclasts in reshaping and replacing portions of the skeleton. They are distinguished from one another in several ways:

1. Modeling involves independent actions of osteoclasts and osteoblasts. Remodeling involves sequential, "coupled" actions by the two types of cells.

---

[3]These terms were coined by Frost. Be aware, however, that many orthopaedists and bone scientists refer to both modeling and remodeling activities, as defined here, as "remodeling."

2. Modeling results in change of the bone's size, shape, or both; remodeling does not usually affect size and shape.
3. The rate of modeling is greatly reduced after skeletal maturity. Remodeling occurs throughout life, although it too is substantially reduced after growth stops.
4. Modeling at a particular site is continuous and prolonged, whereas remodeling is episodic, with each episode having a definite beginning and ending.

Modeling

Modeling is necessary because the longitudinal growth process does not produce a bone of correct geometry, and because each growing child loads his or her skeleton in a somewhat different way, requiring each skeleton to be "customized." Modeling may be resorptive or formative. Examples of modeling include the following.

Metaphyseal modeling to reduce bone diameter: As some long bones grow longer, the diameter of the cylinder of bone created beneath the growth plate must be reduced to create the diaphysis. This change is accomplished by osteoclasts working continuously on the periosteal surface of the metaphysis to cut the shaft down to size (Fig. 2.19, top). This action is most important in bones with a widely flaring metaphysis such as the proximal tibia; in other bones, diametric growth of the physis is coordinated with longitudinal growth in such a way that removal of bone in the lower metaphysis is unnecessary (e.g., metatarsals).

Diaphyseal modeling to increase bone diameter or alter curvature: As a child continues to grow, long bone shafts become larger in diameter, accomplished by slow, continuous addition of bone to the periosteal surface by osteoblasts. Simultaneously, bone is removed from the endosteal surface (Fig. 2.19, middle). In addition, the curvature of long bones is usually adjusted during growth. This is accomplished by increased formation and resorption on the sides of the bone such that the cross section "drifts" sideways relative to the ends of the bone (see Fig. 2.19, bottom). This phenomenon may increase or decrease bone curvature as required in specific bones. It also serves to correct the curvature resulting from a poorly reduced[4] fracture.

Modeling of a flat bone: As a child grows, the bones of the skull must increase in size to accommodate the increasing size of the brain. This enlargement cannot be accomplished simply by addition of bone at the suture lines between the cranial bones because the plate's radius of cur-

---

[4]Physicians say that a fractured bone is "reduced" when the fracture surfaces are placed against one another such that the bone fragments are returned to their original positions and orientations.

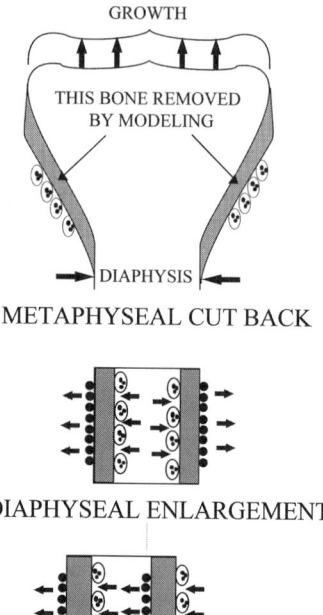

GROWTH

THIS BONE REMOVED
BY MODELING

DIAPHYSIS

METAPHYSEAL CUT BACK

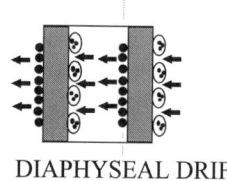

DIAPHYSEAL ENLARGEMENT

DIAPHYSEAL DRIFT

FIGURE 2.19. *Top:* Resorptive modeling beneath the growth plate to form the diaphysis from the metaphysis. *Middle:* Formative (periosteal surface) and resorptive (endosteal surface) modeling to enlarge the diaphysis. *Bottom:* Modeling to "drift" the diaphysis to the left (thereby altering diaphyseal curvature).

vature must also be adjusted. Modeling of these bones (resorption on the inner surface, formation on the outer surface) is required for proportionally correct increases in cranial size. Simultaneously, modeling transforms the facial features of a child into those of an adult.

Remodeling

Remodeling removes a portion of older bone and replaces it with newly formed bone. This process repairs microscopic damage and prevents accumulation of fatigue damage that could lead to fatigue fracture. It has been hypothesized that remodeling also mechanically "fine tunes" the skeleton to increase its mechanical efficiency.

ARF and BMUs

Remodeling is accomplished by teams of osteoclasts and osteoblasts that work together in basic multicellular units, or BMUs. A BMU consists of about 10 osteoclasts and several hundred osteoblasts. There are three principal stages in a BMU's lifetime: *activation, resorption,* and *formation* (ARF). Activation occurs when a chemical or mechanical signal causes osteoclasts to form (by fusion of monocytes) and begin to remove bone somewhere on or in the skeleton. The osteoclasts resorb a volume of bone in the form of a ditch (on bone surfaces) or tunnel (in compact bone) about 200 μm in diameter that progresses along the surface or through the cortex, moving at about 40 μm/day. After the osteoclasts have passed a particular point, osteoblasts are differentiated from mesenchymal cells over a period of several days and begin to replace the resorbed tissue. Formation is much slower than resorption. The resorption period is about 3 weeks in humans; the formation or refilling period is about 3 months. The total remodeling period is about 4 months.

Osteonal Remodeling

When a BMU tunnels through compact bone, it creates a secondary osteon. Figure 2.20 shows the leading portion of an osteonal BMU in longitudinal section, tunneling to the right. A few osteoclasts may be seen on the resorbing surface. At left, osteoblasts and osteoid line the refilling surfaces. The "head" or front end of a BMU contains a capillary "bud" to supply nutrients, and probably to supply the progenitor cells for osteoclasts and osteoblasts. Because osteonal BMUs become isolated deep within the cortex, this vascular supply must be maintained. Therefore, the tunnel cannot be entirely refilled, and each BMU leaves a new Haversian canal in the bone. Within each Haversian canal are two capillaries: a "supply" and a

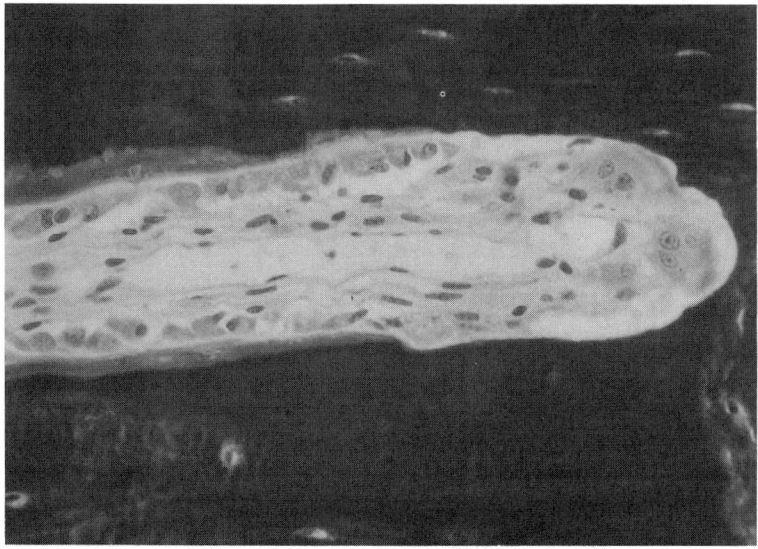

FIGURE 2.20. Photomicrograph of an osteonal basic multicellular unit (BMU). Two multinuclear osteoclasts are visible at *right,* tunneling through the bone; osteoblasts are on bone surfaces to left. Field width, ~1 mm. (Courtesy of Dr. Jenifer Jowsey.)

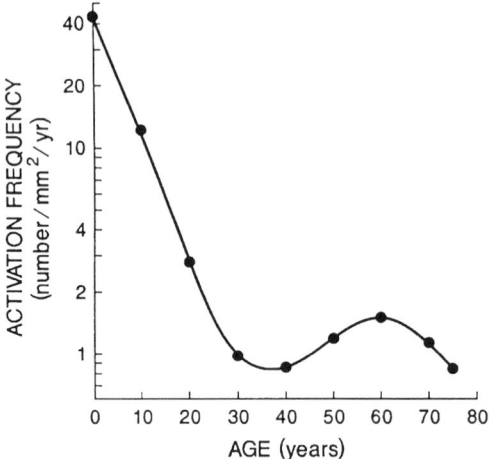

FIGURE 2.21. BMU activation rate vs. age for human ribs. (From data by Frost, 1964b.)

"return" vessel. These vessels connect with the vasculature in the medullary canal or on the periosteal surface.

In human adults, osteonal BMUs replace about 5% of compact bone each year. Figure 2.21 shows how the rate of remodeling changes with age in the human rib. It is very high in children, is reduced in young adults, then rises and falls again in older individuals. (The gender of the individuals in the graph was not reported, but the rise after age 50 probably results from the effects of menopause in women; estrogen inhibits bone remodeling.)

Trabecular Remodeling
Trabeculae are remodeled similarly, except that the BMUs work on their surfaces, digging and refilling trenches. Imagine the lower half of the BMU in Fig. 2.20 moving over the surfaces of the trabeculae seen in Fig. 2.2. (Because a BMU is about the same diameter as a trabecula, it cannot tunnel longitudinally within a trabecular strut without constantly breaking out along its sides.) BMUs replace human adult trabecular bone at a relatively high rate, about 25% each year. In fact, the rate of bone turnover varies widely throughout the skeleton. Table 2.3 demonstrates this using data from various skeletal sites in young adult dogs. It can also be seen that the remodeling rates in dogs are much higher than in humans.

Endosteal and Periosteal Remodeling
In principle, remodeling can also occur on endosteal and periosteal surfaces. Endosteal and periosteal remodeling could be responsible for the observation that long bones expand radially with age in adults as well as children. This expansion would be accomplished by arranging for forma-

TABLE 2.3. Cancellous bone turnover rates in young adult dogs

Site	Turnover rate (%/yr)
Lumbar vertebrae	205
Thoracic vertebrae	167
Cervical vertebrae	121
Mandible	105
Skull	60
Calcaneus	120
Proximal humerus	174
Proximal femur	138
Proximal radius	127
Proximal tibia	112
Carpals	31

After Jee et al., 1991.

tion to exceed resorption on the periosteal surface, and vice versa on the medullary canal surface. However, these effects could be produced more efficiently by modeling, and the issue is unresolved for lack of data.

Bone Structural Units Are Produced by Remodeling
The packet of new bone produced by a BMU is called a *bone structural unit*, whether the remodeling occurs in cortical or trabecular bone or on an endosteal or periosteal surface. Thus, a secondary osteon in cortical bone is a bone structural unit. There is no common name for trabecular bone structural units, although they are sometimes called trabecular osteons or bone packets.

Remodeling and Cartilage
Surfaces covered by cartilage that has not calcified cannot be resorbed by osteoclasts, and cannot serve as a working surface for osteoblasts, so they cannot be modeled or remodeled. The same is considered to be true of surfaces covered by osteoid.

Cellular Events in Modeling and Remodeling

Modeling and remodeling involve some combination of four basic cellular events: migration, mitosis, differentiation, and expression of the mature phenotype (e.g., matrix synthesis by osteoblasts or matrix resorption by osteoclasts). Some of these events are governed by hormones or other chemicals originating outside the skeleton (e.g., parathyroid hormone [PTH] or estrogen). At a local level, most of these complex changes are controlled by proteins called *cytokines* and others called *growth factors*, which are produced by cells present in and around bone surfaces. For example, cell migration may be directed by *chemotaxis*, in which a cell moves itself up a chemical concentration gradient. Bone matrix itself is known to contain more than a dozen cytokines. Other cytokines affecting bone remodeling are synthesized by bone marrow cells.

Skeletal Envelopes and Senile Bone Loss

There are four *skeletal envelopes*, surfaces on which modeling and remodeling occur: *trabecular, endosteal, periosteal*, and *Haversian*. The first two are bathed in marrow, and remodeling that occurs on them is usually characterized by an excess of resorption relative to formation. This endotrabecular deficit of bone replacement during remodeling is the reason for normal, age-related bone loss that occurs in men and women after age 30–35. The periosteal envelope experiences formative modeling or remodeling in which resorption generally removes less bone than formation produces. Such processes partially compensate for involutional (i.e., senile) bone loss. The osteonal, or Haversian, envelope also contributes to involutional bone loss. A Haversian canal must be created for each new osteon, increasing the porosity of compact bone until a level of porosity is attained at which each new BMU removes an existing Haversian canal before creating a new one.

The reason why a deficit in formation occurs during remodeling on surfaces adjacent to marrow is unknown. There is, however, an association between a remodeling deficit and the amount of fat in the marrow (Minaire et al., 1984; Martin et al., 1990).

Regional Acceleratory Phenomenon

A sudden transient increase in the rate of remodeling sometimes occurs in one or more bones. This is called a *regional acceleratory phenomenon*, or RAP. The cause is usually trauma to the bone. When this happens, a transient osteoporosis occurs because there is a lag between resorption and formation in each BMU. The osteoporosis will resolve itself over a remodeling period (about 4 months) as the additional BMUs complete their refilling.

2.9 Fracture Healing

Fracture is the most obvious sign that the mechanical functional capacity of the skeleton has been exceeded. It may be argued that much of skeletal physiology is directed at preventing fracture. Even so, fractures are relatively common, and usually occur because of trauma: an accidental overload substantially exceeds the normal range of loading to which the bone has adapted during its growth and development.

Pathologic fractures are caused by normal loading of a bone weakened by disease. The most common pathologic fractures are caused by osteoporosis in the elderly. Osteoporotic fractures are more common in women than men. Bone tumors are another important cause of pathologic fractures. The tissue in a bone tumor is variable in its mechanical properties, but it will almost always be different than the surrounding bone and therefore serve as a stress concentrator. Removal of a bone tumor obviously also places the patient at risk for fracture. Predicting and preventing pathologic fractures is an important issue in orthopaedic medicine.

Once a fracture occurs, how does it heal? Today, people seek medical attention for the simplest of fractures, but the basic process of fracture healing obviously evolved as a natural phenomenon. Healed fractures are commonly seen in archaeological skeletons of prehistoric people, and in wild animals. Therefore, fracture healing must be a self-controlling process that ordinarily proceeds on its own to reunite the broken bone. The remainder of this chapter is devoted to describing this process.

Basic Concepts

As we have seen, in the embryo and growing child bone is formed directly by osteoblasts (*intramembranous bone formation*) and indirectly by the calcification of cartilage and replacement by bone (*endochondral ossification*). In the healthy adult, the latter process is absent, and bone is formed only by osteoblasts on existing bone surfaces. When a fracture occurs, however, adults as well as children are able to make bone by both intramembranous and endochondral osteogenesis, leading to much more rapid repair of the damaged bone. Unlike skin and other soft tissues, wound healing in bone does not produce a permanent scar. Eventually, the initial repair tissues are replaced by normal lamellar bone. Therefore, while the woven bone and calcified cartilage in the healing fracture can be considered "scar tissue," they are temporary, and the tissues in the healed bone are ultimately indistinguishable from those in the original bone.

A fracture that fails to heal is called a *non-union*. Frequently, non-unions produce a jointlike cartilaginous structure called a *pseudoarthrosis* (literally, 'false joint'); these pathologic joints are lubricated with synovial fluid. A fracture that does not heal in the expected length of time, but which is still progressing, is called a *delayed union*. The length of time required to heal a particular fracture depends primarily on its anatomical site. Ribs, clavicles, and other bones of the torso heal in a few weeks; appendicular bones generally require several months (Table 2.4).

TABLE 2.4. Typical healing times for common fractures

Bone	Typical healing time, weeks
Distal radius	6
Humeral shaft	12
Tibial shaft	18
Femoral neck	24

Important Tissues in Fracture Healing

Two tissues provide the cells responsible for the healing of a typical fracture: the periosteum and the marrow of the injured bone. The periosteum has two layers: a fibrous outer layer, and a highly cellular *cambium* or inner layer. Ordinarily, the osteoblasts of the adult periosteum produce lamellar bone at a relatively slow rate, so that long bones expand very slowly with age. However, even a slight amount of trauma stimulates the periosteum to produce more osteoblasts, which manufacture large amounts of woven bone in a short amount of time. In the case of a fracture, this periosteal woven bone takes the form of trabeculae, which arch over the fracture line and fuse to bridge the gap and make what is known as the *periosteal callus* (Fig. 2.22).

The endosteal surfaces of bones are not actually covered by a membrane. Instead, they are lined with bone lining cells; this is true whether or not the endosteal space contains trabecular bone. In places, the endosteal surface may be experiencing ordinary remodeling, but most of the adult endosteal surface is quiescent. The bone lining cells may be stimulated to

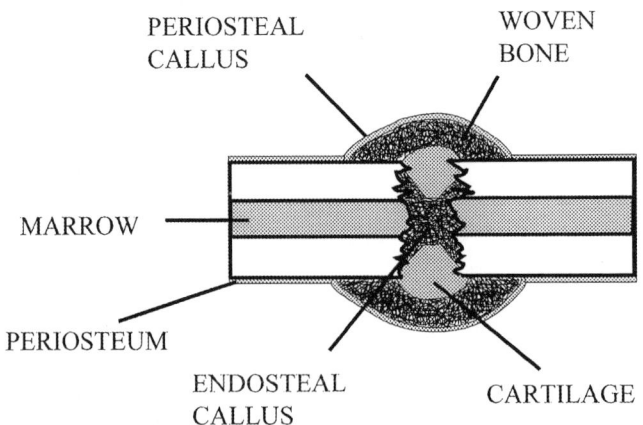

FIGURE 2.22. Diagram of a healing fracture.

initiate bone remodeling by various factors, including mechanical ones, but such remodeling does not play a role in the initial fracture healing response. Instead, trauma to the **marrow** stimulates the formation of the medullary version of the periosteal callus. It has been shown that, even without fracture of the cortex, mechanical disruption of the marrow activates the differentiation of osteoblasts and the formation of woven bone within the medullary canal (Amsel et al., 1969). (This bone formation is a prerequisite to the regeneration of the hematopoietic function of the marrow, and so serves a dual role when fracture occurs.) The periosteal callus tends to be more obvious, but the *medullary callus* is usually considered more important in achieving **initial** union.

Obviously, the jagged ends of the broken bone, as well as other trauma to the fracture site, can be expected to break many blood vessels. The resulting hematoma was once thought to be the primary source of the cells responsible for fracture healing. It is now thought that most of these cells come from the marrow and periosteum.

Three Biological Phases

Healing of a fracture may be divided into inflammatory, reparative, and remodeling phases which, taken together, form a self-driven sequence that leads to a healed fracture.

Inflammatory Phase

This initial phase serves to immobilize the fractured bone and activate the cells responsible for repair. Immobilization is promoted by pain and by swelling, which in many cases serves to hydrostatically splint the fracture. In addition to hematoma formation, this phase is characterized by vasodilation, serum exudation, and infiltration by inflammatory cells. It lasts 3–7 days.

It is important to realize that fracture repair is not accomplished by existing osteoblasts and chondroblasts[5] but depends on the creation of an entirely new work force of cells from a relatively small population of stem cells. The mobilization of this work force requires a variety of chemical mediators and ancillary cell types that are poorly understood. However, it is clear that this process is similar to other biologic cascades. Once set in motion, each step is predicated on what happened earlier, and if a mistake is made, the process may make a wrong turn rather than stopping, with no way to recover the conditions at the point of the error. It follows that many if not most non-unions or delayed unions are caused by a defect in the initial mobilization of the repair process, and once the problem becomes apparent, there is no way to correct it by manipulating the current group of cells. This is consistent with the fact that the most effective treatment for non-unions is to replace the tissues in the

---

[5]Chondroblasts are chondrocytes capable of producing cartilage at a rapid rate.

fracture site with a bone graft, which essentially initiates an entirely new repair process. Alternatively, electrical signals (see following) may be able to start a new cascade.

Interruption of the normal vascular supply to the bone by the fracture also results in death of the osteocytes in the bone matrix. The significance of this is not clear, but it may initiate the remodeling activity that is important in the last stage of fracture healing.

Reparative Phase

In this phase of healing, which lasts about 1 month, the periosteal and medullary calluses are formed by osteoblasts from the periosteum and marrow. The stem cells for medullary osteoblasts are thought to be pluripotential mesenchymal cells in the marrow stroma. These cells are also stem cells for fibroblasts, chondroblasts, and some marrow cells. Mesenchymal cell proliferation and differentiation are accompanied by intense vascular proliferation. The resulting osteoblasts produce woven bone at a high rate.

The fracture site exhibits some degree of endochondral ossification as well as direct bone formation by osteoblasts. This is thought to be related to inadequate blood supply in some portions of the fracture region (see Carter et al., 1988, for citations and discussion). In the absence of adequate vascularization, stem cells are thought to differentiate into chondroblasts rather than osteoblasts. An important contributing factor to inadequate vascularization can be insufficient immobilization of the fracture fragments. Alternatively, the nature of the stresses at a site within the callus may affect the differentiation pathway (Carter et al., 1988). In any event, the chondrogenic regions of the callus, which tend to be near the fracture line, usually calcify and are replaced by lamellar bone, just as the woven bone regions are. If this does not happen, a pseudoarthrosis may develop.

Regardless of how much of the callus is cartilaginous, it takes a week or two for the woven bone or cartilage to accumulate substantial mineral. This early, rather compliant callus is known as the *provisional callus*. As calcification proceeds, the callus material becomes increasingly rigid; it is then referred to as the *bony callus*. The rigidity of the callus is enhanced by the fact that it has a larger cross-sectional area (and moment of inertia, see Section 4.2) than the original bone cortex. This means that even though the **material** in the bony callus is less rigid (and strong) than cortical bone, the rigidity (and strength) of the callus **structure** can equal or exceed that of the intact bone. When this occurs, *bony union* has been achieved, completing the reparative phase.

Remodeling Phase

Once bony union has been achieved, the broken bone is approximately as strong as the intact bone was, but often it has greater mass than the original bone, and is therefore less mechanically efficient. Modeling and remodeling of the fracture site gradually restore the original contour and internal structure of the bone, although radiographic evidence of the fracture may persist for many years. As the bone remodels, it is able to maintain its strength with less material than in the earlier phases of healing, thereby increasing its mechanical efficiency. The medullary and periosteal calluses are removed, and the remaining material (woven bone or calcified cartilage) is replaced by secondary lamellar bone (cortical or trabecular, as the site dictates). If the fracture has healed with incorrect angulation, this may be partially corrected as well by modeling drifts. In general, modeling and remodeling proceed much more rapidly in children than in adults, and this applies to the remodeling phase of fracture healing as well.

In considering the capacity of a malaligned healed fracture to correct itself, there are two important geometric principles to remember. First, an angulation in the plane of motion of the adjacent joint corrects itself better than one out of the joint plane. Second, rotational errors do not correct themselves very well.

Box 2.2 Technical Note
Clinical Assessment of Healing Fractures

It is very important for the physician treating a fracture patient to be able to judge the strength and stiffness of the healing bone. If the cast is removed before the fracture is mechanically competent to bear the loads and moments applied to it, there is risk of refracture or malunion. On the other hand, if the cast is left on too long, there is risk of excessive disuse osteoporosis and contracture of soft tissues about the adjacent joints, not to mention the inconvenience and expense of needless immobilization. The usual clinical methods to assess fracture healing are inspection of radiographs and manipulation to see how stiff the healing fracture feels; both of these are subjective and imprecise. Several groups of investigators have experimented with methods to assess fracture healing in a more quantitative way.

If the fracture has been fixed with an external fixation frame, application of a strain gauge to the fixation frame can allow bone stiffness to be estimated (Burny et al., 1984; Cunningham et al., 1990). A second method for measuring the mechanical characteristics of a healing fracture involves the propagation of sound or stress waves through the tissues. The velocity of sound in an elastic bar is $c = (E/\rho)^{1/2}$ where E is the elastic modulus and ρ is the density of the tissue. The resonant frequency of a vibration in an elastic bar depends on the ratio of c to the length of the bar. When a vibration traveling along the bar reaches a segment having different material properties, a portion of the wave is reflected backward. The magnitudes of the transmitted and reflected waves depend on the ratio of the acoustic impedance, $z = c\,\rho$, of the two materials. These dependencies on the elastic modulus of the tissues are potential means for probing the stiffness of the fracture callus. For example, the velocity of an ultrasonic wave across the fracture site can be measured (Gerlanc et al., 1975). As the fracture heals, the wave velocity increases; healing can be assessed by comparison with the contralateral, uninjured bone (see figure). Alternatively, the resonant frequency can be determined by tapping the fractured bone with a hammer (Collier and Donarski, 1987), or vibrating it over a range of frequencies with a shaker. Another approach is to measure the attenuation of the impulse as it crosses

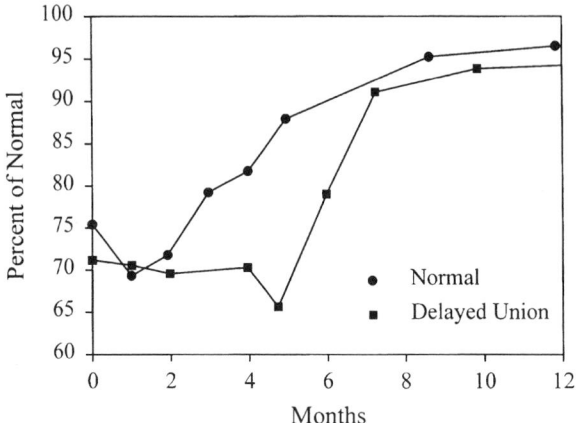

Sound velocity, normalized by the values in the contralateral bone, measured through the tissues in a healing fracture is plotted against healing time. Examples shown represent human tibias with normal (*circles*) and delayed (*squares*) healing. (Replotted from data by Gerlanc et al., 1975.)

the fracture site. These methods are most easily applied to the ulna and tibia, which are subcutaneous along at least one margin. Bones that are fully enveloped in muscle (e.g., the femur) are too inaccessible for these vibration methods.

Although each of these methods is capable of probing the mechanical attributes of a healing fracture, in practice they are too cumbersome and imprecise to have come into general clinical use and their applications have been limited to research. A fast, noninvasive, and easy method to measure bone stiffness in vivo remains to be invented.

Four Biomechanical Stages

A standard way to study fracture healing in experimental animals is to mechanically test the healing bone in torsion and compare the result with the stiffness of an intact bone. When this was done with rabbit tibial fractures (White et al., 1977), four stages of healing were identified (Fig. 2.23).

Stage I: Virtually no stiffness is seen, and failure occurs through the original fracture line. This stage corresponds to the inflammatory and early reparative (i.e., provisional callus) phases of healing. (Days 1–26 in the rabbit tibia.)

Stage II: Substantial stiffness is now encountered, but failure still occurs through the original fracture line. This stage corresponds to the middle of the reparative phase when the callus has become bony but still is not so large or calcified as it will become. (Days 27–49 in the rabbit tibia.)

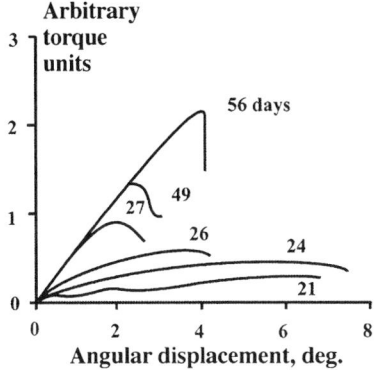

FIGURE 2.23. Torsional stiffness of healing rabbit fractures as a function of days of healing. (After White et al., 1977.)

Stage III: There is no further increase in stiffness, but failure is now partially through the original fracture line and partially through intact bone. (Days 49–56 in the rabbit tibia.)

Stage IV: Failure occurs through the intact bone rather than the callus at the fracture site. (After day 56 in the rabbit tibia.)

Stages III and IV correspond to the later portion of the reparative phase and the early remodeling phase, when the full bony callus exists but its mineral content and strength are still increasing. Another interesting observation from these mechanical studies of healing fractures is that there is typically a slight reduction in the strength of the fracture callus during stage IV of healing. It is reasonable to think that this corresponds to the beginning of the remodeling phase, when the resorptive portion of BMU activity has begun but substantial refilling of the osteonal cavities has yet to occur.

Stability of Fixation
The importance of achieving good stability or immobilization of a fracture depends very much on the anatomical site. For example, ribs heal well even with a great deal of motion—rib non-unions are unknown. On the other hand, tibias and other large bones are very prone to non-union as the result of excessive motion of the fragments. This knowledge must be balanced against the fact that **some** motion is beneficial: there seems to be an inverse relationship between the amount of motion and callus size, up to a point that has not been defined. In the same vein, applying loads across the fracture (i.e., weight-bearing) serves to strengthen the callus, but obviously too much weight can lead to refracture or excessive motion and non-union. There are no clear rules about how much weight-bearing or motion are safe.

Box 2.3 Historical Note
The Rapid Loading Torsion Tester

The torque-deflection curves used to illustrate the four biomechanical stages of fracture healing in Fig. 2.23 were obtained using a device known as the Rapid Loading Torsion Tester. An engineer named Albert Burstein and an orthopaedic surgeon named Victor Frankel developed this apparatus at Case Western Reserve University in the late 1960s. The classic textbook, *Orthopaedic Biomechanics,* that was written by these two investigators was published at about the same time. This book taught a generation of orthopaedic surgeons the application of engineering principles to musculoskeletal function and helped to establish orthopaedic biomechanics as an important discipline.

Burstein and Frankel set out to devise a standard testing apparatus to measure the mechanical strength of whole bone. Knowing that bone strength is dependent on loading rate (a characteristic of viscoelastic materials), the investigators desired an apparatus that could apply a reproducible and rapid loading rate, representative of typical trauma, to a variety of long bones. Other important design criteria included the ability to produce fractures similar to those seen in patients and the ability to subject the entire bone to equally severe loading conditions so as to identify weak points along its length. The device also needed to be inexpensive and easy to use.

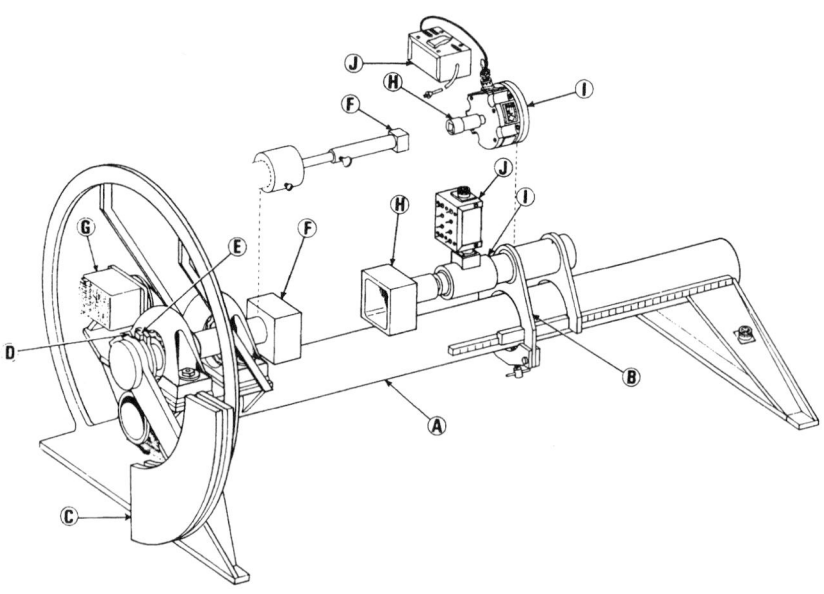

The torsion testing machine: *A,* frame; *B,* tail stock; *C,* pendulum; *D* and *E,* dog clutches on pendulum and rotating shaft, respectively; *F,* rotating grips; *G,* angular deformation transducer; *H,* stationary grips; *I,* torque transducer; *J,* self-calibration system. *Upper drawing* shows a second subsystem used for smaller bones. (Reprinted from *J Biomechanics,* Vol. 4, Burstein, AH, Frankl, VH, A standard test for laboratory animal bone, 155–158, 1971, with permission from Elsevier Science Ltd., The Boulevard, Langford Lane, Kidlington 0X5 1GB, UK.)

Torsional loading best satisfied these requirements. The investigators chose a pendulum system to generate the torsional loads because it could be easily designed to produce a loading rate similar to that estimated for common traumatic fractures (see figure). The pendulum traverses the bottom 30° of its swing in 75 milliseconds (ms), similar to the load duration in fractures caused by skiing or motor vehicle accidents. The pendulum has a mass of 14 kg and a length of 23 cm. The pendulum's energy greatly exceeds the energy required to fracture most bones, so that minimal changes in pendulum velocity or loading rate occur during the fracture event (less than 2% in the case of rabbit bones). Measurement instrumentation consists of a reaction torque sensor, a precision angular position transducer, and an oscilloscope to capture and display the voltage outputs from the two transducers.

To conduct a test, the bone is rigidly mounted in alignment with the rotational axis of the device. The pendulum is dropped from a set height and at the bottom of its swing engages a cam on the load shaft; this imparts an angular deformation to the specimen that is measured by the angular position sensor. The resistive torque in the specimen is simultaneously measured by the torque transducer at the opposite end of the bone. A torque-vs.-angular deformation curve is then constructed, which represents the whole bone's failure characteristics under typical traumatic loading conditions. Common variables derived from the curve include the failure torque and deformation and the energy absorption to failure. Because bones are not perfectly straight cylinders, the derivation of material properties from these data is not usually attempted.

The advent of more advanced material testing machines has reduced the popularity of the Burstein–Frankel apparatus, but, thanks to its superb design, it remains a very legitimate means of testing the structural integrity of whole bones.

Primary Union

If a fracture is rigidly fixed so that the bone ends are closely approximated, healing may occur by osteonal remodeling across the fracture line (Fig. 2.24). This is known as *primary fracture healing*. (Therefore, *secondary healing* is ordinary fracture healing as previously described.) Usually the only way that primary healing can happen is when a plate or screw is used very effectively to internally "fix" (rigidly fasten together the fragments of) the fracture. In these cases, there is almost no periosteal callus.

Posttraumatic Osteoporosis

It is commonly observed clinically that the intact portions of the fractured bone become osteoporotic as healing occurs. This generalized osteopenia of the intact regions, called *posttraumatic osteoporosis* or *posttraumatic bone atrophy*, is caused by two factors. First, in addition to the healing response, the fracture causes a remodeling RAP throughout the bone, so that osteonal BMUs riddle the entire cortex with resorption cavities. The second factor is the removal of mechanical loading from the fractured bone. As is seen later, disuse also activates remodeling. If the fracture is well fixed and sufficient load-bearing is resumed, the resorption spaces will refill and the osteopenia will be transient. To the degree that the bone does not resume load-bearing, the atrophy persists.

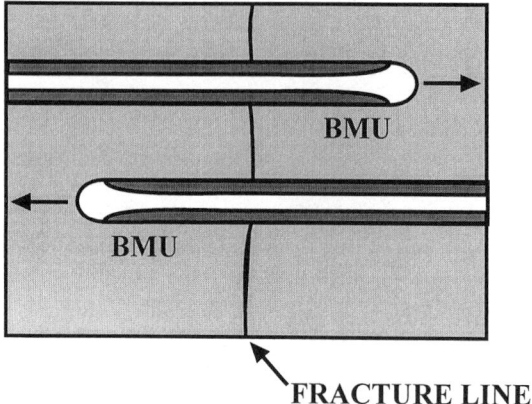

FIGURE 2.24. Sketch illustrating how osteonal BMUs may tunnel across a tightly fixed fracture line or osteotomy (saw cut) interface in cortical bone. This is known as primary healing.

Electrical Phenomena and Fracture Healing

When bones are mechanically deformed, they become electrically polarized via two mechanisms: (1) the ionic fluids in the calcified matrix are forced to move about, creating *streaming potentials* (Eriksson, 1974); and (2) the collagen molecules in the bone tissue are *piezoelectric*, so that they exhibit a dipole moment when strained (Fukada and Yasuda, 1957). The relative roles of these two phenomena in producing *stress-generated potentials* in bone is unclear, but it is well established, for example, that when a bone is bent, the concave surface becomes negatively charged and the convex surface becomes positively charged. Furthermore, when a bone is fractured, the fracture site becomes negatively charged with respect to the remainder of the bone. Experiments have shown that bone formation is enhanced in the vicinity of a cathode, and the optimal current for this effect is about 20 microamperes (μA). For a review of electrical phenomena in bone, see Chapter 4 in Martin and Burr (1989).

Since the 1970s, electrical stimulation has been used as a treatment for non-unions. There are two basic methods of producing such currents. Brighton and Pollack (1985) placed thin wires into the fracture site to carry direct current, with an anode pad on the skin nearby. Bassett and co-worker (1981) passed currents through external coils to induce currents in the tissues noninvasively. Both methods seem to be about 70%–85% successful. They require good immobilization, however, and they do not work in cases of true synovial pseudoarthrosis.

2.10 Summary and Further Reading

This chapter has introduced bone's tissue structure, composition, and cells. We have also described the structure and biology of cartilage, both in regards to articular surfaces and to the growth plate. The physiology of modeling and remodeling were then described. Finally, we described the processes by which broken bones heal. Together with Chapter 1, this chapter lays the foundation for the rest of the book: we now know why the skeleton must support high loads (because of the limitations inherent in muscles) and have a picture of the basic structure and biology of the tissues in the skeleton. In Chapters 3 through 7 we explore more quantitatively the interactions between biology, structure, and function in bones and joints.

Several excellent works can be suggested for further reading. These include *Principles of Bone Biology* (Bilezikian et al., 1996), *The Scientific Basis of Orthopaedics* (Albright and Brand, 1987), *Orthopaedic Basic Science* (Simon, 1994), *Intermediary Organization of the Skeleton* (Frost, 1986), *Bone Formation and Repair* (Brighton et al., 1994), and *Osteoporosis* (Marcus et al., 1996). For more in-depth reading, one may turn to B. K. Hall's series of books titled *Bone*, and an older series by G. F. Bourne titled *The Biochemistry and Physiology of Bone.*

2.11 Exercises

Unlike the exercises at the end of Chapter 1, these are not intended to give you practice in solving problems similar to those illustrated in the text. Instead, they are meant to start you thinking quantitatively about the biological material introduced in this chapter, in preparation for the analytical chapter to follow.

2.1. If a BMU tunnels through cortical bone at a rate of 40 µm/day, how long would it take to tunnel 2 cm through your tibia?

2.2. Estimate the average rate of addition of bone on the diaphyseal surface of your tibia, in micrometers/year, as you grew to skeletal maturity. Palpate your tibia to estimate its present diameter, and assume its diameter was 0.8 cm when you were born. Assume that your tibia's periosteal apposition rate essentially stopped at age 14 if you are a woman or age 16 if you are a man, ignoring any further increase in the diameter of the tibia that occurred after closure of the physes.

2.3. Typically, a newborn infant's tibia is about 6.8 cm long (Jeanty et al., 1984). Estimate the present length of your tibia, then estimate the average rate of growth in the tibial growth plates, in millimeters/year and in micrometers/day. Again, assume that longitudinal growth of the tibia stops at age 14 in females and age 16 in males. How does this compare with the bone apposition rate estimated in Exercise 2.2 and the BMU tunneling rate in Exercise 2.1?

2.4. The rate of growth of a growth plate is equal to the daily chondrocyte production rate for each column multiplied by the height of the maximally hypertrophied cells (Sissons, 1955). Write a paragraph explaining why this is so in a steady-state situation and stating the necessary assumptions.

2.5. Consider the relationship between the rate of growth in a growth plate, the average number of cells being produced each day, and the average amount of hypertrophy each cell experiences. Assume the following. The growth plate consists of many parallel cell columns, all growing at the same rate. The stem cell at the top of each column divides at a rate r_s (divisions/year), replacing itself and producing one new chondrocyte. Each resulting chondrocyte divides, producing two new chondrocytes, which in turn divide. This mitotic sequence continues until N generations of chondrocytes are produced, all of which are aligned in a single column. Each chondrocyte in the column hypertrophies to an ultimate cell height, h. Using the principle of Exercise 2.4, show that the physeal growth rate is $G = h \; r_s \; 2^N$.

2.6. If it takes 2 months to refill an osteonal resorption cavity that is initially 250 μm in diameter, what is the average apposition rate in micrometers/day? Assume the Haversian canal diameter is 50 μm after refilling is complete. Compare this apposition rate with that in Exercise 2.2.

2.7. Imagine a cross section through your femur. Assume that 5% of the bone in this cross section is replaced annually by osteonal remodeling. The osteons average 250 μm in diameter. What would the osteonal BMU activation rate (new osteons/mm^2 per year) have to be to produce the 5% turnover rate?

2.8. In Exercise 2.7, if each osteonal BMU produced a new Haversian canal that was 50 μm in diameter, what would be the rate of increase in bone porosity (mm^2 of canals/mm^2 of cross section/year)? Assume that the new osteons do not obliterate preexisting Haversian canals.

3

Analysis of Bone Remodeling

...there is no philosophy that is not founded upon knowledge of the phenomena, but to get any profit from this knowledge, it is absolutely necessary to be a mathematician.

Letter from Daniel Bernoulli to his nephew, John III Bernoulli,
January 7, 1763.

Good mathematical models don't start with the mathematics but with a deep study of certain natural phenomena.

Stephen Smale, 1930–

3.1 Introduction

The preceding chapter provided a general description of the structure and biology of bone and cartilage. We learned how bones grow, are sculpted by modeling and continuously renewed by remodeling, and repair themselves when fractured. All of these are important processes, but in this chapter we focus on remodeling, for two reasons. First, remodeling is the only one of these processes that occurs throughout one's lifetime. Growth and modeling are essentially restricted to children, and fracture healing is restricted to even more isolated periods of time. Second, because bone remodeling occurs throughout life, it plays a dominant role in determining the structure of most of the tissues in the skeleton, most of the time. Consequently, remodeling is a primary determinant of the mechanical properties of bone, their resistance to fatigue failure, and their ability to function in a changing mechanical environment. Therefore, we need to develop our understanding of bone remodeling in some depth to pursue our study of skeletal tissue mechanics.

As we noted in Chapter 2, the resorption and refilling process by which secondary osteons are formed involves groups of cells functioning as organized units, which Frost named basic multicellular units (BMUs). BMUs are part of what Frost calls the skeletal *intermediary organization*. They operate on bone periosteal, endosteal, and trabecular surfaces as well as within cortical bone, replacing old bone with new bone in discrete packets. In the case of secondary osteons in cortical bone, this is the bone between the cement line and the Haversian canal along the length of an individual osteon. The concept of bone remodeling in quantum units is important because it recognizes a predictable biologic system that is particularly amenable to quantitative analysis. The utility of the quantitative histologic techniques introduced by Frost, and used today in clinical medicine as well as basic science research, is a direct result of these qualities.

After reviewing the methods used to quantify bone remodeling in a histologic section, we shall continue to use mathematical modeling as a tool for learning more about how remodeling works and the effects it has on bone structure. Making mathematical models forces us to be quantitative as we test hypotheses about the unknown. Eventually, these models may lead to a much clearer understanding of skeletal biology and mechanics and the relationship between them. You have the distinction of studying this subject in its infancy, and you should view the mathematical analyses in this chapter as "works in progress," rather than derivations of the kinds of mathematical results that you have encountered in other courses. When asked how they obtained their insights in the world of physics, Newton said "By thinking into them." And Einstein said "I grope." In this chapter, we use mathematical models to think our way into bone remodeling. However, we are still very much in the groping stage, and you should view this chapter in that light. Join in the adventure, look skeptically at what follows, and try to find flaws and fix them!

The A-R-F Sequence

The BMU remodeling sequence follows well-defined phases that always begin with activation (A), followed by resorption (R), followed by formation (F). Figure 3.1 depicts these phases in a schematic dia-

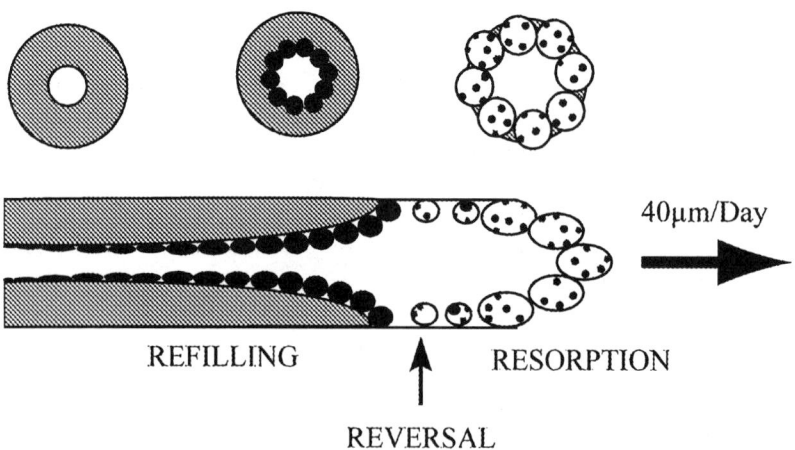

FIGURE 3.1. Schematic diagram of an osteonal BMU. Larger, multinucleated cells to *right* are osteoclasts; smaller cells shown in *black* to *left* are osteoblasts. Cross-sectional views are shown *above*. Not to scale.

gram. This figure also illustrates the relationship between the appearance of an osteonal BMU in longitudinal and cross sections and, in addition, the fact that the duration of events perceived by an observer in a particular bone cross section does not correspond to the duration of events that would be observed by someone moving along the bone with the BMU. For example, the moving observer would say that the resorption phase of the BMU lasts from the time that osteoclasts begin to excavate the tunnel until the tunneling ends, perhaps a centimeter down the bone shaft; this might take 200 days. A stationary observer at some arbitrary cross section along the way would say, however, that the time required for resorption is only the few days required for the osteoclasts to pass through the plane of the section and open a hole corresponding to the osteonal cement line. In virtually all discussion of osteonal BMU dynamics, time periods refer to those observed in cross sections, because it is such sections that are routinely observed and analyzed.

The Six Phases of an Osteon's Lifetime
The A-R-F process may be further divided into six separate phases, which are always sequential in normal bone remodeling.

Activation. Differentiated cells must be recruited from precursor cell populations before any bone resorption or bone formation can occur. It is important to recognize that in the purest sense this process occurs only at the point of origin of the BMU. The osteoclasts produced at this time appear to survive for weeks, and perhaps for the entire duration of the tunneling by the BMU. Thus, the "opening up" of a resorption space on a cross section is not actually "activation" because it does not involve the initial recruitment of osteoclasts, but rather the passage of osteoclasts through a particular plane of observation. The original activation process occurred earlier; in either case, however, the time required is about 3 days.

Resorption. Newly created osteoclasts begin to resorb bone, moving longitudinally at a rate of about 40 μm/day on a so-called cutting cone, an ellipsoidal surface about 200 μm in diameter and about 300 μm long. However, it is difficult to be certain where this region stops and the following region begins, so the length of each is debatable.

Reversal. The transition from osteoclastic to osteoblastic activity takes several days and results in a cylindrical space lying between the resorptive region and the refilling region. The length of this region may vary considerably, depending on the lag between the resorptive and formative phases. In a completed secondary osteon, the cement line coincides with the location of the bone surface during the reversal period; the cement line is also known as the *reversal line*. In humans, about 30 days are required for the resorption and reversal periods combined.

Formation. Osteoblasts appear around the periphery of the tunnel formed by the osteoclasts, and refilling begins. The osteoblasts lay down concentric lamellae at a decreasing rate as refilling progresses, but the average radial closure rate is 1–2 µm/day. Because the osteoclasts and osteoblasts require nourishment, the osteoblasts do not completely refill the tunnel, but leave a central passageway for a vascular loop to support the metabolism of the BMU and bone matrix osteocytes, and to carry calcium and phosphorus to and from the bone when necessary. This passageway is the Haversian canal, and it is typically 40–50 µm in diameter in humans. The formation phase in adult humans averages about 3 months.

Mineralization. Following deposition of the unmineralized organic matrix of bone, or *osteoid*, mineral is deposited within and between the collagen fibers (Fig. 3.2) (Landis, 1995). This process is delayed by a period of time known as the *mineralization lag time*, which is normally about 10 days.

collagen molecules

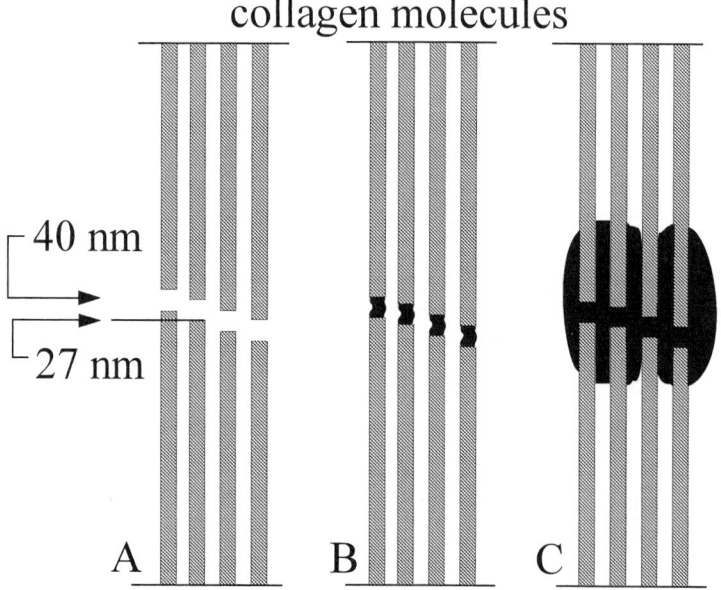

FIGURE 3.2A–C. Schematic diagram of mineralization of collagen matrix. **A** Premineralization; 40-nanometer (nm) gaps between collagen molecules are offset 27 nm in adjacent molecules. **B** Mineral crystals begin to form in the gaps. **C** Mineral crystals grow to fill the spaces between layers of collagen molecules. (Redrawn with permission from *Bone*, Vol. 16, Landis, WJ, The strength of calcified tissue depends in part on the molecular structure and organization of its constituent mineral crystals, 533–544, 1995. Elsevier Science Inc.)

This delay is manifested by a layer of osteoid between the osteoblasts and mineralized bone. Once begun, approximately 60% of the mineralization of osteoid occurs during the first few days (Parfitt, 1983); this is called *primary mineralization*. The remainder of the mineral is added at a decreasing rate for 6 months or so during *secondary mineralization*. The bone in new, incompletely mineralized osteons can exhibit very different mechanical properties from that in older osteons.

Quiescence. After the tunneling and refilling processes are completed, the osteoclasts disappear from the scene, and the osteoblasts become osteocytes or Haversian canal lining cells or disappear. A period of relative inactivity ensues during which the secondary osteon and its associated cells carry on their mechanical, metabolic, and homeostatic functions.

Each of these six phases is location-, magnitude-, and rate specific, so that alterations in the intensity or timing of any one produce an altered histologic morphology characteristic of specific skeletal disorders. For example, an increased lag between the bone formation and bone mineralization stages produces the excessive amounts of osteoid found in *osteomalacia* (a disorder characterized by decreased mineralization of bone).

Osteonal Origins and Trajectories

Osteonal BMUs can originate on any cortical bone surface not covered by cartilage or otherwise inhibited from remodeling. Thus, they can be activated on periosteal or endosteal surfaces, or on the Haversian canal walls of existing primary or secondary osteons. The pathways of osteons may be regular or very irregular. Typically, they tunnel more or less longitudinally through the bone, but spiral slightly around the cortex at an angle of curvature of about 12° and through the cortex at a similar angle (Fig. 3.3). This pathway may be deduced from radial and tangential sections of a long bone or by reassembling tracings of serial sections. Another method is to scrape off the circumferential lamellae of the periosteal surface and stain it with India ink to reveal the Haversian canals. Petrtyl et al. (1996) used this method to show that the osteonal spiral was oppositely directed on the medial and lateral sides of the linea aspera of a human femur.

Cutting cones may dig proximally or distally, and they may be observed passing one another in opposite directions. Also, a BMU may originate at one site and quickly divide to form a pair of BMUs tunneling in opposite directions. The exact direction in which a specific BMU tunnels is probably determined partly by the idiosyncrasies of its osteoclasts and the structure of the bone they encounter and partly by the mechanical strains in the surrounding bone.

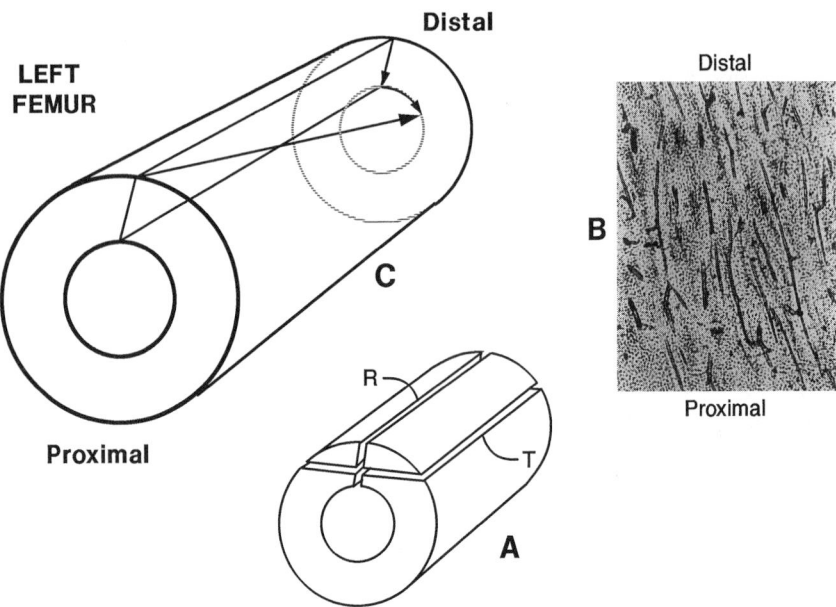

FIGURE 3.3A–C. Directionality of osteons deduced from longitudinal sections cut from a canine femur as shown in **A**. *R*, radial section; *T*, tangential section. Haversian canals in both appear similar to photomicrograph at **B**; they lie at an angle of about 12° to the bone's long axis. **C** shows the osteon path consistent with these sections. In a left femur an osteon originating proximally on the periosteal surface spirals medially through the cortex as it proceeds distally. BMUs activated intracortically apparently tunnel toward the periosteal surface if they are headed proximally, and vice versa. (Note: not all regions of all canine femurs are as well organized as this.)

Remodeling Cycle Duration

Because the wall of the tube of new bone laid down within the excavated tunnel is about 80 μm thick, and the average accretion rate is about 1 μm/day, the time to refill is roughly 80 days. That is, if it were possible to observe a bone cross section in vivo, it would take about 80 days for the initial resorption cavity to be converted into a completed osteon. Before refilling, about 30 days would have passed between the time when the osteoclasts initially began to open the resorption cavity and the appearance of the osteoblasts. Historically, these resorption and refilling times have been called *sigma* times because Frost originally used this symbol to represent them. The total sigma is approximately 4 months in human cortical bone.

Box 3.1 Historical Note
Mathematical Notation

Frost (1964b, 1969) was the first person to develop mathematical models for bone remodeling. Being the first, he was free to choose mathematical symbols for key variables. For example, he used the symbol "sigma" (σ) to represent the time required by each phase of the BMU to be completed at a particular location, M_f to denote the rate at which bone is added to a forming surface, and μ for the rate at which new BMUs are created in a bone.

Subsequently, Frost's mathematical notation was used by other investigators in reporting their experimental data. Eventually, however, concern arose because various investigators would use the same symbol to denote a variable measured or expressed in somewhat different ways. For example, the amount of bone surface on which formation was occurring could be expressed per unit volume of bone (i.e., in three dimensions) or per unit area of histologic cross section (i.e., two dimensions). To eliminate this problem, the research community developed a standard symbology for referring to bone remodeling variables (Parfitt et al., 1987). For example, the area of bone surface covered with osteoid would be called OS in three dimensions or O.Pm in two dimensions. Here, the ".Pm" refers to the fact that surfaces show up as perimeters in a two-dimensional section. The resorption "sigma" became RS.P (resorption period), the reversal period became Rv.P, and the formation sigma became FP (formation period). The total of these three periods is still abbreviated Sg (for "sigma") in the standardized nomenclature because this Frostian term has continued in wide usage.

These abbreviations have proven advantageous for reporting experimental data, but they would be very awkward to use in a mathematical analysis. Therefore, they are not generally used in this book; instead, we use a more convenient mathematical notation. The following table below relates the notation we shall use to the standard notation recommended for bone histomorphometry.

Variable	Units	This text's nomenclature	Standard nomenclature
Resorption period	days	T_R	Rs.P
Reversal period	days	T_I	Rv.P
Formation period	days	T_F	FP
BMU activation frequency	#/mm^2/year	f_a	Ac.f
(Mineral) apposition rate	μm/day	M_F	MAR
Mineralization lag time	days	T_{ml}	Mlt
Osteoid thickness (2-D section)	μm	t_{os}	O.Wi
BMU mean wall thickness	μm	t_{mw}	W.Wi
Bone volume fraction	—	B_v	BV/TV
Bone formation rate	mm^3/mm^3/year	V_F	BFR/BV

3.2 Histomorphometric Measurement of Osteonal Remodeling

Frost demonstrated that histologic sections could be analyzed to quantify bone remodeling. He originally did this for osteonal remodeling in the cortex of the human rib. Since then, his method has been extensively used in animal experiments and adapted for use on human bone biopsies from the iliac crest. The latter is used in preference to the rib because it is more accessible and contains more cancellous bone, where bone remodeling is more active and bone loss appears more quickly. We provide a brief overview of the measurement process used in evaluation of osteonal bone remodeling for three reasons. First, histomorphometric analysis requires the reduction of the mechanics of remodeling and osteon formation to a few well-defined and measurable steps, which leads to a clearer and more integrated understanding of the osteonal remodeling process. Also, this sort of quantitative analysis is an excellent way to identify points of inconsistency in our theories and frame pertinent questions for future experiments. Finally, an understanding of histomorphometric analysis contributes greatly to a clearer perception of the problems associated with diagnosing pathologic states and defining their etiology.

Assumptions

In this simplified approach to osteonal bone histomorphometry, we follow, for the most part, the early rationale of Frost, whose ideas and experimental techniques laid the foundations of this field and still form its backbone. Although thinner sections may be used for cancellous bone, for the purposes of this discussion it can be assumed that the measurements are made on undemineralized cross sections of cortical bone about 100 μm thick. The bone is assumed to have been labeled twice with tetracycline or another vital fluorochrome stain for mineralizing bone. These labels are assumed to be given about 7–14 days apart, with the second label ending at least 1 day before the bone specimen is taken. We refer to the time between labels as T_L, expressed in days. The section is assumed to be stained with tetrachrome, osteochrome, or some other stain that distinguishes osteoid from mineralized bone and also allows the labels to be seen. Thick sections are best observed with epifluorescent (reflected) ultraviolet light to avoid problems associated with having a poorly defined plane of measurement. In the past, the measurements described here required considerable time and ingenuity, using various combinations of stereological principles, photographic enlargement, and microscope eyepiece reticules. Today, a variety of commercial and custom-made computer image analysis systems simplify the work.

The A-R-F sequence, and the fact that osteons generally run parallel to one another, are basic to the relative simplicity of these measurements. Beyond these suppositions, perhaps the most important assumption is that the remodeling is in a *steady state*; that is, new BMUs are being created at a constant rate and are functioning in ways that are not changing with time. Of course, this assumption is never strictly true, given age-related changes in the remodeling process, but compared with the 3 to 4 month period required to complete an osteon, such age-related variations are negligible. Of greater concern are experimental and clinical situations in which the effects of a drug or other variable on remodeling need to be assessed. These situations usually require the investigator to determine *transient* remodeling changes. As the analytical method is based on the existence of a remodeling steady state, many experiments cannot be accurately assessed using conventional histomorphometric techniques. Overcoming this problem presents an important area for future research.

Box 3.2 Technical Note
Bones That Glow

In 1736, John Belchier discovered that developing bone in livestock could be stained in vivo by feeding the animals madder, a dye derived from the Eurasian herb *Rubia tinctorum*. This happened when he observed red streaks in the bones of some pork that he was having for dinner. He investigated and found that the pig had been fed waste plant material from a dye shop. During the ensuing two centuries Belchier's discovery provided impetus for several qualitative studies of bone growth and fracture healing. In the late 1700s, using madder-fed pigs, the famous English surgeon-anatomist John Hunter was the first to grasp the significance of what we now call modeling in the growth of the skeleton. Hunter's experiments demonstrated that bones grow radially by endosteal resorption and periosteal apposition. (The prevailing view had been that bones expanded outward as a consequence of new bone deposition within the cortex.) In the 1800s, synthetic alizarin replaced the natural dye and was found to be equally effective as a bone label. These dyes bind to hydroxyapatite as it is being formed, and therefore preferentially stain the interface between osteoid and mineralized bone. More recently, water-soluble versions of such dyes have been administered by injection, producing more sharply defined labels. However, these dyes would not be considered safe to administer to humans.

The modern use of such labels in human bone histomorphometry began in the 1950s when Milch et al. (1958) reported that the common antibiotic tetracycline was incorporated into actively mineralizing bone surfaces and fluoresced yellow when histologic sections were observed in ultraviolet light. Along with Milch, Frost (1960c), Harris (1960), and others recognized the significance of this finding: for the first time, the dynamics of human bone modeling and remodeling could be safely studied. Soon, tetracycline-based histologic analysis of bone remodeling was being widely used to study bone physiology. Fluorochrome labeling techniques are now routinely used to study the effectiveness of new drugs aimed at preventing osteoporosis and other metabolic bone diseases, new techniques and treatments intended to increase the rate of fracture healing, and the responses of bone to total joint replacements and other implants.

Alizarin and tetracycline are not the only in vivo labels available. Calcein green, calcein blue, and xylenol orange are other compounds used to label mineralizing bone in vivo. It turns out that all these compounds fluoresce, and they are collectively termed *fluorochromes*. Sequential administration of different fluorochromes, each fluorescing a different color, allows the researcher to track the time course of a skeletal process such as fracture healing.

Obviously, bone specimens intended for fluorescent histomorphometric analysis must be processed with the mineral intact. This is contrary to standard histologic procedures, which demineralize bone so that it can be cut with a knifelike instrument (called a microtome), just as soft tissues are. The simplest (but most tedious) means of making a section of undemineralized bone is to saw off a slab approximately 1 mm thick, usually using a diamond-impregnated saw, and then grind it down between two pieces of wet sandpaper until it is about 100 μm thick. Much of Frost's pioneering work was done using this simple method (Frost, 1958). More modern methods embed the bone in plastics or resins with material properties similar to the bone tissue. It is then possible for a skilled technician to cut trabecular bone sections a few micrometers thick (or cortical cross sections 30–50 μm thick) using special knives or saws.

Histomorphometric analyses of labeled bones are performed using *epifluorescent* microscopes. The term "epi-" refers to the fact that the ultraviolet light is reflected off the surface of the section, which renders a sharper plane of focus than would transmitted light. Filters are used to enhance fluorescence and enable polychrome analysis of a single image. Fluorescence decays with time, so measurement techniques must be efficient. As outlined in the text, quantitative analysis requires measuring several histologic features, only a few of which involve fluorescent labels. The bone tissue must be stained to identify features such as osteoid and cement lines. If these stains interfere with the observation of fluorochrome labels, separate sections must be made for quantifying labels.

Measurements

A series of measurements is usually made in a standardized sequence:

1. The area of cortical bone in the section is first measured so that results can be given per unit area of cross section measured.
2. The section is scanned to find the osteonal resorption cavities or spaces that it contains. These spaces are identified by their scalloped surfaces (R in Fig. 3.4) and lack of osteoid or fluorochrome label, and their number per unit area is N_R.
3. The perimeter of each resorption space is measured, and the mean perimeter is calculated (S_R).
4. The section is scanned for its refilling BMUs, identified by theirosteoid seams (F in Fig. 3.4); the number per unit area of bone is called N_F.
5. The mean perimeter of the osteoid seams is determined and called S_F.

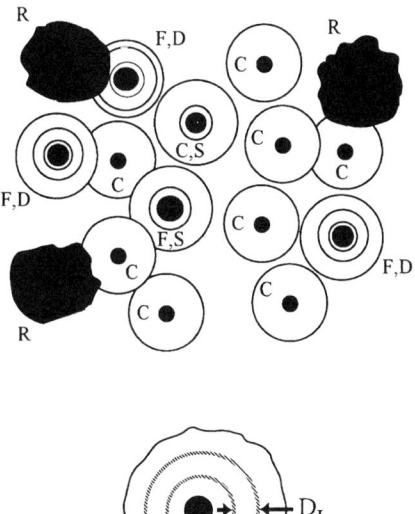

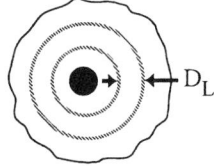

FIGURE 3.4. *Upper:* Schematic diagram of BMUs in a cross section of cortical bone. *Black*, irregularly shaped BMUs are resorption spaces (*R*). BMUs with large canals are refilling; they have either a single label (*F,S*) or a double label (*F,D*). Completed osteons have small canals (*C*); some of these have a single label (*C,S*). *Lower:* A single BMU in the last stages of refilling is depicted. It has a double label; the distance between the labels is measured as shown (*D*$_L$).

6. The mean distance (in micrometers) between the two labels[1] in all the doubly labeled BMUs is determined (Fig. 3.4) and called D_L.
7. The section is scanned for osteons that have been completed, as indicated by the absence of osteoid seams (*C* in Fig. 3.4). The "mean wall thickness" (t_{mw}, in micrometers) of these completed osteons is found by averaging the distance from the Haversian canal wall to the cement line.

Calculating BMU-Level Results

The *mineral apposition rate* is calculated as

$$M_F = D_L/T_L \text{ (in micrometers/day)} \tag{3.1}$$

---

[1]Some of the BMUs in the figure are labeled with an "S" as well to indicate that they have only a single label. This is explained later under "Label escape error."

The word "mineral" is attached to this variable because it is measured from fluorochrome labels, which lie at the junction between the osteoid "seam" (i.e., layer) and the layer of bone experiencing primary mineralization. In a steady-state situation, the mineral apposition rate is the same as the rate at which osteoblasts are laying down osteoid. Therefore, it is common to shorten the term to "apposition rate."

The time that elapses between the production of osteoid by the osteoblasts and its initial mineralization is the *mineralization lag time*, which is symbolized here as T_{ml}. Histologically, this delay is represented by the thickness of the osteoid seam, t_{os}, which would be zero if the new bone mineralized immediately. The mineralization lag time may be determined by measuring the average t_{os} for the refilling BMUs on the section and dividing the result by the apposition rate:

$$T_{ml} = t_{os}/M_F \text{ (in days)} \qquad (3.2)$$

The time required to refill a typical BMU is then logically assumed to be the mean wall thickness, t_{mw}, divided by the apposition rate:

$$T_F = t_{mw}/M_F \text{ (in days)} \qquad (3.3)$$

(There is a fundamental problem in this part of the analysis. The measurements are supposed to reflect the current situation, but measuring completed osteons means that the results will be skewed toward the conditions that existed several months or years ago. One way to handle this problem would be to limit the t_{mw} measurements to completed osteons that microradiographs reveal to be incompletely mineralized [see Fig. 2.3]. Another method would be to measure t_{mw} in those few osteons bearing at least one of the labels given for the analysis but no osteoid seam.)

The calculation of the activation frequency of new BMUs relies heavily on the steady-state assumption. The times for each phase of BMU development are assumed to be constant, so that the number of BMUs activated each day is the same as the number entering the refilling phase. Use is also made of the fact that the size of any **constant** population is equal to the product of the birthrate and the average longevity of its individuals. Applying this to refilling BMUs yields

$$N_F = f_a T_F \qquad (3.4)$$

That is, the population of refilling BMUs per millimeter squared of bone section is equal to the birthrate (activation frequency, f_a) of refilling BMUs multiplied by their longevity, or the time required for refilling, T_F. Rearranging this equation, one has

$$f_a = N_F/T_F \text{ (BMU/mm}^2 \text{ per day)} \qquad (3.5)$$

for the BMU activation frequency.

Calculating Tissue-Level Results

The *bone formation rate*, V_F, is defined as the amount of new bone made per unit time, expressed as a fraction of the area or volume of bone being considered. For example, if within each cubic millimeter of bone in a human femur the remodeling process formed a volume of bone equal to 0.05 mm^3 in 1 year, the bone formation rate would be 0.05 mm^3/mm^3 per year. Because fractions of area measured on a bone cross-section are equivalent to volume fractions in the same region, this variable may be computed from the histologic measurements already described:

$$V_F = 0.365 \times 10^{-3} \ M_F \ S_F \ N_F \ (\text{mm}^2/\text{mm}^2 \text{ per year}) \qquad (3.6)$$

In this equation, 0.365×10^{-3} serves to convert days to years and μm^2 to mm^2. As the amount of bone resorbed during osteonal remodeling is normally only slightly greater than that formed (the difference in osteonal bone resulting from the creation of new Haversian canals), the bone formation rate is often taken to be the same thing as the bone "turnover" or replacement rate.

There is no straightforward way to measure the osteoclastic equivalent of M_F, which one may call the *erosion rate*, M_R, but it may be approximated if the rate at which bone is being lost can be determined. Suppose, for example, that biopsies are obtained before and after an experimental treatment, and measurements of porosity indicate that 0.01 mm^2/mm^2 of bone was lost during the course of a year. This time rate of change of porosity must equal the difference between the bone resorption rate and the bone formation rate:

$$dp/dt = V_R - V_F \qquad (3.7)$$

Writing an expression for osteonal resorption that is exactly analogous to that for formation, one has

$$V_R = 0.365 \times 10^{-3} \ M_R \ S_R \ N_R \ (\text{mm}^2/\text{mm}^2 \text{ per year}) \qquad (3.8)$$

Because V_R can be determined from Eq. 3.7, and S_R and N_R were obtained from the previous histomorphometry, it is clear that this equation can be solved for M_R.

Table 3.1 shows typical values for a number of these histomorphometric variables as they have been measured in the ribs of humans, dogs, and monkeys. Clearly, there is variation between species and with age, so that numbers such as $T_F = 90$ days, which we cited earlier as typical in humans, must be used with discretion.

TABLE 3.1. Bone remodeling data for rib osteonal BMUs

Variable	Dogs[a]	Rhesus monkeys[b]	Humans by age group, years[c]		
Age, years	1.5–3.5	"Adolescent"	1–9	30–39	70–89
N_F, #/mm^2	3.09 ± 1.75	4.03 ± 1.64	2.7 ± 1.0	0.21 ± 0.54	0.66 ± 0.50
N_R, #/mm^2	0.58 ±. 034	1.96 ± 1.15	2.0 ± 0.62	0.25 ± 0.33	0.84 ± 0.46
S_F, μm	120 ± 30	200 ± 30	280 ± 40	250 ± 140	300 ± 72
M_F, μm/day	0.95 ± 0.26	1.33 ± 0.23	1.40 ± 0.46	1.1 ± 0.36	0.72 ± 0.20
T_F, days	117 ± 110	54 ± 11	51 ± 47	73 ± 58	109 ± 55
f_a, #/yr/mm^2	15.5 ± 11.8	27.9 ± 13	19 ± 9	1.1 ± 0.5	2.2 ± 1.1
t_{mw}, μm	57 ± 6	63 ± 2	68 ± 4	70 ± 5	66 ± 7
V_F, %/yr	11.4 ± 9.4	35.9 ± 18.6	38 ± 24	1.8 ± 0.6	4.4 ± 1.5

BMU, basic multicellular unit.

[a]A. Villanueva, personal communication.

[b]Schock et al., 1972.

[c]Frost, 1969.

Label Escape Error

When two fluorochrome labels are given and the bone is examined afterward, some of the labeled osteons only have one label. One reason for this
is that some osteons began to refill after the first label was given, and others stopped refilling before the second label was given. Frost (1983)
worked out the correction for this using the graph shown in Fig. 3.5. In this
figure, time is represented on the horizontal axis, and the two vertical lines
(T_1 and T_2) represent label times. The time between labels is $T_L = T_2 - T_1$.

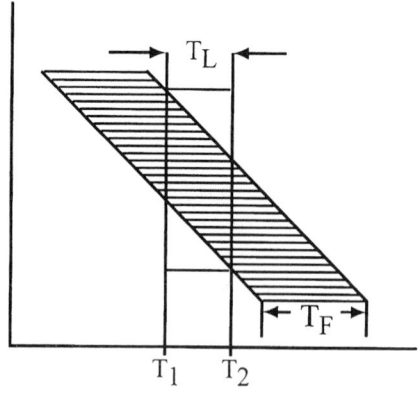

FIGURE 3.5. Frost's "ladder diagram" for deriving the label escape equation.
(Reproduced with permission from Frost, 1983.)

The diagonally distributed horizontal lines represent BMUs that began refilling at the time indicated by their left end and completed refilling at the time indicated by their right end. Thus, the lengths of the BMU lines are all T_F. Doubly labeled BMUs have both vertical lines through them and singly labeled BMUs have only one vertical line. Frost (1983) showed that the ratio of the number of singly labeled osteons (N_{1L}) to those doubly labeled (N_{2L}) should be

$$N_{1L}/N_{2L} = 2\ T_L\ /\ (T_F - T_L) \tag{3.9}$$

and the "true" number of refilling osteons (N_F) should be taken as

$$N_F = {}^1\!/\!_2\ N_{1L} + N_{2L} \tag{3.10}$$

Demonstration of this relationship using similar triangles (Fig. 3.5) is left as Exercise 3.1.

There is experimental evidence that the N_{1L}/N_{2L} ratio is larger than predicted by Eq. 3.9 because some (or all) osteons pause during refilling. Such pauses may be exacerbated by age or mechanical disuse. This is an important detail of remodeling, which has been further analyzed (Martin, 1989) but remains to be fully understood.

True Vs. Histologic BMU Activation Frequency

As noted, activation frequency has two definitions. The first is the number of BMUs created per unit volume of cortex per unit time (e.g., BMUs/mm^3 per year). We refer to this definition as the *true activation frequency*, or f_{aT}. Typically, BMUs are "born" when bone lining cells on the surface of a Haversian canal are stimulated to recruit osteoclasts to the site. These osteoclasts dig into the adjacent bone matrix, and in a few days create a tunnel aligned with the long axis of the bone. The BMU continues to tunnel through the bone until it reaches the periosteal or endosteal surface or is otherwise caused to stop.

The second definition of activation frequency is the one we previously discussed; it arises when osteonal remodeling is quantified histomorphometrically. In this context, "activation" is not defined as the **original** creation of a BMU at a specific site along the bone, but as the **appearance in the cross section** of a functioning BMU in a particular stage of activity. As we have seen, the standard method of calculating activation frequency is to determine the population of refilling BMUs/mm^2 of section and divide this number by the refilling time (i.e., the "lifetime" of the refilling stage). This yields the "birthrate" of refilling BMUs in the section, that is, the rate at which the "front ends" of refilling regions enter the section. If all BMUs are similar with respect to the lengths of their resorbing and refilling regions, this "refilling region birthrate" is identical to that for the resorbing region, which is in turn identical to the rate at which BMUs themselves arrive at the cross section and begin to create a resorption cavity (Frost, 1964b). In other words, this definition involves the "creation" of **sections** of BMUs within the two-dimensional world of a histologic section, as opposed to the creation of the entire BMU in the three-dimensional world of the bone. We refer to the variable measured in this two-dimensional way as the *histologic activation frequency*, f_a. Its units are typically given as BMUs/mm^2 per year. How is f_a, which is routinely measurable, related to f_{aT}, which is not, but quantifies the biologically controlled event?

Consider a region of cortical bone having cross-sectional area A and its long axis oriented in the *x*-direction (Fig. 3.6). If f_H is measured in a histologic section, Σ, made perpendicular to the bone at $x = 0$, we wish to know its relationship to f_{aT} in the bone as a whole. Let v be the mean velocity at which BMUs tunnel through the cortex, and let R be the mean BMU range, that is, the average distance that they dig before ceasing their activity. Now, those BMUs just arriving at Σ originated at various locations, x_i, in the cortex, and at various times, T_i, in the past. The latter times may be called the ages of the BMUs arriving at $x = 0$. If all traveled at the same average velocity, then the originating distances and the ages are proportional to one another such that

$$x_i = vT_i \tag{3.11}$$

Let $f_{aT}(x,t)$ be the rate at which BMUs are being created at time t and position x along the bone. The number of BMUs originating within an infin-

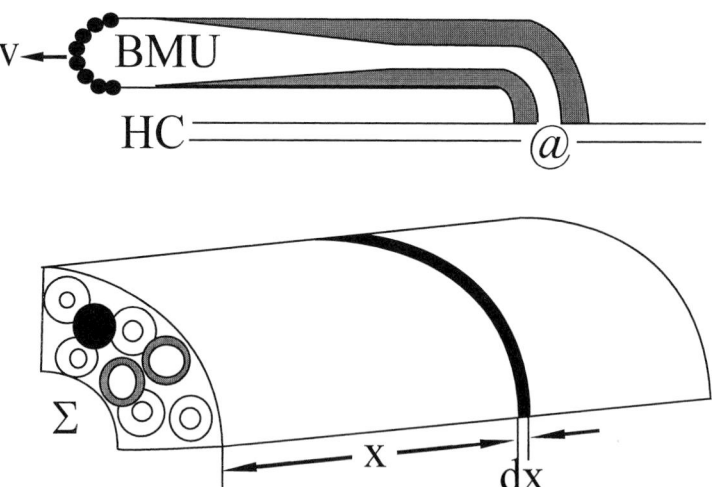

FIGURE 3.6. *Upper:* An osteonal BMU has been activated on the wall of an existing Haversian canal (*HC*) at the site marked @, and is tunneling to the left with velocity v. *Lower:* A quadrant of cortex having cross-sectional area A is shown. A histologic section Σ at the left end is histomorphometrically analyzed to obtain f_a. BMU resorption cavities arrive at Σ after having originated at a distance x, within a zone dx. (Reprinted from *Bone*, Vol. 15, 547–549, 1994, with permission from Elsevier Science.)

itesimal section dx at time t in the past is

$$dN = f_{aT}(x,t)\ A\ dx \tag{3.12}$$

where x is the location of dx. The number of BMUs that originated in dx arriving at Σ during each unit of time is

$$dN = f_{aT}(x,T)\ A\ dx \tag{3.13}$$

where $T = x/v$ is the time required for a BMU to travel from dx to Σ. The total number of BMUs/mm^2 arriving at Σ during each unit of time is obtained by integrating Eq. 3.13 from $x = 0$ to $x = R$, the maximum distance from which a BMU could travel to reach Σ:

$$f_a = N/A = \int_0^R f_{aT}(x,T)\ A\ dx \tag{3.14}$$

If f_{aT} varies in time and space, its histologic analog must be found by integrating Eq. 3.14. Let us now assume that f_{aT} is the same throughout the bone, varying only with time. Then substituting $dx = vdt$, and letting $L = R/v$ be the lifetime of a BMU's tunneling activity, one has

$$f_a = \int_0^L f_{aT}(t)\ v\ dt \tag{3.15}$$

If f_{aT} is also constant in time, one has

$$f_a = f_{aT}\ v\ L \tag{3.16}$$

or

$$f_a = f_{aT}\ R \tag{3.17}$$

That is, in a steady-state situation, the two-dimensional, histologically measured activation frequency is equal to the three-dimensional, true activation frequency multiplied by the mean BMU range.

This derivation has only considered BMUs arriving at Σ from one direction. When BMUs arriving from the other direction are considered, it might seem that Eq. 3.17 would yield values of f_a that are too small by half. That is not the case, however, because it can be assumed that half the BMUs represented by f_{aT} will tunnel away from Σ. Thus, the effects of ignoring both these details cancel one another. Another detail should be mentioned as well. As noted, sometimes activation results in a "double-headed" BMU, which arises at one site but forms two cutting cones that tunnel in opposite directions. For the purposes of this analysis, and in terms of the amount of remodeling that results from such incidents, these events should be counted as two activations.

If f_a and f_{aT} are related through the BMU range, R, the next question is, what is the value of R? This is not an easy question to answer. Because of the difficulty in sorting out anastomoses and Volkmann's canals, it has been practically impossible to trace the paths of BMUs from their point of origin to their terminus. However, Cohen and Harris (1958) observed that "In approximately one centimeter of the length of the femur, an osteon and its descendants coursed from periosteal lamellae to endosteum." Tappen (1977) was able to trace forming osteons back 5–6 mm from the cutting cone toward their origin. These observations suggest that a BMU may be expected to tunnel several millimeters before it ends. In that case, the true activation frequency, in BMUs/mm^3 per year, would be several times smaller than the histologically measured value, in BMUs/mm^2 per year.

In some early work, activation frequency was measured as just described, obtaining values of f_a, but the data were reported as f_{aT}, that is, as BMUs activated annually per **cubic millimeter** rather than per **square millimeter**. Other authors would simply report f_a. This may seem to be a small distinction, but it is important for two reasons. First, the true activation frequency is probably only a fraction of that measured. This is not important when comparing groups in the same experiment, but it could be very important elsewhere, such as when three-dimensional mathematical models are constructed for remodeling phenomena. Second, ignoring the distinction deemphasizes important details of remodeling dynamics that are rarely discussed. Activation of BMUs at one site is only the beginning of an ongoing process that reaches out in both space and time to generate effects which may appear unrelated to the initial event. If BMUs travel several millimeters after they are activated, the lifetime of a BMU in a bone, as opposed to its lifetime in a section, is considerably longer than Frost's sigma. This means that transients caused by changes in activation frequency may take longer to play themselves out than one sigma. For example, activation of remodeling at a screwhole in the middle of a long bone sends waves of BMUs traveling up and down the diaphysis, so that its whole length is affected for months afterward (Martin, 1987). Other events that alter activation frequency can be expected to produce effects of similar scope, and it is important to know their scale in time and space.

The application of these results to cancellous bone is unclear. While investigators describe trabecular BMUs as migrating over trabecular surfaces, no information appears to be available concerning how far an individual BMU may wander before it merges with another, divides into two, or "dies."

3.3 Remodeling Details

Activation: What Initiates New BMUs?

Osteonal BMUs may originate on periosteal or endosteal bone surfaces, or along the walls of Haversian canals. Various activation signals are thought to be present in the skeleton, including chemical, mechanical, and electrical signals, but just how or where they affect cells, and what sorts of intermediate messengers may exist, are questions still being resolved. In Chapter 5 we shall see that activation may stem from fatigue damage, and of course there are many chemical candidates for the role of initiating new BMUs in both osteonal and trabecular bone.

The Rodan Theory

Rodan (1992) developed a theory of BMU activation based on the hypothesis that bone lining cells are responsible for initiating new BMUs. That is not to say that lining cells differentiate to osteoclasts, but that they may be the target of various activation signals, responding in ways that lead to the appearance of osteoclasts. This idea springs from the observation that osteoblasts-turned-lining cells possess receptors for parathyroid hormone and $1\text{-}\alpha,25\text{-}(OH)_2$ vitamin D_3, but osteoclasts and their precursors apparently do not. Furthermore, lining cells respond to such signals in ways consistent with osteoclast activation. First, they change shape, contracting so as to expose the bone surface to the osteoclasts. Also, in cell culture mechanical and electrical stimuli increase cyclic adenosine monophosphate (cAMP) activity in osteoblast-like cells, which stimulates them to release prostaglandins, a potent intercellular messenger. Thus, the Rodan theory provides an interesting way for osteoclasts to respond to hormonal and physical signals to which they are not sensitive. It is also a theory that appeals to a sense of symmetry and simplicity; it provides a cytologic loop in which the final cell in the remodeling sequence becomes the initiator of the next remodeling cycle.

Hormonal Activation

Many hormones affect bone remodeling both directly and indirectly; these include parathyroid hormone (PTH), glucocorticoids, the active form of vitamin D (i.e., $1\text{-}\alpha,25\text{-}(OH)_2$ vitamin D_3), and calcitonin. The first two are usually thought of as leading to bone loss, and the last two as producing increased bone mass. In truth, however, we currently know very little about the effects of these hormones on the bone remodeling system, because so much of the available information has been obtained in tissue or cell culture experiments. Clearly, such experiments distort or destroy many characteristics of the skeletal intermediate organization in general and of BMUs in particular. Furthermore, many in vivo experiments have observed the effects of a hormone soon after it was administered, thereby missing effects related to the refilling stage of the BMUs activated by the treatment. Finally, much of the information obtained from experiments and clinical biopsies is of limited usefulness because it is difficult to sort out direct and indirect effects. For example, in dogs, serum vitamin D is elevated when serum estrogen is reduced by loss of ovarian function (Snow et al., 1984), but it is unclear which of these changes is primarily responsible for the accompanying elevation of trabecular BMU activation (Dannucci et al., 1987).

Activation Is Prime

Whatever causes activation of osteonal BMUs in a given situation, the body's ability to vary activation is much greater than the ability to control the numbers of osteoblasts or osteoclasts within the BMU or the rates at which they work. Figure 3.7 shows this by comparing the correlations between bone formation rate and apposition rate (M_F, a measure of osteoblast activity, top graph), and activation frequency (f_a, bottom graph). Clearly, the correlation between bone formation rate and f_a is much higher than that between M_F and formation rate. From this we deduce that changes in activation frequency have a greater effect on the amount of bone formed (or turned over) each year. Thus, effective (i.e., reasonably rapid and precise) responses to a situation calling for decreased bone mass, for example, cannot be achieved simply by decreasing osteoblast activity or stimulating osteoclasts; activation of many more BMUs is required. It turns out that activation is both the most important bone remodeling parameter and the one we know the least about, because it cannot be directly studied in vitro or in vivo.

Resorption: Out with the Old

Bone resorption is primarily the function of osteoclasts, which are mobile, multinucleated cells closely related to macrophages. Osteoclasts attach themselves to a bone surface, form a peripheral seal, and break down the bone within the sealed area by means of enzymes and other chemicals. The final stages of the dissolution of the bone tissue appear to be carried out by mononuclear cells following along behind the osteoclast. Nevertheless, osteoclasts are the primary resorbing cells within the BMU. Although osteoclasts exhibit considerable mobility on trabecular bone surfaces, apparently ranging over resorption territories several times their contact area, in osteonal cutting cones they seem to dig straight ahead in a tightly bunched configuration.

How Fast Is an Osteoclast?

In 1972, Jaworski and Lok measured the longitudinal advancement rate of osteonal cutting cones in dog ribs using double tetracycline labels (Fig. 3.8). By measuring the distance between the leading end of the two labels (x) and dividing by the interlabel time, T_L, they determined that the region where the "refilling cone" began to mineralize advanced longitudinally some 39 ± 14 µm/day. Assuming the BMU's geometry does not change as it moves through the bone, the cutting cone should have a similar rate of advancement. The rate at which the osteoclasts dig **radially** can be deter-

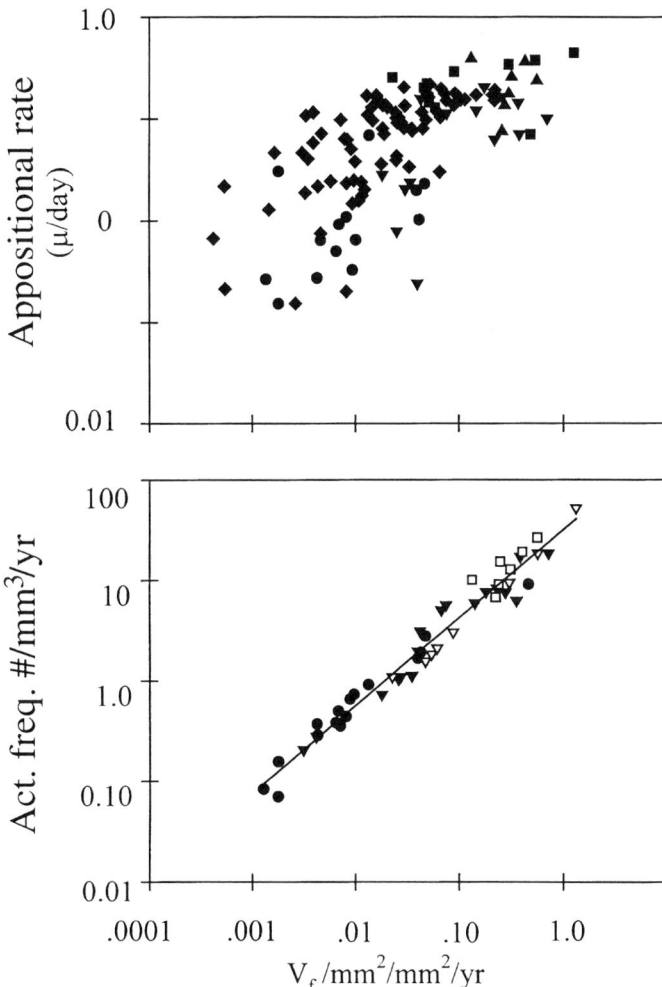

FIGURE 3.7. Graphs of apposition rate (*top*) and activation frequency (*bottom*) vs. bone formation rate (BFR) in human rib specimens. Note the much tighter correlation in the lower plot. (Reproduced with permission from Villanueva and Frost, 1970.)

mined from the longitudinal advancement rate and geometry of the cutting cone. As seen in Fig. 3.1, in longitudinal section the cutting cone is ellipsoidal rather than hemispherical, with (in dog ribs) a major (longitudinal) radius of about 300 μm and a minor radius (i.e., the cement line radius) of 100 μm. Thus, the time to open up the resorption space is 300/39 or about 8 days, and the rate of radial erosion is 100/8 or 12 μm/day. It is not clear just how the osteoclasts move in the process of opening the tunnel. One

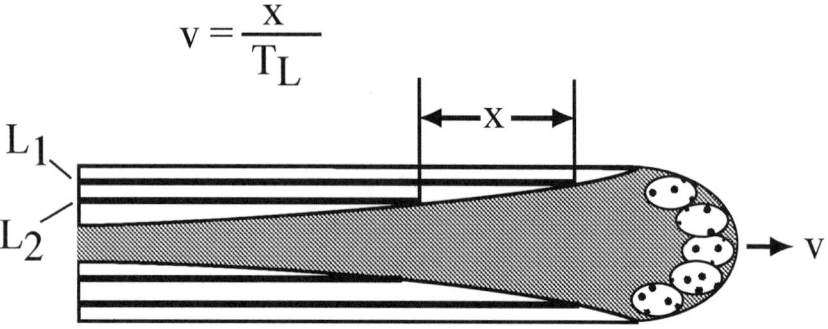

$$V = \frac{x}{T_L}$$

FIGURE 3.8. Schematic diagram shows measurement of osteonal BMU advancement rate in a longitudinal section. x is the distance between the leading edges of labels L_1 and L_2 given T_L days apart.

can imagine each cell starting at the apex of the cutting cone and digging radially, or all the cells digging straight ahead. So little is known of the dynamics of cells in the osteonal BMU that neither possibility has much evidence in its favor, but radial motions would seem to require that the cells either have short lifetimes or recirculate.

How Long Does an Osteoclast Live?
The average lifetime of osteoclasts is a rather arbitrary concept, for they appear to consist of an evolving syncytium of mononuclear cells that have fused. This was learned as a result of experiments in which dogs were injected with tritiated thymidine and the labeled cell nuclei traced (Jaworski and Hooper, 1980; Jaworski et al., 1981). When this was done, the number of labeled osteoclast nuclei increased and then decreased, demonstrating that nuclei enter and leave the cell while it persists in its activity. The average life span of a labeled nucleus in canine osteoclasts is 11 days.

The Phagocytic Connection
New cells are apparently added to and removed from the osteoclast as time goes by. These pre- and postosteoclasts are thought to be *mononuclear phagocytes* (MNPs), cells of hematopoietic origin that circulate in the blood and are attracted to sites of injury or infection. There, groups of them fuse to form macrophages that engulf, digest, and eliminate foreign debris and damaged tissue. The osteoclast is also a composite of MNPs. The branching point between macrophages and osteoclasts seems to occur early in the phylogeny of the MNP. There are complex relationships between bone cells and hematopoietic cells, and strong arguments for close connections between inflammatory responses and bone remodeling. This idea recalls the intense activation of BMUs precipitated by trauma or stress, known as a regional acceleratory phenomenon, or **RAP**. It is possi-

ble that this phenomenon is related to inflammatory responses mediated by cells of the immune system.

MNP-like cells are observed in the reversal region, "dressing" the walls of the resorption cavity and possibly creating the cement line over this surface (Baron et al., 1984). Perhaps MNPs in the reversal region are "post-osteoclasts" in the sense that they are phagocytes that have reemerged from osteoclasts in the cutting cone. There is evidence that by-products of osteoclasis, such as collagen fragments, lysosomal enzymes, and glycosaminoglycans, are chemotactic for MNPs, causing these cells to approach the cavity wall and begin preparing the surface to receive osteoid. There is also evidence that osteoblast precursors are attracted and induced to differentiate and begin laying down bone by similar chemotactic factors near the cement line.

Refilling: In with the New

In some respects, the refilling process is more amenable to analysis than the resorption process, partly because newly mineralizing bone surfaces can be labeled with fluorochrome markers. The basic processes of bone formation within BMUs are clear, but many questions remain about the details.

Osteoblast Origins

Osteoblasts are mononuclear cells that are smaller and less mobile than osteoclasts. As previously noted, they apparently originate from stem cells in the mesenchyme (the embryonic tissue from which all connective tissues are derived). Thus, although bone formation follows resorption in bone remodeling, labeling BMU cell nuclei has excluded the possibility that osteoblasts are differentiated from osteoclasts or their remnants. Cells from many tissues (e.g., spleen) may be induced to differentiate into osteoblasts and form bone by continuous exposure to certain cytokines or tissues (e.g., urinary bladder epithelium), but only cells from bone marrow and the periosteum **spontaneously** produce osteoblasts.

An interesting observation by Coccia et al. (1980) underscores the different origins of osteoblasts and osteoclasts. After a man's marrow was transplanted into a female osteopetrosis[2] patient, her osteoclasts were found to contain the man's Y chromosome, but her osteoblasts did not. This implies that the two cell types have different origins; it also suggests that the osteoblast precursors in marrow do not survive transplantation. There are two principal types of cells in marrow: hematopoietic and stro-

---

[2]Osteopetrosis (literally, "stone-bones") is a disease caused by an inability to produce sufficient osteoclasts. Thus, normal resorption and remodeling cannot occur, and the patient's bones are much harder and more dense than usual. It can be treated by transplanting marrow containing osteoclast pre-cursor cells from a normal person to the patient.

mal cells. Owen (1980) proposed that hematopoietic cells are the source of osteoclasts, and that the mesenchymal stem cells which are the source of marrow's stromal cells are the source of osteoblasts. Presumably, mesenchymal cells are also required for growth of the vascular system within the osteon. As mesenchymal cells do not circulate, it seems that osteoblast precursors must come from a self-perpetuating pool of mesenchymal cells that follows the cutting cone from its point of origin.

In any event, assuming that osteoblast precursor cells are present in the reversal region, there must be mitogenic and chemotactic factors to orchestrate their conversion to functioning osteoblasts on the refilling surface. The products of bone dissolution by osteoclasts could perform this function and ultimately attract differentiated osteoblasts to the bone surface. Both monocytes and osteoblast-like cells are attracted to some portion of the disassembled collagen molecule.

Mechanical loading appears to be required for osteoblasts to differentiate from their precursors. Using a simple yet elegant experimental approach, Roberts et al. (1982) characterized the stages of osteoblast differentiation in terms of nuclear size and used mathematical models to reveal a differentiation pathway through four precursor cell types to the functional osteoblast (Fig. 3.9). They found that, in the rat periodontal ligament,[3] the differentiation process is keyed to five light–dark cycles, so that the entire process requires 60 h or 2.5 days. Furthermore, the transition from stem cells to committed osteoprogenitors (type A cells) to preosteoblasts (type C and D cells) requires the presence of normal levels of mechanical stress (or strain) in the tissues; this has been confirmed in rats

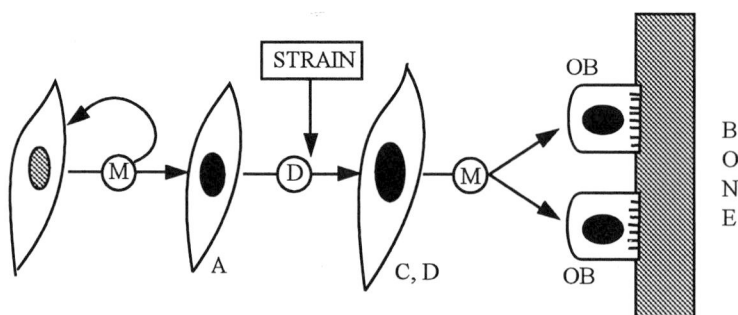

FIGURE 3.9. Diagram shows osteoblast precursor cells differentiating through intermediate stages. M, mitosis; D, differentiation; OB, osteoblasts. (After Roberts et al., *American Journal of Anatomy*, 1984, with permission from Wiley-Liss, Inc., a subsidiary of John Wiley and Sons, Inc.)

---

[3]The periodontal ligament is the tissue that supports each tooth in its socket. It is the interface between tooth and bone, and is the source of cells that excavate or fill in bone when teeth are moved naturally or by an orthodontist.

subjected to either simulated or actual weightlessness (Barou et al., 1996). In an osteonal BMU advancing at 40 μm/day, the cutting cone would move forward 100 μm during the time required for osteoblasts to go through the differentiation process. Allowing some time for events preliminary to those observed by Roberts, these observations are entirely consistent with the fact that reversal regions are typically 100–200 μm long. Abnormally long reversal regions could mean that the osteoblast differentiation process has been compromised, but the effects of disuse on osteoblast differentiation in **osteonal** BMUs have not yet been observed.

Labeling with tritiated thymidine has revealed that osteonal osteoblasts derive from spindle-shaped cells located near the vascular sinus in the resorption cavity, behind the advancing osteoclasts (Jaworski and Hooper, 1980). These spindle cells are believed to constantly divide to maintain their population while half the new cells differentiate into osteoblasts, a process consistent with Roberts' observations in periodontal ligament and trabecular bone. Furthermore, the differentiation process appears to carry each subsequent group of precursors further from the capillary bud. Once created, osteoblasts begin work as cube-shaped cells occupying an area of bone surface about 15 μm square, and become shorter, flatter, and extend over more bone surface as they work. Ultimately, some of them end up embedded in bone as osteocytes, and others as lining cells, each covering perhaps 300 μm² of the finished bone surface.

Jaworski and Hooper (1980) also showed that osteoblasts do not move along the filling cone, but remain stationary with respect to the surrounding bone, falling further and further behind the cutting cone. Because the BMU is advancing about 40 μm/day, and each osteoblast occupies about 15 μm of the closing cone's length, simple geometry suggests that about three new rings of cells must be created each day. If the radius of the resorption cavity is initially 100 μm, its circumference is about 600 μm, and each new ring could contain some 40 osteoblasts. Thus, about 120 osteoblasts must be produced daily to maintain the filling cone.

One or Many Generations?

Is the initial ring of osteoblasts responsible for the complete refilling job at its section of the osteon, or do additional generations of osteoblasts follow to complete the work? If the former possibility obtains, then the lifetime of some osteoblasts must reach T_F, or about 3 months. Alternatively, several generations of osteoblasts could do the job, each living a fraction of this time.

Frost (1960b) described "resting" osteonal osteoid seams based on a qualitative change in their histologic appearance when stained with basic fuchsin. He associated these characteristics with lack of tetracycline uptake but normal osteoid thickness, suggesting that *both* the elaboration of osteoid by osteoblasts and its mineralization ceased simultaneously. This could occur because one generation of osteoblasts has reached the end of its lifetime and another was being recruited to resume the work. If multiple generations of osteoblasts refill the osteon, both resting osteoid seams and lamellar structure could result from the transition from one generation of osteoblasts to another. The exact nature of lamellar structure is controversial, but if one merely accepts the view that completed Haversian systems contain a number of discrete layers of bone between the cement line and the canal, then the boundaries between these layers may somehow be associated with interruptions or pauses in the refilling process.

Johnson (1964) suggested that refilling discontinuities involve the replacement of generations of osteoblasts rather than pauses in the activity of the original cells; existing data may be used to test this hypothesis. Using polarized light, Kelin and Frost (1964) counted the numbers of lamellae in 350 osteons in normal human ribs; they counted bright layers, which were separated by dark layers. They found 7.9 ± 0.5 lamellar **pairs** per osteon. The mean wall thickness of an osteon is typically 80 μm. If this wall contains eight lamellar pairs, the average pair thickness is 10 μm. For a mean apposition rate of 1 μm/day, the time required to form the first lamellar pair is 10 days; this can be regarded as a **maximum** time, as we show that the initial accretion rate is normally much greater than 1 μm/day. Thus, the 11-day period of the Jaworski–Hooper experiment should have seen the labeled osteoblasts well into formation of the second lamellar pair, suggesting that lamellar structure is not correlated with separate generations of osteoblasts. On the other hand, this observation certainly does not prove that refilling is completed by only one generation of cells, and it leaves open the possibility that lamellar structure is associated with mere pauses in osteoblastic activity.

Tam and his co-workers (1980) showed that apposition rate measurements in rats yield results that are inversely proportional to the time between labels when this time is greater than 72 h. This result could be explained by the occurrence of 3-day rest periods and longer work periods during refilling. As shown in Fig. 3.10, interlabel times longer than 3 days (e.g., *A* and *B*) could produce double-labeled BMUs that had rested between labels. These BMUs would underestimate apposition rate because they would overestimate the time actually devoted to making bone. Double-labeled BMUs obtained with interlabel times shorter than 3 days (e.g., *C* and *D* in Fig. 3.10) could not have rested, and would thus give accurate results. A 3-day rest would be long enough to include Roberts' 60-h osteoblast differentiation time. Thus, the "rest period" could be either a "sabbatical" for a persistent population of osteoblasts, or a time in which a new generation of osteoblasts is produced.

Apposition Rate Diminishment During Refilling

There has been some controversy regarding the uniformity of the rate at which osteonal BMUs refill. As refilling is initiated, it appears that there is rapid bone formation that "plasters over" the cusps of Howship's lacunae

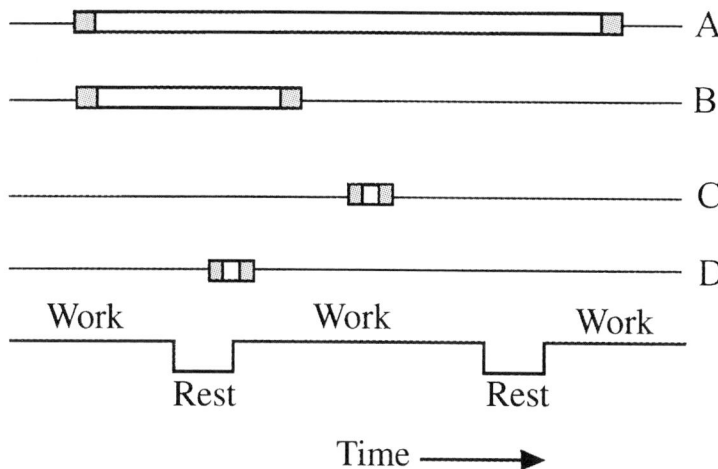

FIGURE 3.10. Schematic diagram shows possible relationships between fluo-rochrome labels and putative osteoblast work-rest cycles in a refilling BMU. Labels of varying intervals are shown as *shaded blocks* at ends of label intervals on time lines A–D. Assumed work and rest periods for bone formation are shown at *bottom*.

and fills in other eccentricities in the resorption cavity surface. Vincent (1955) originally observed that the rate of refilling slows down in propor-tion to the diameter of the remaining cavity. Other investigators suggested that the apposition rate was constant over most of the refilling period, or depended on the diameter of the original osteonal resorption cavity rather than on the age of the osteoblasts or the time since refilling began.

Manson and Waters (1965) performed an analysis that resolved the issue. They postulated that the apposition rate, M_F, is directly proportional to the radius, R, of the refilling cavity. That is,

$$M_F = dR/dt = -k\,R \qquad (3.18)$$

where k is a constant and the minus sign occurs because dR/dt is directed opposite to the radius vector. The solution to this equation is

$$R = R_c\,e^{-kt} \qquad (3.19)$$

where R_c is the cement line radius. Figure 3.11 shows this exponential model's prediction about how refilling would proceed. T_F can be found by solving for t when R is the Haversian canal radius, R_H. One obtains

$$T_F = -\ln(R_H/R_c)/k \qquad (3.\ 20)$$

This expression is fundamentally different from the conventional histomor-phometric formula,

$$T_F = (R_c - R_H)/M_F \qquad (3.21)$$

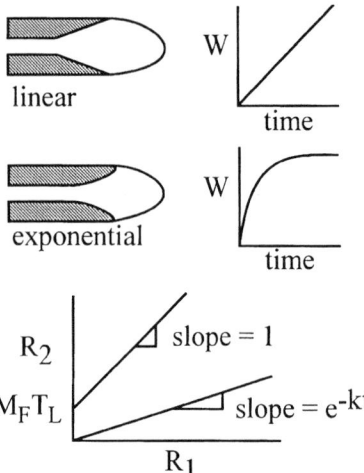

FIGURE 3.11. Linear and exponential models for refilling compared. *Upper left:* The shapes of the refilling surfaces in longitudinal profile. *Upper right:* Osteon wall thickness, *W*, as a function of time at a single longitudinal location. *Bottom:* Plots of the radii of double fluorochrome labels have different slopes and intercepts in the two models.

which assumes a constant apposition rate.

To test the model and obtain values for the constant k, Manson and Waters planned an experiment in which fluorochrome labels would be given at times T_1 and T_2, producing rings of radii R_1 and R_2, respectively, in those osteons that were doubly labeled. Consider now the result of plotting R_2 vs. R_1. **If the apposition rate were constant,** one would have

$$R_2 = R_1 + M_F T_L \tag{3.22}$$

where $T_L = T_2 - T_1$ is the interlabel interval. The plot of R_2 vs. R_1 should produce a straight line with a slope of 1.0 and an intercept proportional to M_F and the interlabel time, T_L. From their exponential model, Manson and Waters obtained a distinctly different expression:

$$R_2 = R_1 \exp(-kT_L) \tag{3.23}$$

which indicates the plot should pass through the origin and have a slope determined by k and T_L. Upon executing their experiment in cats of two different ages, and analyzing several different bones, they found that their data clearly fit Eq. 3.23 much better than Eq. 3.22 (Table 3.2). The intercepts of the R_2 vs. R_1 plots were nearly zero, and the slopes were significantly different from unity. Thus, the data confirmed the hypothesis that the apposition rate declines exponentially during refilling.

TABLE 3.2. Apposition rate experiments

Age of cat, years	0.75	2.5
Osteons analyzed	101	169
Interlabel time, days	14	28
Mean slope	0.539	0.436
Mean correlation coefficient	.933	.943
k, days^{-1}	0.0441	0.0296

Data from Manson and Waters, 1965.

Computation of T_F

Although the formula for computing T_F that arises from the Manson–Waters theory is different from the conventional histomorphometric formula, if the **mean** apposition rate is used in the conventional formula (Eq. 3.3 or Eq. 3.21) the correct result is nevertheless obtained. The mean apposition rate is related to k by the expression

$$M_{av} = k\,(R_c - R_H)/\ln(R_c/R_H) \tag{3.24}$$

Table 3.3 shows values of initial, final, and mean apposition rate, and T_F for four possible values of k, assuming that the cement line and Haversian canal radii are 100 and 20 µm, respectively.

The Equal Lamellar Work Areas Principle

The fact that osteonal refilling proceeds at an exponentially decreasing rate leads to an important geometric characteristic of the filling cone. As we have seen, there are typically eight lamellar pairs in the osteonal wall, and their thickness is remarkably uniform. In an idealized filling cone, nonlinear refilling leads to a horn-shaped rather than a truly conical surface (Fig. 3.12). Note that the appositional surface of each lamella increases as one proceeds from right to left in the figure (e.g., $x_2 > x_1$ and $s_2 > s_1$). Because the apposition rate declines during refilling, the inner lamellae not only take longer to form than the outer lamellae, but they also present more surface on which the cells can work. In fact, it can easily be shown that this longitudinal extension of the lamellar work surface just compensates the

TABLE 3.3. Apposition rate modeling: effect of k on other variables

k, day^{-1}	M_F, µm/day Initial	Final	Mean	T_F, days
0.01	1.0	0.2	0.5	160
0.02	2.0	0.4	1.0	80
0.03	3.0	0.6	1.5	53
0.04	4.0	0.8	2.0	40

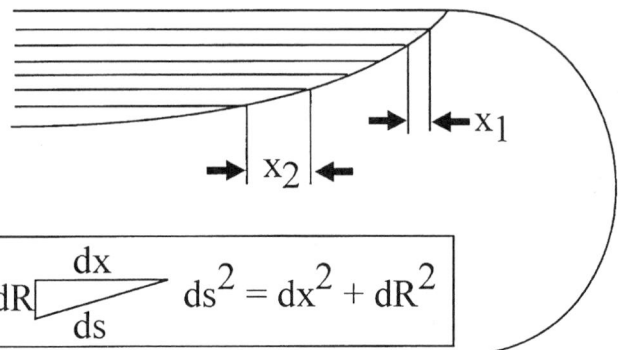

FIGURE 3.12. Effect of nonlinear refilling on exposed surface of individual lamellae. Inner lamellae have a broader surface on which formation can occur, but osteoblasts at this point lay down bone more slowly. *Inset:* Relationship between *ds*, *dx*, and *dR*.

loss of work area resulting from the contracting circumference. To see this, consider the calculation of the surface, dS, of a single lamellar work region bound by longitudinal positions x_1 and x_2. Because $dS = 2 \pi R(x) \, ds$ and $ds^2 = dx^2 + dR^2$, the surface integral is

$$S = 2 \pi \int_{x_1}^{x_2} R(x) \sqrt{dx^2 + dR^2} = 2 \pi \int_{x_1}^{x_2} R(x) \sqrt{1 + (dR/dx)^2} \, dx \quad (3.25)$$

where x is the longitudinal coordinate and R is the radius of the filling cavity. For exponentially declining refilling,

$$R = R_c \, e^{-kt} \quad (3.26)$$

This expression can be converted to one for $R(x)$ by using the longitudinal advancement rate, v, of the osteon. That is, substituting

$$t = x/v \quad (3.27)$$

into Eq. 3.26 gives

$$R(x) = R_c \, e^{-kx/v}, \quad dR/dx = -(kR_c/v) \, e^{-kx/v} \quad (3.28)$$

and substituting these expressions into the integral yields

$$S = 2 \int_{x_1}^{x_2} R_c \, e^{-kx/v} \sqrt{1 - (kR_c/v)^2 \, e^{-2kx/v}} \, dx \quad (3.29)$$

It can be shown that the square root term is almost equal to 1 for any value of x and typical values of k (0.02/day), R_c (100 μm), and v (40 μm/day). Making such an approximation and performing the integration,

$$S = 2\ R_c\ (v/k)\ [\exp(-kx_1/v) - \exp(-kx_2/v)] \qquad (3.30)$$

Note that because $R_1 = R_c \exp(-kx_1/v)$ and $R_2 = R_c \exp(-kx_2/v)$, the term in brackets can be replaced by $(R_1 - R_2)/R_c$. Therefore

$$S = 2\ v\ (R_1 - R_2)/k \qquad (3.31)$$

Now $(R_1 - R_2)$ is just the lamellar pair thickness, which is nearly constant for all lamellae. Thus, the lamellar work surface, S, turns out to be equal to a constant function of k, v, and lamellar thickness.

Equation 3.31 may be used to estimate the number of osteoblasts that work on each lamellar pair. Letting T_F = 90 days, R_c = 100 μm, and R_H = 20 μm, Eq. 3.20 yields k = 0.0179/day. Then if v = 40 μm/day, and the lamellar pair thickness is 10 μm, S = 2 × 3.14 × 40 × 10/0.0179 = 140,000 μm^2. Osteoblasts that are rapidly forming bone have a "footprint" on the bone surface which is about 10 μm in diameter; those which are slowly forming bone are about 20 μm across (Holtrop, 1990). Assuming, then, that the average osteoblast occupies approximately 15 × 15 = 225 μm^2, the number of cells working on each lamella is 140,000/225 = 622.

Why Does the Apposition Rate Slow Down?
We might postulate four mechanisms for declining apposition rates during refilling. (1) A single generation of osteoblasts refills the cavity and its cells slow down, or spend more time resting as the refilling proceeds, from either internal or external factors. (2) Several generations of osteoblasts refill the cavity, and each generation becomes more indolent, owing to genetic or environmental factors. (3) Several generations of osteoblasts refill the cavity, and the time required for differentiation of each new generation is increased, causing increasingly longer "rest periods." (4) Both (2) and (3) occur.

Manson and Waters asserted that osteoblastic activity is enhanced when the cells are further from the blood vessels in the osteonal cavity. They did not offer a rationale for this assertion, which seems contrary to a straightforward notion of cell nutrition via diffusion from a capillary. It is, however, consistent with another observation by Roberts et al. (1987) concerning osteoblast differentiation. When a periodontal blood vessel is too close to the bone-forming surface, osteoblast precursors are obliged to differentiate along a path that initially takes them away from the bone. They subsequently circle back to the bone surface. On the basis of this behavior, it could be postulated that as the osteonal resorption cavity is refilled, osteoblast precursors find it increasingly difficult to get enough "elbow room" between the blood vessel and the bone surface, and this affects their rate of differentiation or their subsequent ability to function.

This mechanism, based on spatial factors within the osteonal cavity, is external to the osteoblast; alternatively, it is possible that the decreased rate of osteoid synthesis is caused by factors internal to the cell. This distinction is important because internal actors would be likely to affect osteoblasts in osteonal and surface (i.e., periosteal, endosteal, or trabecular) BMUs alike, but external factors such as cell crowding may be peculiar to osteonal osteoblasts.

Martin and co-workers (1987) developed a model that assumes that variations in apposition rate are caused by factors within the osteoblast, and compared the abilities of the three available models (linear refilling, slowing from external factors, slowing from internal factors) to describe apposition rates in osteonal and trabecular bone. For osteons, the internal effects model yields essentially the same result as the Manson–Walters model. However, applying the internal effects model to trabecular bone by measuring R_1 and R_2 from the cement line,

$$R_2 = R_1 \exp(-kT_L) + C\,[1 - \exp(-kT_L)] \tag{3.32}$$

where k and C are constants. In this case, the slope term is similar to that in the Manson–Waters theory, but now there is also an intercept term. Thus, the three apposition rate models yield distinctly different predictions about R_1 vs. R_2 plots, as shown in Table 3.4.

These theoretical models were tested against R_1–R_2 data from both osteonal and trabecular BMUs in adult dogs. As shown in Table 3.5, it was found that *osteonal* apposition rates had slopes significantly less than 1 and intercepts close to 0, but *trabecular* BMUs had slopes very close to 1 and intercepts nearly equal to $M_F T_L$. Thus, apposition rates were found to be

TABLE 3.4. Apposition rate models: predictions about R_1 vs. R_2 plots

	Slope	Intercept
Linear refilling	1	$M_F T_L$
External factors	<1	0
Internal factors		
Osteonal BMUs	<1	0
Trabecular BMUs	<1	C (1 − slope)

TABLE 3.5. Apposition rates in osteonal and trabecular canine BMUs

	No. of dogs	No. of BMUs	Correlation coefficient	Slope	Intercept	$M_F T_L$
Osteonal BMUs						
Rib	7	88	.97 ± .02	.77 ± .06	−5.3 ± 1.4	15
Humerus	5	140	.95 ± .02	.70 ± .12	−1.5 ± 3.3	14
Trabecular BMUs						
Ilium	5	67	.99 ± .03	1.02 ± .05	9.6 ± 1.7	8

Data from Martin et al., 1987.

constant in trabecular BMUs but not in osteons. If internal factors were responsible for the nonlinear refilling in osteons, trabecular osteoblasts should be similarly affected; as they were not, it appears that external factors are responsible for reducing the rate of bone apposition as osteons refill.

In thinking about the external factors that could cause this effect, we may reiterate our previous suggestion that decreasing space between the cavity wall and the blood vessel interferes with osteoblast differentiation. Another possibility has to do with Roberts' observation that the differentiation of osteoblasts is stress dependent. One would expect to find a stress concentration at the front end of a BMU, with diminishing stresses along the refilling region. Perhaps this diminishment of bone strain interferes with osteoblast production. This idea is clarified in Chapter 6, where we describe a theory relating wall strains in the osteonal BMU to cell function and directional control.

Osteoblast Accounting
Another aspect of refilling that appears to be related to both lamellar structure and cell lifetime is the conversion of osteoblasts to osteocytes. As shown in Fig. 2.12, osteocytes are formed when osteoblasts bury themselves in osteoid. Jaworski and Hooper found that the first osteocytes appeared 9 days into the refilling period; this places the first osteocytes near the junction between the first and second lamellar pairs.

It has been argued that because the working surface contracts as osteonal refilling proceeds, some osteoblasts should be forced out into the new bone matrix or back into the resorption cavity (Parfitt, 1983). Pursuing this assumption quantitatively, it appears that the number of displaced osteoblasts would be much greater than the number of osteocytes and bone lining cells in a unit length of completed osteon. We first consider what fraction of the osteoblasts become osteocytes using the scheme of cell dynamics presented here. During the formation of each lamellar pair, 1 of every 28 cells gets trapped and becomes an osteocyte. This figure is deduced as follows. Consider the ith lamellar pair, which consists of a "pipe" of bone having an outer radius R_o and an inner radius R_i. Segments of the length of this pipe may be thought of as being formed by successive rings of osteoblasts arranged around the BMU cavity. If $b = 15$ μm is the breadth of the work area of an individual osteoblast, then the volume of such a segment of bone formed by a single ring of osteoblasts is

$$V = \pi b \, (R_o^2 - R_i^2) \qquad (3.33)$$

The average density of osteocytes in cortical bone is 12,000–20,000/mm^3 (Boyde, 1972; Frost, 1960d; Mullender et al., 1996). Let q be such an osteocyte density expressed as cells/μm^3 and let its average value be 16×10^{-6}. Then the expected number of osteocytes in the pipe segment of volume V is

$$N_c = q \, V = \pi b \, q \, (R_o^2 - R_i^2) \qquad (3.34)$$

The average circumference of the work surface during formation of the pipe segment is

$$C = 2 \pi (R_o + R_i)/2 \qquad (3.35)$$

Assuming there is an osteoblast every b μm along this circumference, the average number of cells in the osteoblast ring would be

$$N_b = \pi (R_o + R_i)/b \qquad (3.36)$$

Combining and simplifying Eqs. 3.34 and 3.36, the ratio of the number of osteocytes buried in the pipe segment to the number of osteoblasts that formed the segment is

$$N_c/N_b = q \, b^2 \, (R_o - R_i) \qquad (3.37)$$

Note that the term in parentheses is just the lamellar pair width, which can be assumed to be about 10 μm in all cases. Therefore, the fraction of osteoblasts converted to osteocytes is also constant for all lamellae, and works out to be $N_c/N_b = (16 \times 10^{-6})(15)^2(10) = 0.036 = 1/28$.

Assuming the cement line radius was 100 μm, and the osteoblasts averaged 10 μm in diameter at the start of refilling, the original ring of osteoblasts would have contained about $2\pi \times 100/10 = 63$ cells. Then only a couple of these cells could be expected to end up as osteocytes. If the radius of the completed Haversian canal is 20 μm, and it is covered by bone lining cells having a diameter of about 20 μm, there would be about 2π or 6 lining cells left from the original ring of 63 osteoblasts. That leaves about 55 osteoblasts unaccounted for, assuming that a single generation of cells accomplishes the refilling. One interesting possibility is that these cells are "surfing" on the advancing wave of new bone, that is, translating forward in the wake of the advancing resorption cavity, which would reduce the numbers of new osteoblasts that have to be produced. Alternatively, some of these cells may be forced off the shrinking refilling surface and die, or revert to another type of cell. What actually happens is unknown.

3.4 Long-Term Effects of Osteonal Remodeling: Implications for the Aging Skeleton

Activation frequency in the human rib declines from very high levels in childhood to a minimum at about age 35 years, and then rises and falls again at the end of an average lifespan (see Fig. 2.21). Unfortunately, such data are not available for long bones such as the femur or for trabecular bone. It is assumed, however, that this curve is generally descriptive of f_a vs. age, although the magnitudes of the values would be different in other bones.

Anyone who has studied dynamic systems theory will be struck by the similarity of this curve to that describing the behavior of a damped oscillation. One way that such oscillations can occur is when a feedback system is "overdriven," or operated with too much feedback. Whether the variation in activation rate has anything to do with an overdriven regulatory system remains to be learned, but it is certain that the shape and scale of this curve have a number of important consequences.

One of the most obvious of these consequences is that the rate of addition of new osteons, and the rate of turnover of cortical bone, is much higher in children than in adults. Most cortical bone is created by modeling processes associated with growth. The high values of f_a in children ensure that most of this bone is rapidly converted to secondary osteonal bone. One may postulate that later, during adulthood, f_a is maintained at some minimal level by systemic factors and is increased only under conditions of mechanical or metabolic need.

It appears that the difference between the values of f_a in children and adults is greater than the difference in levels of activity or mechanical usage in these two groups. The elevated activation rate is generally assumed to be a growth hormone-driven phenomenon, but part of the elevation may be caused by age-related variations in material properties of the bone matrix. The rapid growth and high rate of remodeling of children's bone causes it to be undermineralized. This deficit results in a relatively low elastic modulus, producing higher strains for a given load or level of stress. Nunamaker et al. (1990) have shown that peak strains in growing horses approach 5000 microstrain but are reduced to less than 3000 microstrain after maturity. Strain levels this high may cause increased fatigue damage and thus increased activation of new BMUs.

After declining to a minimum at about age 35, the BMU activation rate increases again to about twice the minimal value in men and women in their sixties. This increase tends to amplify any deficit in remodeling at the BMU level that occurs in adults as well as that which may be associated with aging itself. The final decline in f_a in people in their eighties and nineties tends to assuage the effects of senile remodeling deficits.

Osteonal Overlapping

Osteon accumulation is limited by the fact that osteons soon begin to overlap one another as more and more are added. When an osteon overlaps another, it is not prevented from "obliterating" (i.e., refilling) its predecessor's Haversian canal. While in general new osteons create new canals, they may also destroy an old canal. Age-related changes in porosity therefore are affected by the probability that old canals will be obliterated.

The analysis of osteon accumulation and overlapping is a geometric problem not unlike the one astrogeologists have encountered in trying to estimate planetary age from the numbers and overlapping of craters seen on the surface. Physical anthropologists have used osteon population density to estimate the age at death of skeletal remains. A mathematical theory can help to clarify the nature of senile osteoporosis in cortical bone. It can also allow one to use a simple census of osteons to estimate activation and bone turnover rates—information that normally can only be obtained from vital labeling of living animals.

The Random Remodeling Assumption

Because osteons are quite long in comparison to their diameter, their overlapping must be determined as much by the "steering" of the BMU as by its point of origin. It has been hypothesized that osteonal orientation is controlled by the principal strain direction, but it is also reasonable to think that it is affected by variations in the material properties or chemical composition of the bone which appears before the advancing osteoclasts. That is, regions of increased mineral content, patches of osteoid, and canals could conceivably alter the path. As all these factors may be expected to be at least a little different for the new BMU than for its predecessors, one would expect the steering of an osteon to be different from its nearby older neighbors, leading to quasirandom positioning of each new osteon. This "local randomness" does not imply that new osteons are randomly positioned with respect to the whole cross section but only with respect to local organization. Such local randomness is an underlying assumption in most analyses of the effects of osteonal remodeling on such tissue properties as porosity and mean tissue age.

Mathematical Theories

The problem of determining the average age of a region of bone (tissue age, or mean skeletal age) is very closely linked to that of osteon accumulation, and has been discussed in some detail by Hattner and Frost (1963). Tissue age depends on the rate of remodeling and the degree to which new osteons obliterate those previously formed. We now examine the effects of the accumulation of osteons inside a cortical bone cross section using probability theory.

Suppose that within an area A_T of cross section (Fig. 3.13) a new resorption space is equally likely to appear at any location. The area of existing Haversian canals that this resorption space can be expected to obliterate is, on the average,

$$A = A_o\, p \tag{3.38}$$

where A_o is the area of the resorption space and p is the porosity (total area of Haversian canals/total area, $p = A_V/A_T$) of the region. The resorption space will refill to leave a Haversian canal of area A_H. The net change in void area to be expected is, when averaged over many such incidents,

$$dA_V = A_H - A_o\, p \tag{3.39}$$

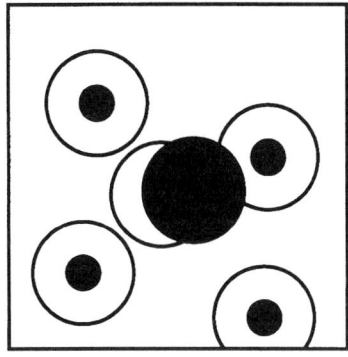

FIGURE 3.13. A new resorption space (*black circle*) appears at a random location in a field of osteons in a cross section having total area A_T.

The number of new osteons added to the section per square millimeter per year is just the activation frequency, f_a. If one defines an *osteon area ratio*, f, as A_H/A_o, then in an interval of time dt one has

$$dA_V = (f - p) f_a A_o A_T \, dt \qquad (3.40)$$

where A_T is the total area of the region. Dividing by $A_T dt$, one has

$$dp/dt = (f - p) f_a A_o \qquad (3.41)$$

where $dp/dt = dA_V/A_T dt$ is the time rate of change of porosity resulting from Haversian canals. If f, f_a, and A_o can be assumed to be constant, Eq. 3.41 has the simple exponential solution

$$p = f - (f - p_0) \exp(-f_a A_o t) \qquad (3.42)$$

where p_0 is the initial porosity. When this equation is plotted, it is apparent that as t becomes large, the porosity of the bone does not increase indefinitely, but asymptotically approaches f, which may be viewed as the porosity of each osteon. The equation may be used to study the ways in which osteon dimensions and activation frequency influence the rate of increase of porosity.

Equation 3.42 can be converted to a formula for the number of Haversian canals per unit area (N) by substituting $f = A_H/A_o$ and $p = NA_H$. The result is

$$N = 1/A_o - (1/A_o - N_o) \exp(-f_a A_o t) \qquad (3.43)$$

or, if the initial osteon density, N_o, is negligible,

$$N = [1 - \exp(-f_a A_o t)]/A_o \qquad (3.44)$$

This variable also is limited in its growth, and one can see that the maximum number of osteons/mm^2 that can be packed into the bone is equal to the reciprocal of the osteonal area, A_o. (In Chapter 5 it will be seen that this reciprocity between osteon size and packing density has significance in the resistance of bone to crack propagation.) The initial rate at which osteon density increases turns out to be a function only of activation frequency; subsequently, smaller osteons cause N to grow at a faster rate.

Also of interest is the fractional area of a section converted to secondary osteons or their fragments after a given period of time. This fraction, F_r, can be calculated as follows. Suppose a single resorption space of area A_o appears in a field of area A_T of which $A_r = F_r A_T$ has already been remodeled. The expected portion of A_r that would be removed by the new resorption space is

$$A = A_o F_r \tag{3.45}$$

The expected increase in A_r is then

$$dA_r = (1 - F_r) A_o \tag{3.46}$$

In an increment of time dt the number of new resorption spaces appearing on the section would be $f_a A_T \, dt$. Over this period,

$$dA_r = (1 - F_r) A_o A_T f_a \, dt \tag{3.47}$$

Because $dA_r = dF_r A_T$, one may rewrite this as

$$dF_r/dt = (1 - F_r) f_a A_o \tag{3.48}$$

Assuming that F_r was initially zero, this differential equation has the solution

$$F_r = 1 - \exp(-f_a A_o t) \tag{3.49}$$

where t is the time during which the remodeling has occurred. F_r asymptotically approaches 100%, with the initial rate of increase being equal to the product of activation frequency and osteon size.

The fraction, F_p, of the cross-sectional area occupied by primary (i.e., unremodeled) bone would be (since $F_p = 1 - F_r$)

$$F_p = F_{p0} \exp(-f_a A_o t) \tag{3.50}$$

where F_{p0} is the initial value of F_p. Naturally, this variable decreases at the same rate that F_r increases, eventually reaching zero.

This kind of analysis shows the essential ways in which mechanically important variables such as porosity and osteon density depend on remodeling parameters such as osteon dimensions and activation frequency. If the models can be substantiated by comparison with experimental data, they have the advantage that, unlike empirical correlations between the variables (e.g., polynomial curves), they lead to an understanding of the mechanisms underlying the phenomena.

The derivations of these equations assumed that A_o, A_H, and activation frequency are constant. Are these assumptions valid? A_H and A_o do not vary much over the course of a lifetime, but we have seen that the activation frequency can vary greatly, contrary to the assumption made in deriving the equations. Figure 2.21 showed that f_a may be 20–30 times greater in childhood than in adults, and a substantial increase also occurs as adults age. This problem can be treated by using computer models that incorporate age-corrected values of f_a into the afore-mentioned integrals. On the other hand, Eq. 3.42, 3.43, 3.49, and 3.50 are useful in showing the basic exponential nature of osteonal accumulation, and they may be reasonably accurate in those regions of cortical bone that are created in late childhood, so that the variations in f_a are relatively small over an extended period of time. In general, the outer cortex of most long bones is formed in the latter stages of growth, and many bones "drift" in a particular direction so that much of the adult cross section contains little or no bone from early childhood. For example, most of the adult rib cross section is created after age 12 years (Wu et al., 1970). By comparing the foregoing equations to experimental data, one can find out how this kind of theory holds up in the face of age variations in f_a and A_o.

There are two important applications for these relationships. First, the increase in osteon numbers with age means that age-at-death may be predictable on the basis of the osteon population density or the fraction of the cortex occupied by secondary osteons. Second, the activation frequency may be estimated from the rate at which these variables increase with age. One may also compare the values of these variables with those determined by histomorphometric methods (which have their own limitations, as we have discussed.)

Comparison of the Theoretical Results with Experimental Data

Kerley (1965) pioneered the use of osteon population density measurements to determine age from skeletal remains. Figure 3.14 contains osteon density data that he published for the midshaft of the femur. These data are means for four sites evenly spaced around the outer half of the cortex. The best least-squares regression fit of these data to Eq. 3.44 yields the curve shown in Figure 3.14. The form of the theoretical curve matches the data very well, and the parameters of the curve allow one to deduce the mean values of A_o and f_a. (The radius of the osteon, R_o, is calculated from A_o assuming the osteons are circular.) The results of this analysis are shown in the upper part of Table 3.6, along with results for the tibia and fibula obtained from other graphs in Kerley's paper.[4] Note that the fibula is similar to the femur with respect to osteon

---

[4]In his 1965 paper, Kerley underestimated his microsopic field area (Kerley and Ubelaker, 1978). We have corrected for this in Table 3.6, but not in Figure 3.14, where the ordinate values should be multiplied by 0.60. Figure 3.15 needs no correction as it represents fractional area.

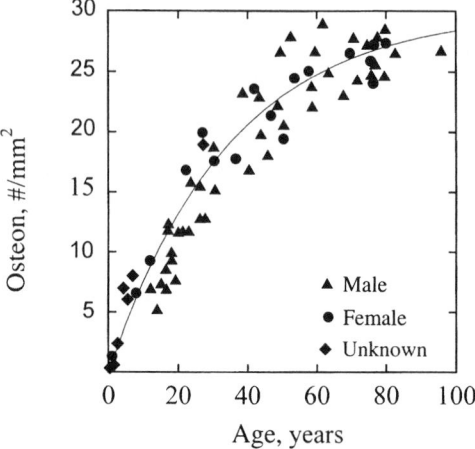

FIGURE 3.14. Osteon density vs. age for a sample of human femurs. (Reproduced with permission from Microscopic determination of age in human bone, Kerley, ER, *American J Physical Anthropology,* 1965, by Wiley-Liss, Inc., a subsidiary of John Wiley & Sons.)

size and activation frequency, but the tibia seems to have smaller osteons and a reduced rate of activation.

Kerley also published graphs relating age to the percent of cross-sectional area occupied by primary bone. Fitting Eq. 3.50 to these graphs yields a second estimate of f_aA_o for the same set of samples. Figure 3.15 shows the femoral example, and the second set of f_aA_o values appears in the bottom of Table 3.6. The form of Eq. 3.50 again matches the data, and the two estimates of f_aA_o for the femur are quite similar. The separate estimates of f_aA_o for the other two bones do not agree as well, but this may be partly because they are based on less than half as many data points. Also, while the points in the F_p vs. age graphs suggest an F_{p0} of 1, the regression concludes that the cross section started with a significant amount of secondary bone. This conclusion probably reflects the inaccuracy resulting from assuming f_a is

TABLE 3.6. Bone remodeling parameters calculated from experimental data

Parameter	Femur	Fibula	Tibia
Osteon density data			
f_aA_o, #/yr	0.029	0.026	0.018
f_a, #/mm^2/yr	0.53	0.46	0.38
R_o, μm	133	137	122
Fraction primary bone data			
f_aA_o, #/yr	0.031	0.036	0.026
F_{p0}	0.70	0.55	0.60

From Kerley (1965). Corrected according to data by Kerley and Ubelaker (1978).

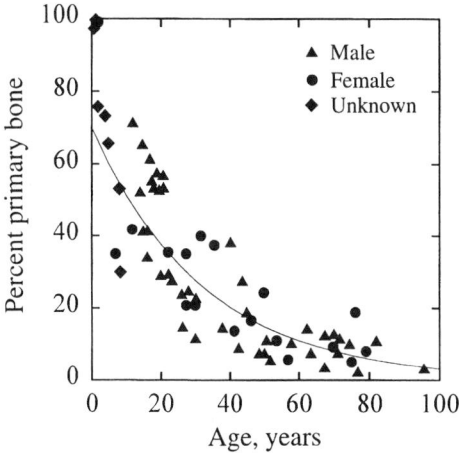

FIGURE 3.15. Percent of cortex that is primary bone vs. age for a sample of human femurs. (Reproduced with permission from Microscopic determination of age in human bone, Kerley, ER, *American J Physical Anthropology*, 1965, by Wiley-Liss, Inc., a subsidiary of John Wiley & Sons.)

constant when it is not. Even so, the consistently lower k of the tibia suggests that it remodels differently than the femur and fibula.

Others have independently reached similar conclusions. Evans and Bang (1967) correlated mechanical properties with microscopic structure in the same three bones. They, too, found that the fibula has the largest osteons and the tibia has the smallest. As one would expect from this, the fibula had the fewest osteons/mm^2, but the femur rather than the tibia has the greatest osteon density. The tibia, however, has the greatest number of osteon fragments/mm^2, suggesting a difference in the way osteons are positioned in these two bones. That is, new osteons in the tibia seem to be more likely to overlap existing osteons than those in the femur, creating more and smaller osteon fragments between whole osteons. With aging, the femur exhibited both the greatest increase in osteon density and the highest f_a estimate (see Table 3.6).

As we have noted, Kerley obtained the data in Figs. 3.14 and 3.15 for the purpose of finding the age-at-death of archeological bone specimens. Others have pursued this goal, sometimes correlating age with several different histomorphometric variables (Ahlqvist and Damsten, 1969; Bouvier and Ubelaker, 1977; Thompson, 1979; Stout and Gehlert, 1980; Ubelaker, 1989). The diminishing slope of the graphs in these figures suggests that determining the age-at-death from the amount of remodeling that has occurred will be less accurate in older individuals, and this is what Walker et al. (1994) found. In addition to the slope problem, one cannot be sure that the bone age reflects the person's age-at-death, as was noted. Because the diaphyses of adults expand radially with age, the oldest bone near the endosteal surface is preferentially removed; thus, the average age of the remaining bone may be considerably younger than the person. These limitations serve to illustrate the complex relationship between aging and remodeling in cortical bone, which depends on many other factors as well, including nutrition, activity, genetics, body size, and personal behavior (e.g., smoking).

The activation frequencies in Table 3.6 are all lower than those for adult rib shown in Table 3.1. There is a paucity of histomorphometrically measured values (i.e., using tetracycline labels) for the activation frequency of osteonal BMUs in bones other than the rib. One reason for this is that f_a is so low in cortical bone of the extremities of adults that a great many sections must be examined to find enough remodeling sites for a valid measurement. The theory presented here offers a way to estimate f_a for such low-turnover regions and to compare remodeling in one bone with that in another.

Another Theoretical Approach to Bone Remodeling

Polig and Jee (1990) applied stochastic renewal theory to the problem of determining the relationship between bone turnover and mean skeletal age. Mathematicians have developed the theory of stochastic processes in such a way that it is directly applicable to bone remodeling. By taking advantage of this, some very useful insights and relationships are quickly established. For example, it can be shown that mean skeletal age must be greater than half the average lifetime of any particular modicum of bone tissue. How much greater depends on the degree to which remodeling is random. The authors also showed how, if a single fluorochrome label is given and the amount of this label remaining in the bone is plotted as a function of time, the parameters of bone turnover and mean skeletal age can be obtained from the graph.

Box 3.3 Technical Note
Calculating Activation Frequency Without In Vivo Labels

Suppose that you dug up a mastodon bone in your garden and wanted to know know what its average bone formation rate had been. Could you estimate this without benefit of double tetracycline labeling? It is possible to do this by counting the number of osteons/mm^2 in a section and assuming that the total osteons created is the product of the activation frequency and the age of the bone (Wu et al., 1970). However, you have to find a way to estimate the number of osteons that were created but have since been obliterated by more recent remodeling, so that you can add them to the currently visible osteons to get the grand total. This is shown in the accompanying graph of osteon density (N) vs. bone age (t). As in Fig. 3.14, the curved line represents the visible osteons at any t; it has an asymptote at N_A. The straight line represents the total number of osteons created, $N_T(t)$. The difference between the two lines is the number of osteons obliterated. Frost (1987a) developed a simple algorithm for finding N_T when you have measured N and know the asymptotic value, N_A. First, calculate $\alpha = N/N_A$. Then compute $\beta = (1 - \alpha^{3.5})^{-1}$, and $N_T = \beta N$. If you know the age of the bone, t, the activation frequency is estimated as $f_a = N_T/t$ and the bone formation rate as $V_F = f_a A_o$, where A_o is the mean osteon area.

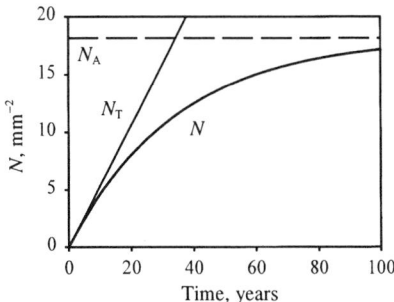

Theoretical plot of N (osteons/mm^2) vs. t (time) for a bone cross section having a constant activation frequency. The curve $N(t)$ is that obtained from Eq. 3.44. It has an asymptote at $N = N_A = 1/A_o$. The line $N_T(t)$ represents the total number of osteons/mm2 that have been created and would be seen if new osteons never obliterated old ones. Its slope is the activation frequency, f_a. The number of observed osteons, N, is always less than N_T and N_A.

Stout and Paine (1994) compared this method's results to measurements made with double tetracycline labels in human ribs. They found that Frost's method gave similar but somewhat lower results, especially in people more than 40 years old. Similar techniques have been used to estimate remodeling rates in human patients (Wu et al., 1970), nonhuman primates (Burr, 1992), and modern archeological populations (Burr et al., 1990), as well as in fossils of extinct animals (Wu et al., 1970) or of Neandertal and early *Homo sapiens* (Abbott et al., 1996).

Frost's algorithm is empirical, based on the asymptotic nature of $N(t)$ as seen in graphs like those of Kerley (1965). One may derive an alternative procedure on the basis of our analysis of randomly overlapping osteonal remodeling. According to Eq. 3.44, the asymptote of the $N(t)$ curve is $N_A = 1/A_o$. Thus, N_A can be actually be obtained by measuring A_o in the specimen of interest, assuming that this area does not change appreciably with age. Solving Eq. 3.44 for f_a, one obtains

$$f_a = -\ln(1 - A_o N)/A_o t$$

This equation can be solved for the activation frequency if A_o and N are measured directly from the section and the age of the bone, t, is known. Frost's algorithm and this approach give similar results (see Exercise 3.10).

An important difficulty in calculating activation frequency using either of these methods is that the chronological age of an individual often exceeds the age of the bone (t) because of drifts that occur during growth. Based on fluorochrome labeling studies, Wu et al. (1970) calculated an adjustment of 12.5 years for the middle of the sixth rib to account for the effective "birth" date of the compact bone seen in an adult; i.e., t = person's age − 12.5 years. The correction for other human bones is unknown, and the data necessary to calculate it are not available. Therefore, studies using these techniques often assume the bone age is the subject's age or some other constant; this makes comparison between studies difficult, but maintains consistency within a study.

3.5 Summary and Further Reading

In this chapter we have tried to gain a deeper understanding of bone remodeling. In the process, we have introduced you to several attempts to mathematically model this phenomenon and its effects. We have seen how Frost originally formulated the remodeling process in cortical bone so that it could be histomorphometrically quantified. We have probed deeper into some nooks and crannies of remodeling, trying to pin down details of the refilling process, for example. We concluded by examining the ways in which osteons could accumulate and change bone structure over a lifetime.

We must remember that our mathematical approach was based on the assumption that the morphology and physiology of the osteonal BMU can be expressed quantitatively, even though the histologic picture is one of considerable variation. The reason that this variability does not negate the usefulness of mathematical analysis is that, to be useful, the result must only correctly describe the average behavior of the system. The truth of this is aptly illustrated in the case of the histomorphometric measurement of remodeling dynamics and in the work of Manson and Waters. In both cases, conceptual models of osteonal remodeling that assume simple geometric and physiologic features are found to be very successful in describing the average behavior of large numbers of BMUs. It does not matter, for example, that the osteonal apposition rate is not exactly an exponential function, but seems to have an especially rapid early phase. This phenomenon is not apparent when the experimental data are compared to an exponentially varying apposition rate, so it may be regarded as a second-order effect. Second-order effects and histologic variability mark the limitations of remodeling theories, but do not prevent them from being useful tools for understanding physiology and disease.

Further reading should begin with Frost's 1964 book, *Mathematical Elements of Lamellar Bone Remodeling*, which laid the groundwork for this subject. *Bone Histomorphometry: Techniques and Interpretation*, edited by Recker (1983), contains several chapters on the details of quantitative histomorphometric analysis, as well as a chapter by Parfitt that remains an excellent resource for the details of bone remodeling. Two mathematical analyses of bone remodeling by Polig and Jee (1987, 1990) illustrate how a physicist and anatomist can very productively collaborate in this area. In addition, several investigators have published computer simulations of bone remodeling: Krumme et al. (1984), Kimmel (1985), Martin (1985), and Reeve (1986). Turner et al. (1993b) mathematically analyzed the effects of remodeling on fluoride uptake by the skeleton. Weinhold et al. (1994) examined bone remodeling as a dynamic system capable of oscillation. Finally, more details and examples of the approaches found in this chapter may be found in papers by Martin (1984, 1986, 1989, 1991).

3.6 Exercises

3.1. Derive Eq. 3.9 using the principle of similar triangles in Fig. 3.5. Explain how Eq. 3.10 is represented on the diagram.

The following exercises are intended to start you thinking in terms that can be used to mathematically model the consequences of remodeling processes which are varying in time. To do this, it is necessary to keep track of how many BMUs are currently in each stage of remodeling. These populations depend not on the current activation frequency but on what f_a was at some time in the past. For example, BMUs currently in the resorbing phase were activated or "born" between the present time and T_R days ago. BMUs in the reversal phase were born before that, and refilling BMUs were born even earlier than that. Also, keep in mind that we are considering the remodeling going on in a representative, two-dimensional cross-section of cortical bone, e.g., Fig. 3.4. In this context, "activation" of a BMU occurs when its leading osteoclasts first break the plane of the section, creating a new resorption cavity. It helps to imagine that you can observe such a cross section in your own femur as BMUs tunnel across it, excavating and refilling their tunnels.

3.2. Draw a horizontal line representing time. Place a tick mark representing the present time at the left end of the line, *so that past time is measured to the right*. Let the time scale be 1 cm/week. Assuming the BMU resorption period is T_R = 21 days = 3 weeks, mark the region of the line containing the "birthdays" (activation days) of all BMUs that are currently resorbing bone. Assuming the reversal period is T_I = 7 days = 1 week, mark the time span during which all BMUs presently in the reversal phase were activated. Finally, assuming the refilling phase lasts T_F = 56 days = 8 weeks, show the time span during which all currently refilling BMUs were "born."

3.3. Using the time line of Exercise 3.2, construct an ordinate axis to represent activation frequency, f_a. Plot the following BMU activation history: constant at 2 BMU per mm^2/week until 8 weeks ago, when it suddenly increased to 4 BMU per mm^2/week. It remained at that level until 2 weeks ago, when it rapidly decreased back to the original value. Show graphically the numbers of BMUs/mm^2 in the three phases of activity (N_R, N_I, and N_F, respectively).

3.4. Using the notation given in Exercises 3.2 and 3.3, write integral equations to represent N_R, N_I, and N_F as functions of $f_a(t)$, where t is time. Be sure to specify the integration limits, and continue to let past time be represented by positive numbers.

3.5. Now think about the cross-sectional view of a single osteon. Assume that the mean diameter of an osteonal cement line is 200 μm and that of the mean Haversian canal is 40 μm. Using these numbers, calculate Q_C, the mean rate at which bone is resorbed in each osteon during its resorption phase, in μm²/day, and Q_B, the mean rate at which bone is replaced during refilling of each osteon, in μm²/day. You will use these values in the next exercise. (B and C stand for 'blasts and 'clasts, respectively.)

3.6. Calculate N_R and N_F assuming the activation frequency has been decreasing linearly at a rate of 5 BMU per mm²/year and is presently 1 BMU per mm²/year. Use the integration methods you practiced in the earlier exercises, and the same values for T_R and T_F as before. Then calculate something called the "bone balance":

$$Q = Q_B N_F - Q_C N_R$$

using Q_B and Q_C from Exercise 3.5. Based on the equation, what are the units and meaning of $Q_B N_F$, $Q_C N_R$, and Q? Compare this bone balance with that which exists when the activation frequency is constant at 1 BMU per mm²/year, assuming everything else remains the same. Why is one number positive and the other negative? Discuss the relevance of this calculation to the porosity of cortical bone in children and adults.

3.7. Suppose a tetracycline label is given for 1 week. Write an equation for the volume fraction of the bone that has been labeled by the end of the week. (Remember, volume fraction is the same as the area fraction in our representative cross section. Also, assume that none of the labeled bone has been resorbed during the week.) Use data for a 30- to 39-year-old human from Table 3.1 to calculate the volume fraction. Now, derive an equation for the time rate of decrease of the labeled bone thereafter, assuming it is only removed by remodeling and that the new osteons are randomly placed in the section. Use the concept of Fig. 3.13 and Eq. 3.38, only now let the Haversian canal areas become labeled bone. Calculate the time it would take for half the labeled bone to be removed, assuming again the subject is a 30- to 39-year-old human. What would this "half-life" be for a child?

3.8. Discuss the implications of Exercise 3.7 for an adult who has been caught in a nuclear reactor accident and ingested a significant quantity of strontium-90, which is radioactive with a half-life of 28 years and is incorporated into bone mineral in a manner similar to tetracycline. Quantitatively explore the possibilities for speeding up the elimination of the strontium from the victim's skeleton using what you have learned in this chapter.

3.9. Show that the initial slope of the N vs. t curve represented by Eq. 3.44 and Fig. 3.14 is the activation frequency, f_a, and that the asymptote is $1/A_o$.

3.10. In Box 3.3, two methods were described for determining average activation frequency (f_a) from the numbers of secondary osteons visible in a cortical cross section from a bone of known age. Compare the results of these two methods using the following assumptions. Let the age of the bone be $t = 40$ years and the observable osteon density be $N = 40$ mm^{-2}. Let the asymptotic value of N be $N_A = 50$ mm^{-2}. Calculate α, β, and f_a using Frost's algorithm. Then calculate f_a using Eq. 3.44 as described in the box. For consistency, assume that $A_o = 1/N_A$. Stout and Paine (1994) found that Frost's method underestimated f_a compared to conventional histomorphometric measurements. Does it also yield a lower f_a than Eq. 3.44?

4

Mechanical Properties of Bone

Strength of bones is dependent upon the material, the microscopic structure and the shape of the whole bone.

A.A. Rauber, 1876, as quoted in Roesler, 1987.

4.1 Introduction

We have now learned much about the variable structure of the material inside bones. Previously, we learned that the muscle forces acting on the skeleton are quite large relative to the forces exerted on it by the outside world. It is now time to bring these two aspects of skeletal tissue mechanics together: to learn how strong bones are relative to the loads they must bear, and how the mechanical properties of bones depend on their structure.

The mechanical properties of bones are governed by the same principles as those of manmade load-bearing structures. However, the ability of vertebrates to adapt their bone structure to the imposed loading, in terms of both the evolution of a genetic plan for the skeleton and the ability to acutely modify this plan by cellular activity, results in a structure that is highly complex and, when healthy, exquisitely efficient. The purpose of this chapter is to explain how the various attributes of whole bones combine to determine their *monotonic* strength and stiffness. (The term monotonic refers to the behavior of a structure loaded steadily to failure, in distinction to repetitive or *fatigue* loading, which is considered in Chapter 5.)

4.2 Fundamentals of Solid Mechanics

To lay a foundation for subsequent discussion of the mechanical properties of bone, this section briefly reviews the mathematical formulation of the mechanical properties of materials and structures. Students trained in engineering may well skip this section; those with no background in this area can find other biologically oriented treatments of this material in an excellent review by Turner and Burr (1993) and books by Currey (1984), Frost (1973a), and Wainwright et al. (1976).

Strength and Stiffness of a Structure

To determine the mechanical properties of a structure—or, more typically, a major part of a structure, such as a beam in a bridge or the shaft of a femur—engineers conduct a *load-deformation* test. The structure is loaded in a manner appropriate for its actual function—tension, compression, torsion, or bending—and its deformation is recorded as a function of the applied load. Figure 4.1 shows such a test for a femur shaft loaded in compression. Typically, load and deformation are linearly proportional for some time, until the *propotional limit* is reached; then, the slope of the load-deformation curve is reduced. Usually, if the structure is unloaded after the proportional limit has been passed, it has a permanent deformation, as shown by the dashed line in the figure. The point on the load-deformation curve where this begins to happen is known as the *yield point* because the structure begins to "yield" or *plastically* deform. Before the yield point the deformation is *elastic*, meaning that the structure will return to its original shape when the load is removed. If the load continues to increase, an *ultimate* load is reached; in some materials (e.g., metals), this may not be the point at which a catastrophic failure (or fracture) occurs. The point marking the latter event is known as the *failure point*. (In cortical bone, the ultimate and failure loads are usually identical.) The *strength* of the structure may be defined as the load at the yield or failure points, or as the ultimate load, depending on the

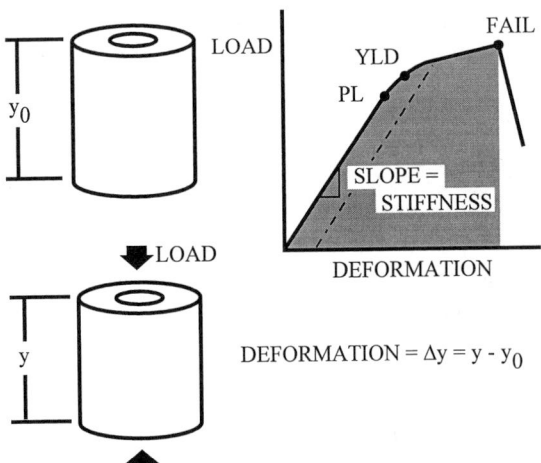

FIGURE 4.1. A load-deformation test for a segment of cortical bone is schematically depicted. The unloaded bone segment, of length y_0, is shown at *upper left*. The loaded bone segment, shortened to length y, is shown at *lower left*. The load-deformation curve at *right* is used to characterize the mechanical properties of the bone segment's structure. *PL*, proportional limit; *YLD*, yield point; *FAIL*, failure (fracture) point. The slope of the initial, linear portion of the curve is the stiffness of the structure. The *shaded area* under the curve is the energy-to-failure. The *dashed line* shows how unloading between the failure and yield points results in a residual deformation.

circumstances. There is no absolute definition of strength, but if it is not otherwise specified, it may be taken as the load at the failure point.

The *stiffness* or *rigidity* of the structure is the load required to deform it a given amount. This is the same thing as the slope of the load-deformation curve. The slope is usually measured within the proportional range, where it is constant. *Compliance* is the reciprocal of stiffness, and is a measure of the ease of deforming the structure. The deformations at the yield and failure points are measures of the extent to which the structure may be deformed.

The *work* or *energy* required to yield or fracture the structure is another important mechanical property that is sometimes referred to as *toughness*. This characteristic may be determined by measuring the area under the load-deformation curve out to the yield or failure point. *Hysteresis* refers to the area between the dashed line and the load curve in Fig. 4.1; it represents the energy lost when a structure is loaded to a point short of failure and then unloaded.

Stress and Strain

The mechanical properties of a structure depend on both its geometry and the properties of the material inside. To determine the properties of the material, it is necessary to "normalize out" the geometric effects. To do this, engineers have defined *stress* and *strain*. In its simplest definition, stress is load per unit area, as shown at upper left in Fig. 4.2. The more general def-

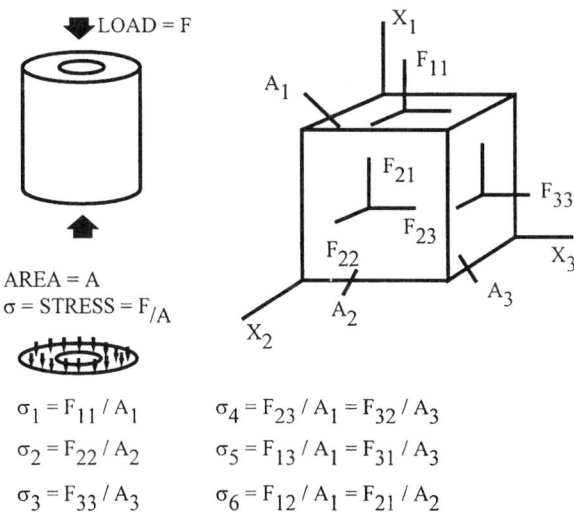

$$\sigma_1 = F_{11} / A_1 \qquad \sigma_4 = F_{23} / A_1 = F_{32} / A_3$$
$$\sigma_2 = F_{22} / A_2 \qquad \sigma_5 = F_{13} / A_1 = F_{31} / A_3$$
$$\sigma_3 = F_{33} / A_3 \qquad \sigma_6 = F_{12} / A_1 = F_{21} / A_2$$

FIGURE 4.2. Simple and tensoral definitions of stress. The *cube at right* represents a free-body diagram of a small region within the loaded bone segment at *left*. F represents the normal and shear forces exerted on this region by the stresses in the surrounding bone. When divided by the areas of the cube's faces (A_1, A_2, A_3) , these forces become the stress tensor components at the cube's location.

inition of stress is also shown in this figure. If one imagines a small cube of material at a point inside a loaded structure, one can see that each surface of the cube has three loads acting on it: a *normal* load perpendicular to the surface, and two *shear* loads parallel to the surface. If these three loads are divided by the area of the cube's face, they become the three stresses associated with the direction of that face. The cube has six faces, but because there must be equal forces on opposite faces for stability, it turns out that there are $3 \times 3 = 9$ *stress components* at each point in a loaded material. The first three of these are normal stress components, and the other six are shear stresses. However, some of the latter are equal to one another, so that the number of stress components is ultimately reduced to six: three normal and three shear. We represent these by the symbol σ_i, where $i = 1$ to 6. Stress has more components than a vector (which has only three); such mathematical quantities are known as tensors. Stress is usually measured in newtons/m^2 (N/m^2), which is the same thing as pascals (abbreviated Pa). A typical failure stress in bone would be 150 MPa (megapascals).

Strain is another tensor quantity that engineers define to be able to discuss the deformation of a material without regard to its structural geometry. In its simplest definition, strain is the fractional change in dimension of a loaded body; Fig. 4.3 illustrates normal and shear strains. One can show that there are, in general, six strain tensor components, three normal and three shear, which describe the deformation of a cube of material at any point in a loaded structure. We represent these by the symbol ε_j, where $j = 1$ to 6. Strain is a dimensionless ratio, but it is common to measure it in *microstrain*, or 10^{-6} mm/mm. For example, a strain of 0.02 mm/mm would be the same as 2% strain or 20,000 *microstrain* (abbreviated $\mu\varepsilon$); this would be a dangerously large strain in a bone.

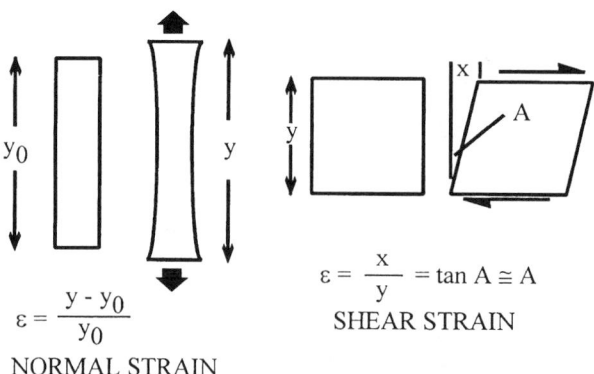

$$\varepsilon = \frac{y - y_0}{y_0}$$

NORMAL STRAIN

$$\varepsilon = \frac{x}{y} = \tan A \cong A$$

SHEAR STRAIN

FIGURE 4.3. Normal *(left)* and shear *(right)* strains are defined. A represents the angle through which the side of the shear specimen is rotated. Strains are dimensionless; shear strains may be expressed as radians.

When a structure is loaded in tension, it becomes longer in the load direction, but it usually "waists" or becomes narrower in the transverse directions. (Of course, the opposite phenomenon occurs in a compressed specimen.) The ratio of the transverse strain component to the longitudinal strain component is known as *Poisson's ratio*. For geometric reasons, this ratio must be less than 0.5 in ordinary materials. For incompressible materials (e.g., liquids), Poisson's ratio is 0.5. Typical values for solids, including bone, are about 0.3. (For another biological material, cork, Poisson's ratio is nearly 0. This makes it easier to push a cork into a wine bottle, but that obviously has nothing to do with why the cork material grows with this unusual property.)

Principal Directions

Perhaps it is intuitively clear that the stresses associated with a cube in an arbitrarily loaded structure depend not only on the location of the cube but also on its orientation. In general each face of the cube experiences both normal and shear stresses, and the relative magnitudes of these vary if the cube is placed in different orientations. It happens, however, that it is **always** possible to find an orientation such that all the shear stresses become zero. These three mutually perpendicular directions are called the *principal stress directions*. A similar set of *principal strain directions* exists for each point in a loaded structure. In simple materials, such as steel, these two principal direction sets are identical, but in complex materials like bone they may differ.

Strength and Stiffness of a Material

A load-deformation curve may be converted to a stress-strain curve by using appropriate formulas to change load to stress and deformation to strain. For example, in a compressed bone shaft, the load is divided by the cross-sectional area of the bone cortex to obtain stress, and the deformation is divided by the original length of the shaft segment to obtain strain. After this is done, all the strength terms previously defined for the structure (yield, ultimate, and failure strengths) can be redefined for the **material** in terms of stresses. Toughness is still the area under the curve, but it now has units of work or energy per unit volume of material.

Stress and strain are related by Hooke's law, which says they are linearly proportional to one another. In its simplest form, this law relates one component of stress to its corresponding strain component (e.g., stress and strain in longitudinal tension), and it applies only to the linear portion of the stress-strain curve between the origin and the proportional limit. The slope of this line is the stiffness of the **material**. For normal loading (tension or compression), the material stiffness is known as the *elastic modulus* or *Young's modulus*. For shear loading, the material stiffness is known as the *shear modulus*. These moduli have the same units as stress, and values for bone are typically of the order of gigapascals (10^9 pascals), abbreviated GPa.

Box 4.1 Historical Note
Robert Hooke and Thomas Young

C E I I I N O S S S T T U V

Thus did Robert Hooke (1635–1703) record his law of elasticity as a Latin anagram in 1676. (Can you solve it?) He did this as a result of his well-known quarrel with Isaac Newton, whom Hooke thought had stolen some of his ideas regarding planetary mechanics and other matters. This situation arose from the fact that Hooke held the position of Curator of Experiments in the Royal Society, of which Newton was a contemporaneous member. It was Hooke's responsibility to present experimental demonstrations at the society's meetings, and these experiments had often directly demonstrated aspects of what we now know as Newton's laws, well before Newton published them. Newton, often described as a reclusive and self-absorbed person, refused to acknowledge Hooke's contributions to these concepts in his *Principia* until the second edition, after Hooke's death.

Hooke was a leading scientist of his day. His first published work was the *Micrographia* in 1665, containing his microscopic studies and introducing the term "cell" to biology. He served as Robert Boyle's assistant for several years, and invented the air pump used in the experiments leading to Boyle's law. (Some historians believe that this law also was rightfully Hooke's.) Subsequently, Hooke pursued many experiments related to respiratory physiology, involving both himself and animals as experimental subjects. In the next decade Hooke turned his attention to clock-making. The impetus for this was the need to determine longitude at sea. Because pendulum clocks are inaccurate aboard ship, timekeeping based on the resonant frequency of a spring became of interest. The resulting experiments led Hooke to discover the linear relationship between applied force and extension of a spring: *ut tensio sic vis* (as is the deformation, so is the force). Interestingly, he saw this law as an extension of Boyle's law, which expresses the "springiness" of air.

Young's modulus is named for Thomas Young (1773–1829), the physician-physicist who coined the term "modulus of elasticity" and made other important contributions to mechanics. Unlike Hooke, Young had no desire to guard his scientific insights, but his inability to write clearly concealed his ideas about elasticity almost as effectively as Hooke's anagram: "The modulus of elasticity of any substance is a column of the same substance, capable of producing a pressure upon its base which is to the weight causing a certain degree of compression as the length of the substance is to the diminution of its length."

Such unintentional obfuscation aside, Young was, like Hooke, a person of enormous breadth of interest and accomplishment. He is perhaps best known for his work in optics, having established both the wave nature of light and, by means of sometimes gruesome and painful experiments on his own eyes, the foundations of physiological optics. In addition to his practice of medicine and his work in physics, Young helped decipher the Rosetta Stone and wrote 61 articles for the newly founded *Encyclopaedia Britannica*. Much of his scientific work was published anonymously because he was afraid that his medical practice would suffer if the public knew that he was a scientist!

Generalized Hooke's Law: Anisotropy

These simple versions of Hooke's law do not suffice for more complex materials, so it is necessary to write a *generalized Hooke's law*. To do this, it is assumed that each of the six stress components is proportional to all six of the strain components. That is,

$$
\begin{aligned}
\sigma_1 &= C_{11}\,\varepsilon_1 + C_{12}\,\varepsilon_2 + C_{13}\,\varepsilon_3 + C_{14}\,\varepsilon_4 + C_{15}\,\varepsilon_5 + C_{16}\,\varepsilon_6 \\
\sigma_2 &= C_{21}\,\varepsilon_1 + C_{22}\,\varepsilon_2 + C_{23}\,\varepsilon_3 + C_{24}\,\varepsilon_4 + C_{25}\,\varepsilon_5 + C_{26}\,\varepsilon_6 \\
\sigma_3 &= C_{31}\,\varepsilon_1 + C_{32}\,\varepsilon_2 + C_{33}\,\varepsilon_3 + C_{34}\,\varepsilon_4 + C_{35}\,\varepsilon_5 + C_{36}\,\varepsilon_6 \\
\sigma_4 &= C_{41}\,\varepsilon_1 + C_{42}\,\varepsilon_2 + C_{43}\,\varepsilon_3 + C_{44}\,\varepsilon_4 + C_{45}\,\varepsilon_5 + C_{46}\,\varepsilon_6 \\
\sigma_5 &= C_{51}\,\varepsilon_1 + C_{52}\,\varepsilon_2 + C_{53}\,\varepsilon_3 + C_{54}\,\varepsilon_4 + C_{55}\,\varepsilon_5 + C_{56}\,\varepsilon_6 \\
\sigma_6 &= C_{61}\,\varepsilon_1 + C_{62}\,\varepsilon_2 + C_{63}\,\varepsilon_3 + C_{64}\,\varepsilon_4 + C_{65}\,\varepsilon_5 + C_{66}\,\varepsilon_6
\end{aligned} \tag{4.1a}
$$

This may be written in compressed notation as

$$
\sigma_i = C_{ij}\,\varepsilon_j \tag{4.1b}
$$

where both i and j assume the values 1–6 and j is understood to be summed over these values. The 36 constants of proportionality C_{ij} are known as the elastic constants. It turns out that $C_{12} = C_{21}$, etc., so that the matrix of C_{ij} constants is symmetric, with only 21 independent values, and has the form:

$$
\begin{bmatrix}
C_{11} & C_{12} & C_{13} & C_{14} & C_{15} & C_{16} \\
 & C_{22} & C_{23} & C_{24} & C_{25} & C_{26} \\
 & & C_{33} & C_{34} & C_{35} & C_{36} \\
 & & & C_{44} & C_{45} & C_{46} \\
 & & & & C_{55} & C_{56} \\
 & & & & & C_{66}
\end{bmatrix}
$$

Fortunately, this number is further reduced for most materials. In *isotropic* materials (which have no "grain" but behave the same in every direction, e.g., steel), there are only two independent elastic constants, and these related to the elastic modulus, E, and shear modulus, G, which are sometimes known as *technical constants*. Poisson's ratio, ν, is considered to be a third technical constant. For isotropic materials, because there are only two elastic constants, there are only two independent technical constants as well, and $E = 2G(1 + \nu)$. For materials like bone, which have a grain and thus are anisotropic, the number of elastic constants is between 2 and 21, and the technical constants have different values in different directions. For example, the elastic modulus in tension of a long bone's diaphysis is greater longitudinally than it is across the shaft.

Most anisotropic materials are not completely so, but have some symmetry to their internal structure. Two common types of limited anisotropy happen to be found in bone, wood, and some other biologic materials. The mechanical properties of *orthotropic* materials are different in three perpendicular directions (e.g., longitudinal, radial, and circumferential). Their matrix of elastic constants has nine independent values and the following form:

$$
\begin{bmatrix}
C_{11} & C_{12} & C_{13} & 0 & 0 & 0 \\
 & C_{22} & C_{23} & 0 & 0 & 0 \\
 & & C_{33} & 0 & 0 & 0 \\
 & & & C_{44} & 0 & 0 \\
 & & & & C_{55} & 0 \\
 & & & & & C_{66}
\end{bmatrix}
$$

Transversely isotropic materials behave similarly in every direction about a single axis of symmetry. They have five independent constants and a C-matrix of the form:

$$
\begin{bmatrix}
C_{11} & C_{12} & C_{13} & 0 & 0 & 0 \\
 & C_{11} & C_{13} & 0 & 0 & 0 \\
 & & C_{33} & 0 & 0 & 0 \\
 & & & C_{44} & 0 & 0 \\
 & & & & C_{44} & 0 \\
 & & & & & C_{55}
\end{bmatrix}
$$

where $C_{55} = 2(C_{11} - C_{12})$. Later in this chapter we show which kinds of bone exhibit these forms of stiffness anisotropy. (Bone is also anisotropic in its strength properties, but there is no straightforward theory covering this behavior.) A much more complete and extremely lucid explanation of anisotropy and elasticity theory may be found in the treatise, *Theory of Elasticity of an Anisotropic Elastic Body*, by Lekhnitskii (English translation, 1963).

4.3 Determinants of the Strength of a Whole Bone

Mechanics

Generally speaking, for a particular mode and rate of loading, the strength (or stiffness) of a structure is a product of (1) its shape and size, and (2) the strength (or stiffness) of the material within. For example, in compression, tension, or shear, the stress in a specimen of cross-sectional area A and bearing a load L is

$$\sigma = L/A \tag{4.2}$$

When this stress reaches the failure stress for the material, the structure will fail. Therefore, by simply rearranging this equation and using the subscript f to represent the failure condition, one obtains an equation for the compressive load that would fail a long bone diaphysis,

$$L_f = A\,\sigma_f \tag{4.3}$$

where L_f is the failure load, and σ_f is the material strength in the appropriate mode of loading. In this case, the shape and size factor is simply the cross sectional area, A. Analogous equations describe the strength of the diaphysis in other modes of loading. When a perfectly cylindrical structure is twisted by a torque T, the shear stress varies linearly with distance, r, from the center, and is given by

$$\sigma_s = T\, r/J \tag{4.4}$$

where J is the *polar moment of inertia* (Fig. 4.4). Rearranging this equation, one has, for the failure torque of a cylindrical long bone diaphysis loaded in torsion,

$$T_f = (J/r)\, \sigma_{sf} \tag{4.5}$$

where r is now the periosteal radius, and σ_{sf} is the torsional (or shear) strength of the bone material. The ratio J/r may be called the torsional "shape and size factor." When a structure is bent by a moment M, there is tension on one side and compression on the other; the boundary between these regions is called the *neutral plane* or *axis*. The tensile and compressive stresses are given by

$$\sigma = M\, c/I \tag{4.6}$$

where I is the *cross-sectional moment of inertia* (Fig. 4.4) and c is the dis-

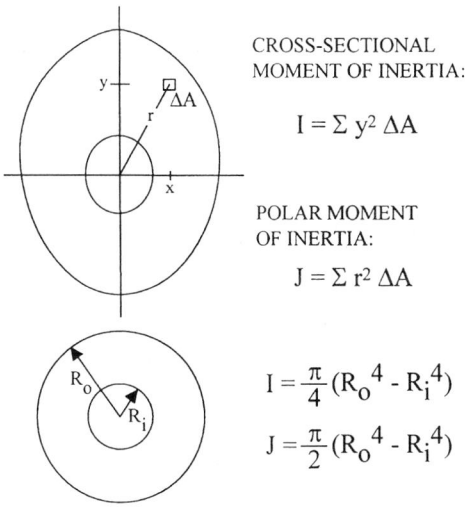

CROSS-SECTIONAL MOMENT OF INERTIA:

$$I = \Sigma\, y^2\, \Delta A$$

POLAR MOMENT OF INERTIA:

$$J = \Sigma\, r^2\, \Delta A$$

$$I = \frac{\pi}{4}(R_o^4 - R_i^4)$$

$$J = \frac{\pi}{2}(R_o^4 - R_i^4)$$

FIGURE 4.4. *Upper:* Cross-sectional and polar moments of inertia are defined for a bone cross section. The cross section is divided into many small elements of area, ΔA. The x-axis is assumed to be the neutral axis for bending. *Lower:* Equations for I and J of a cylindrical tube.

tance from the point in question to the neutral plane. The failure moment is

$$M_f = (I/c) \ \sigma_{bf} \tag{4.7}$$

where σ_{bf} is the failure stress for bending and c is now the distance to the outermost fibers of the bone. Here, the shape and size factor is I/c.

Cross-sectional and polar moments of inertia are defined in Fig. 4.4. Consider that the cross-section of the bone cortex is divided into elements of area ΔA, of which one is shown. J is the sum of the products of each ΔA and the square of its distance from the centroid. I is similarly defined, except that the distance is measured from the neutral axis, which divides the tensile and compressive areas of a bent structure.

Equations of similar form would describe the stiffness of a bone: the primary difference is that a shape and size term is multiplied by the stiffness of the material (the elastic modulus in tension, compression, or bending; the shear modulus in shear or torsion).

Examples

Suppose a person's femoral diaphysis is idealized as a circular cylinder with a periosteal diameter of 3 cm and an endosteal diameter of 1.2 cm. Calculate the diaphyseal failure load for uniaxial compression and the failure moments for bending and torsion.

Compression:
Cross-sectional area = $A = \pi(R_p^2 - R_e^2) = \pi(0.015^2 - 0.006^2) = 5.94 \times 10^{-4}$ m^2.
From Table 4.1, use the ultimate stress in compression, $\sigma_f = 195$ MPa.
Then from Eq. 4.3 one has
$$L_f = A \ \sigma_f = (5.94 \times 10^{-4}) \ (195 \times 10^6) = 1.16 \times 10^5 \text{ N}$$
For a 70-kg person, this is about 170 × body weight.

Torsion:
Polar moment of inertia = $J = \pi(R_p^4 - R_e^4)/2 = \pi(0.015^4 - 0.006^4)/2$
$$J = 7.75 \times 10^{-8} \text{ m}^4$$
From Table 4.1, use the ultimate shear stress, $\sigma_{sf} = 69$ MPa.
Then from Eq. 4.5 one has, letting $r = R_p = 0.015$ m,
$$T_f = (J/r) \ \sigma_{sf} = (7.75 \times 10^{-8}/0.015) \ (69 \times 10^6) = 356 \text{ N-m}$$

Bending:
Cross-sectional moment of inertia = $I = \pi(R_p^4 - R_e^4)/4 = \pi(0.015^4 - 0.006^4)/4$
$$= 3.87 \times 10^{-8} \text{ m}^4$$
From Table 4.1, use the ultimate stress in bending, $\sigma_{bf} = 208.6$ MPa.
Then from Eq. 4.7 one has, letting $c = R_p = 0.015$ m,
$$M_f = (I/c) \ s_{bf} = (3.87 \times 10^{-8}/0.015) \ (208.6 \times 10^6) = 539 \text{ N-m}.$$
If the femur were 42 cm long and supported at its ends, a 5130-N load in the middle would produce this moment. This would be about 7 times the body weight of a 70-kg person.

TABLE 4.1. Typical mechanical properties for cortical bone

Property	Human	Bovine
Elastic modulus, GPa		
Longitudinal	17.4	20.4
Transverse	9.6	11.7
Bending	14.8[a]	19.9[b]
Shear modulus, GPa	3.51	4.14
Poisson's ratio	0.39	0.36
Tensile yield stress, MPa		
Longitudinal	115	141
Transverse	—	—
Compressive yield stress, MPa		
Longitudinal	182	196
Transverse	121	150
Shear yield stress, MPa	54	57
Tensile ultimate stress, MPa		
Longitudinal	133	156
Transverse	51	50
Compressive ultimate stress, MPa		
Longitudinal	195	237
Transverse	133	178
Shear ultimate stress, MPa	69	73
Bending ultimate stress, MPa	208.6[a]	223.8[b]
Tensile ultimate strain		
Longitudinal	0.0293	0.0072
Transverse	0.0324	0.0067
Compressive ultimate strain		
Longitudinal	0.0220	0.0253
Transverse	0.0462	0.0517
Shear ultimate strain	0.33	0.39
Bending ultimate strain	—	0.0178[b]

Human data are for adult femur and tibia; bovine data are for femur.
Data compiled from Cowin (1989), pp. 102, 103, 111–113, except as indicated.
[a]From Currey and Butler (1975); adult femur, three-point bending.
[b]From Martin and Boardman (1993); tibia, three-point bending.

It should be emphasized that the foregoing equations are restricted to fairly simple geometries of the cross section and load application. Bones may frequently depart from these conditions. For example, torsion of a bone having a cross section that departs significantly from axial symmetry, or bending in which the load vector does not pass through the centroid of the cross section, would not accurately follow these descriptions. Even so, it generally remains possible to separate the effects of material properties (which depend on histologic structure) from those of the shape and size of the structure (which depend on anatomy). That is the important point insofar as this discussion is concerned.

Another limitation of this approach is that it assumes that skeletal "shape and size" characteristics include only macroscopic anatomical features, and not microscopic features such as trabecular or osteonal architecture. It is assumed that the mechanical effects of microscopic features can be embodied in the material properties. This strategy, used in *continuum mechanics*, is consistent with the approach taken when *finite element modeling* (see Box 4.3) is used to study skeletal stresses. The gross anatomy of the bone is modeled by the "shape" of the model, and variations in microscopic architecture are satisfied by assigning different material properties to individual regions (elements) within the model. The alternative is to attempt to formulate the shape and size of the microscopic as well as macroscopic structure. Not only would this be exceedingly difficult, but one would have to determine the material properties of the tissue within these small structures. The approach favored here allows one to define material properties on a macroscopic scale, combining the mechanical effects of variations in microscopic architecture with those in microscopic material properties. This approach is also consistent with the common practice of measuring the material properties of bone using specimens small enough to define an anatomical region, but large compared to the size and distribution of microscopic features. For example, in the case of trabecular bone, test specimens are made large enough to contain perhaps 10–50 trabeculae (Brown and Ferguson, 1980). For further discussion of this problem, see Fung (1977) and Harrigan et al. (1988). We return to this subject when we discuss the mechanical properties of trabecular bone later in this chapter.

Box 4.2 Clinical Note
Mechanical Significance of Bone Fracture Patterns

When a person comes to the hospital with a fracture, the biomechanically knowledgeable physician can often deduce the kind of loading that caused the break from the pattern of the fracture surfaces. To do this, one must keep two things in mind. First, the ultimate stress for bone differs in various modes of loading: it is strongest in compression, weakest in shear, and intermediate in tensile failure (see Table 4.1). Secondly, the stress inside a bone depends on orientation as well as position. If the cube in Figure 4.2 is rotated to some new orientation, the stresses on its surfaces will change. Just as the cube can be rotated to minimize the shear stresses on its surfaces, there is another orientation that will maximize the shear stresses. Because the ultimate stress in compression is much higher than that in shear, when a bone is loaded in compression it tends to fail along the planes which carry the highest shear stresses, not those having the highest compressive stresses. These planes are typically at an angle of 45° to the load direction.

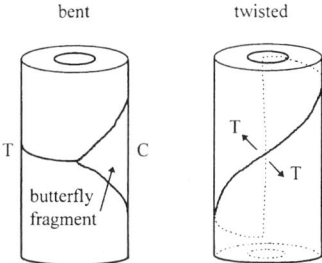

Left: In a bent bone, a transverse crack originates on the tensile side *(T).* On the compressed side *(C)* the fracture surface often follows a shear plane in one or two directions, producing a wedge-shaped fragment. *Right:* In a twisted bone, tensile stresses produce a spiral crack that winds around the cortex, and the bone breaks when the crack's ends are connected by a longitudinal fissure.

When a bone is bent, failure often begins on the tensile side and propagates across the shaft because the tensile stresses on a cross section exceed the strength before the shear stresses on a 45° plane do. When the crack enters the compressed tissue, it again tends to travel along the 45° planes of maximum shear, causing 'butterfly' fragments or other types of *comminution.* When a bone is loaded in torsion the shear stress created by the twisting is across the fibers. In this case, compressive and tensile stresses lie at a 45° angle to the fibers. The tensile stress tends to separate the fibers, producing a fracture surface that spirals around the bone at a 45° angle. The crack then returns to its starting point along a longitudinal shear stress plane. Clinically, each of these fracture patterns is common, indicating that loading of long bones can occur by compression, bending, torsion, or a combination of these.

Mechanical Failure of Whole Bones

In considering the mechanical requirements of whole bones, several conflicting demands must be balanced. In normal daily life, the skeleton provides struts (bones) that the muscles can work against to create motion. The stiffer these struts are, the more efficient is the action of the muscles. If bones are less stiff, then part of the mechanical output of the muscles is used in deforming the bones, rather than creating useful motion. This wastes metabolic energy. In this context, bones of greater stiffness are desirable. On the other hand, if the individual is in a situation where there is risk of bone fracture from a fall or a blow, less stiff (more compliant) bones would be preferred. This is because, for a given strength, the more compliant structure is able to absorb a greater amount of energy before reaching its failure load (Fig. 4.5). Another consideration is weight: stronger bones weigh more and require more metabolic energy to carry around. The mechanical characteristics of bones, therefore, represent a compromise between the need for stiffness to make muscle actions efficient, the need for compliance to absorb energy and avoid fracture, and the need for minimal skeletal weight.

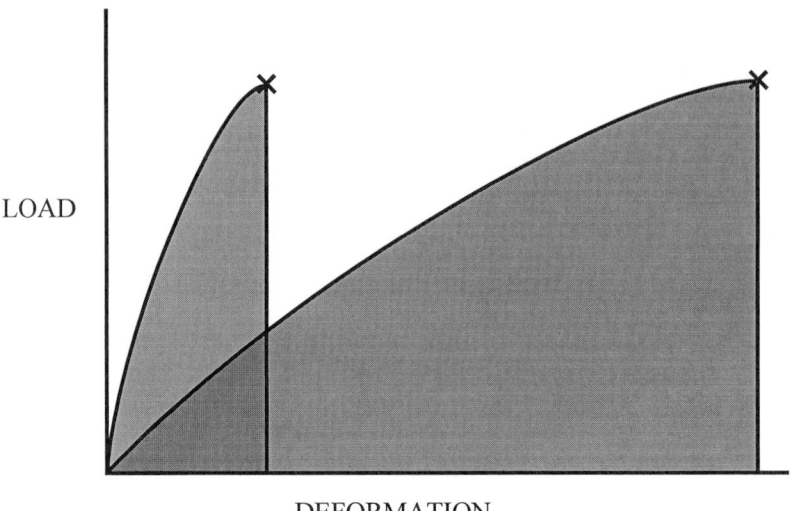

LOAD

DEFORMATION

FIGURE 4.5. If two bones have the same strength (failure load), the more compliant one (at *right*) requires more energy to break it.

Different bones have different functions, and the best solution to the conflicting mechanical demands need not be the same for every bone. The bones of the inner ear need to be extremely stiff to efficiently transmit sound waves; the fact that they could absorb very little energy before fracture is unimportant because they are protected from extrinsic loads. At the other extreme, the material in the bones of children is much more compliant than those of adults. Currey and Butler (1975) found the femoral bone of 2- to 8-year-old children was 68% as stiff and required 45% more energy to break in bending than that of 26- to 48-year-old adults. This increased compliance decreases the efficiency of children's muscle action, but it confers more protection against falling and crashing into things.

The energy required to break a bone is small compared with the energy encountered in daily activities. For example, the energy required to break the shaft of an adult tibia or femur is only about 15 joules (J). The energy released when a 70-kg person falls to the ground from a standing position is about 500 joules. The ability to absorb this energy using eccentric muscle contractions and deformations of soft tissues usually prevents fractures from occurring in such a fall. Soft tissue deformation absorbs energy during an accident in two ways. First, loads applied perpendicular to the tissues compress them and propagate stress waves up and down the body. The tissues effective in this way are skin, fat, and muscle (plus fur on animals and clothing on humans). Second, tissues such as fascia, tendons, ligaments, joint capsules, and contracted muscle brace bones against bending by supporting part of the tensile forces. These tissues absorb energy as they are stretched, but they must be nearly as stiff as the bone to be effective. If these energy-absorbing mechanisms are impaired by surprise, restrictions, or incapacity, fracture occurs easily. The propensity of elderly persons to fracture their bones reflects not only the weakness of their bones but also their increased risk of falling and the inability of their muscles, ligaments, and soft tissues to adequately absorb the applied energy, which instead is transmitted to the bones.

Box 4.3 Technical Note
Finite Element Modeling

Finite element analysis, or modeling, is a process by which stresses and strains can be computed in a structure of irregular shape and varying internal properties. The name comes from the process in which the volume of the structure (e.g., the proximal end of a human femur, as shown in the figure) is divided into *elements*, each having a prescribed geometry and a set of mechanical properties (e.g., elastic modulus and Poisson's ratio). The model also includes *boundary conditions* that prescribe the forces applied to, or constraints on, portions of the structure. For example, in the case of a human femur, forces may be applied to some of the elements on the articular surface of the head and along the greater trochanter where the abductor muscles act. In addition, the elements at the distal end of the model may be constrained from moving in one or more directions. A computer is used to solve the equations that govern the ways in which the elements deform and the ways that they interact with one another as they deform. This method of analysis allows many problems to be solved that are far too complicated for a "closed form" solution (i.e., a single equation, such as Eq. 4.6, to give an simple example), either because of their irregular geometry or the variability of the material properties inside. Many commercial programs have been developed for finite element analysis, and the method has become very important in orthopaedics because of the irregular shapes of bones, their varying material properties, and the interest in studying the stresses produced when a metal implant is implanted in a bone.

More detailed explanations of finite element modeling of the skeleton may be found in Hart (1989) and Beaupre and Carter (1992).

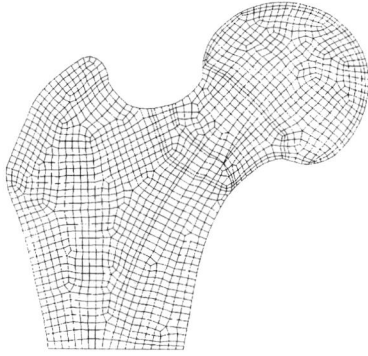

Typical "mesh" of elements for a finite element model of the proximal human femur. (Courtesy of S. Hazelwood.)

Relationship to Modeling and Remodeling

It happens that the two parts into which structural strength may be factored coincide with modeling and remodeling. Shape and size changes are effected on external (i.e., periosteal and endosteal) bone surfaces, where there is more modeling than remodeling during growth, and little capacity for change in adults (Frost, 1986). On the other hand, changes in material properties are effected on internal (i.e., Haversian and trabecular) bone surfaces. Internal bone surfaces are governed primarily by remodeling throughout life (Frost, 1986).

Thus, the gross anatomy (i.e., the shape and size) of a bone is in large part determined during growth. Modeling serves to sculpt the bone to suit both the genetic plan and the demands of current mechanical usage. In adults, modeling is reduced in rate and extent, so that alterations of bone geometry require much more time (or a mechanical stimulus large enough to produce woven bone). This restriction is important with respect to joints as well as bones. An adult who greatly increases the loads on his skeleton is, in principle, more likely to acquire osteoarthritis than a juvenile because his bones have insufficient capacity to alter their metaphyseal size so as to reduce joint stresses (Frost, 1994).

The mechanical properties of the bone material are matched to mechanical usage by remodeling. Unlike modeling, remodeling rates can be substantially increased in adults by various events, including changes in mechanical loading. That means that in adults, substantial changes in skeletal loading must be accommodated primarily by changing bone material properties rather than bone geometry. (The same principles apply with respect to the latter stages of fracture healing, in which the callus is replaced by lamellar bone and errors in the geometry of the healed bone are corrected. Internal replacement by lamellar bone occurs fairly rapidly in both children and adults, but sculpting the surface back to its normal geometry takes much longer in adults.) This limitation is significant because it is practically impossible to substantially increase the strength or stiffness of cortical bone by remodeling, there being no large voids to be filled. Therefore, in adults it is much easier to reduce cortical bone strength than increase it.

In many bone diseases, the lesion primarily affects remodeling and therefore the bone material, and the skeleton frequently tries to adjust the shape and size factor to compensate. For example, senile or idiopathic osteoporosis involves loss of trabecular bone mass in the spine, which reduces the failure stress of the vertebral bodies. However, in men there is a concurrent increase in vertebral diameter that prevents the structural strength from being reduced to the same extent as the material strength (Mosekilde and Mosekilde, 1990). This form of compensation does not occur in women, for reasons that are not clear. A similar phe-

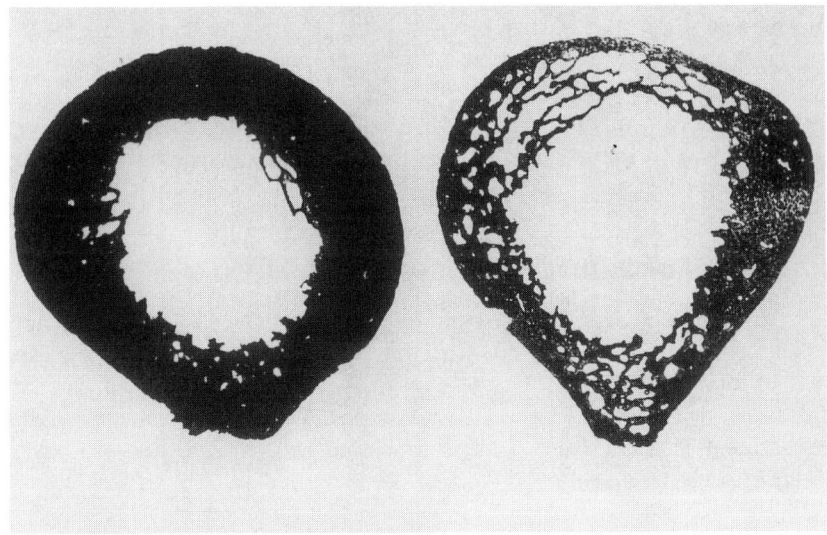

FIGURE 4.6. Midshaft cross sections of femur from a woman in her thirties *(left)* and from a woman in her nineties *(right)*.

nomenon occurs in long bone diaphyses (Martin and Atkinson, 1977; Martin et al., 1980; Ruff and Hayes, 1988) and the neck of the femur (Beck et al., 1993). These adaptive functions often continue even during serious pathologic states, resulting in the strongest structure that circumstances allow. For example, in Chapter 6 we shall see how a human femur drastically alters its cross-sectional geometry (by modeling of the periosteal and endosteal surfaces) to compensate for large bending stresses arising from curvatures produced by rickets. Another example is shown in Fig. 4.6, which shows silhouettes of young (left) and old (right) femur midshaft cross sections. The neutral axis for bending during gait is approximately from the top (anterior) to the bottom (posterior) of the sections. Note that the increased porosity associated with aging is largely confined to endosteal regions near the neutral axis, where the resulting reduction in the bone's failure stress has minimal effect on torsional and bending strength.

4.4 Material Properties of Cortical Bone

It has been clearly demonstrated that osteonal remodeling of compact bone reduces its flexural fatigue strength and resistance to tension (Carter and Spengler, 1978). Compression, shear, and bending strengths are also reduced. This section explores the reasons for this and otherwise examines the mechanical properties of osteonal bone. Osteons should affect the mechanical properties of the cortex in several ways: by the replacement of highly mineralized bone matrix with less calcified material, by increasing cortical porosity, possibly by altering collagen fiber orientation, and by the introduction of cement line interfaces that have unique mechanical properties. We first consider the mechanical properties of single osteons, and then consider cortical bone as a composite of numerous such structural units.

Properties of Individual Secondary Osteons

In a remarkable series of papers beginning in 1964, Antonio Ascenzi, Ermanno Bonucci, and their colleagues presented a great deal of experimental information about the relationship between structure and mechanical function in individual secondary osteons. Taken together, their papers constitute the major part of our knowledge in this area, and we review them in some detail.

Tensile Properties

In their initial work, Ascenzi and Bonucci (1964) examined the tensile properties of the secondary osteon. Undecalcified longitudinal sections, 20–50 µm thick, were prepared from bovine and human femurs. A sharp needle was used to dissect free the lamina between the Haversian canal and the cement line on one side of an individual osteon. This procedure produced an approximately rectangular segment of osteonal wall about 50 µm wide and 100–300 µm long. Figure 4.7 shows how the osteonal lamel-

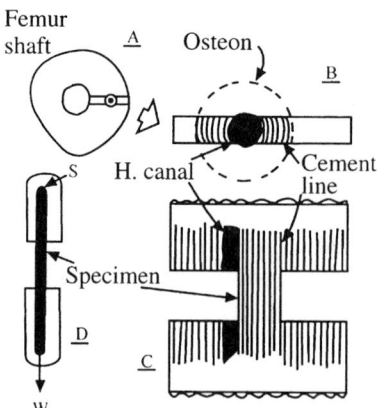

FIGURE 4.7. Diagram showing tension test of osteonal wall segment. A, End view of test slab removed from femur cross section. B, Relationship of a single osteon's geometry to the slab in an enlarged end view. C, Side view of central portion of the test slab. Material has been removed to leave only the lamellae on the right side of the osteon in place. D, The test slab supported at S and pulled in tension by weight W. (Drawn from description by Ascenzi and Bonucci, 1964, 1967.)

lae formed an approximately rectilinear system within the specimens. In their initial experiments, these osteonal segments were simply broken in quasistatic tension, with only the failure load recorded. Later, load-deformation curves were recorded when a microwave extensometer had been developed to measure the extremely small deformations encountered at this scale (Ascenzi and Bonucci, 1967).

The specimens were categorized as either longitudinal (type L) or alternating (type A) in their lamellar structure (c or b, respectively, in Fig. 2.5). Microradiography was used to classify the degree of mineralization as either low or fully calcified. Both dry and wet bone were tested, but only the wet bone data are considered here. Figure 4.8 shows typical curves for secondary osteons with different lamellar orientations and degrees of mineralization. It was found that:

- The tensile strength of osteon wall segments was about 114 MPa, a figure that lies in the midrange of data for larger test specimens composed of primary and secondary bone.

- Type L osteons had greater tensile strength and lower ultimate deformations than did type A osteons mineralized to the same degree.

- The degree of calcification of type A osteons had little effect on the tensile strength of their wall segments. The less mineralized type L osteons were more plastic and deformable than highly mineralized type L osteons, which were extremely stiff. The properties of poorly mineral-

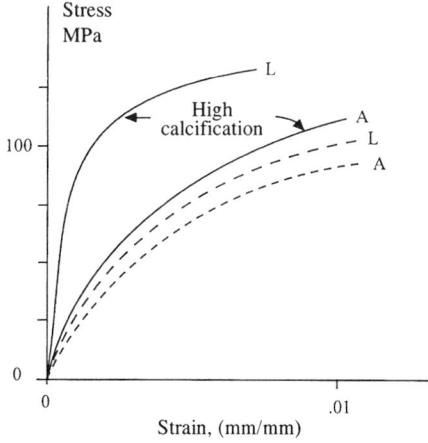

FIGURE 4.8. Typical stress-strain curves for osteon segments in tension. *Solid curves,* fully calcified osteons; *dashed curves,* osteons in the initial stages of calcification. *L,* type L osteon; *A,* type A osteon (see Fig. 2.5). (Redrawn from The tensile properties of single osteons, Ascenzi, A, Bonucci, E, *Anatomical Record,* 1967. With permission of Wiley-Liss, Inc., a subsidiary of John Wiley & Sons, Inc.)

ized type L osteons and all type A osteons (regardless of how well mineralized) were nearly the same.

No differences were found between human and ox osteons, but the authors remarked on the considerable variation in the results for apparently similar osteons. No specific effect of osteocyte lacunae could be mathematically formulated, but examination of the published photographs of failed specimens indicates that lacunae and other irregularities in the lamellar structure may have played an important role in the results. The specimens did not have the reduced central cross section that is preferred in tensile specimens. This factor surely led to stress concentrations at the points where the specimens were gripped, and the published photographs show breakage at these points.

Compressive Properties

Next, Ascenzi and Bonucci (1968) studied the compressive properties of single osteons using specimens from a human femur. They had by now developed a technique for cutting cylindrical segments from osteons contained in cross sections 500 μm thick. These minute cylinders were 180–200 μm in diameter and contained the Haversian canal and all but the outermost laminae of the osteon. The specimens were classified as to mineralization and lamellar structure as before, except that now osteons of circumferential structure (type T in Fig. 2.5) were also studied. In this case, they found that type T osteons were strongest and stiffest. Type A osteons were stronger than type L, but there was no difference in their elastic moduli. As before, the more fully mineralized osteons were more elastic, stiff, and brittle, regardless of their type. The specimens uniformly exhibited fracture planes at an angle of 30°–35° to the longitudinal direction, regardless of the lamellar structure or degree of calcification. Thus, the failure mode appeared to be shearing, regardless of the orientation of the collagen fibers.

Shear Properties

The next experiments examined the shear properties of osteons (Ascenzi and Bonucci, 1972). A 160-μm-diameter cylindrical punch was used to push out the central lamellae from osteons in sections 300 μm thick. The section was supported from below by a plate containing an 800-μm-diameter circular hole, so that the test region involved the cement line. It should also be noted that the test involved significant bending as well as longitudinal shear (Fig. 4.9). The osteons did not fail at the cement line, but along the margins of the punch. Because there would have been a stress concentration at this boundary, one ought not necessarily accept Ascenzi and Bonucci's assertion that this proves that the cement line is stronger than the interstitial lamellae, but their results are not consistent with a weak cement line. Type L osteons were the weakest, and types A and T were of equal strength. Again, strength and stiffness increased with calcification in each

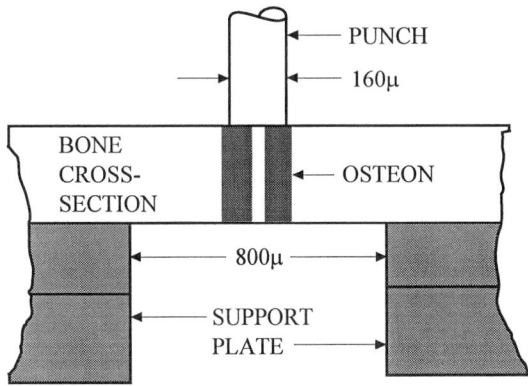

FIGURE 4.9. Diagram of relative dimensions of the specimen and test apparatus used in Ascenzi and Bonucci's (1972) shear experiments. Note that the 300-μm-thick bone section must have been subjected to significant bending loads.

type. The usefulness of the data from this experiment is limited by the fact that the shear test arrangement departed significantly from "simple shearing"; that is, the lateral distance between the edge of the punch and that of the support plate hole was relatively large. Unless one is willing to make a custom apparatus for each osteon, this condition must be compromised.

There is another report of a similar experiment. Frost et al. (1961) arranged a simple shear test without regard to individual osteons or cement lines. Using 70-μm-thick sections, a 900-μm-diameter punch, and a "slightly larger" hole in the support plate, they pushed out areas containing several osteons and osteon fragments. The average stress for longitudinal shearing of bone from a human femur was about 74 MPa, considerably above the range reported by Ascenzi and Bonucci.

Summary

The results of Ascenzi and Bonucci's experiments are summarized in Table 4.2. Variations in the mechanical properties of osteons with different collagen fiber orientations suggest that they are individually adapted to locally enhance the ability of bone to support a particular type of stress. The lamellae of type L osteons are stronger in tension, but it is relatively easy for their inner lamellae to slip longitudinally relative to the outer lamellae. This observation fits nicely with the frequently reported tendency of osteonal bone tested in longitudinal tension to show "osteon pull-out" (Moyle et al., 1978; Moyle and Bowden, 1984). Because type A and type T osteons are weaker in tension, but shear less readily in the longitudinal direction, it is not clear whether macroscopic specimens in which these osteonal types predominate should be weaker or stronger in tension.

TABLE 4.2. Mechanical properties of different types of osteons

Mode of loading	Ultimate stress, MPa	Elastic modulus, GPa	Ultimate strain, %
Tension			
Type L	114 ± 17	11.7 ± 5.8	6.8 ± 2.9
Type A	94 ± 15	5.5 ± 2.6	10.3 ± 4.0
Type T	—	—	—
Compression			
Type L	110 ± 10	6.3 ± 1.8	2.5 ± 0.4
Type A	134 ± 9	7.4 ± 1.6	2.1 ± 0.5
Type T	164 ± 12	9.3 ± 1.6	1.9 ± 0.3
Shear			
Type L	46 ± 7	3.3 ± 0.5[a]	4.9 ± 1.1[b]
Type A	55 ± 3	4.1 ± 0.4[a]	4.6 ± 0.6[b]
Type T	57 ± 6	4.1 ± 0.4[a]	4.6 ± 0.6[b]

[a]The shear modulus is defined as the slope of the linear portion of the stress-strain curve.
[b]The strain in the shear test is defined as the ratio of punch advancement to section thickness, expressed as a percent.

Effects of Osteons on Bone Mechanical Properties

We first consider the effects of osteons per se, that is, the simple addition of osteons to primary bone, on mechanical properties. Subsequently, the effects of intraosteonal variations in collagen fiber orientation, that is, of adding type L, A, or T osteons, are examined.

Walmsley and Smith (1957) suggested that Haversian systems weakened cortical bone. Soon after, Currey (1959) found that completely remodeled cow bone was 35% weaker in tension than primary bone. This reduction in strength was attributed to the relative deficiency of mineral in more recently formed Haversian systems and the additional porosity caused by new Haversian canals. The former concept was reinforced when Hert and co-workers (1965) found that **fully mineralized** osteonal cow bone was as strong in compression as primary cow bone. A decade later, Reilly and Burstein (1974) tested Haversian and primary cow bone in both tension and compression, and found that osteonal remodeling decreased the strength about 20% in both modes of loading. There was, however, no change in the elastic modulus. A year later Reilly and Burstein (1975) presented additional data showing that osteonal remodeling weakened cow bone about 11% in tension and about 9% in shear.

Margel-Robertson and Smith (1978) studied the effects of Haversian remodeling on the 3-point bending properties of bovine femur and the compressive properties of porcine mandible. Although they did not present their data in enough detail to confirm statistically significant effects, they reported that remodeling reduced strength in both tests. Their bending data are unique and support the hypothesis that remodeling reduces the strength of cortical bone in all modes of loading.

Carter and colleagues (Carter et al., 1976; Carter and Hayes, 1977b) found that osteonal remodeling weakened bovine bone with respect to both monotonic and fatigue loading. However, their analysis indicated that this was not caused simply by reduced bone density (that is, increased porosity and reduced mineralization). The osteonal bone matrix seemed to contain another source of weakness. This additional factor may be the cement line. Burr et al. (1988) found that the cement line contained 85%–90% of the calcium and phosphorus found in adjacent bone lamellae. They also found increased ground substance in the cement line. These observations suggest that the cement line may be more compliant and viscoelastic than other components of the bone matrix.

Cortical bone exhibits substantial amounts of plastic deformation when loaded to failure in either tension or compression. The amount of plastic deformation is greatest when the bone is loaded parallel to the collagen fibers, in which case it typically exceeds the elastic deformation (Burstein et al., 1972). Osteonal bone exhibits more plastic deformation than primary bone (Reilly et al., 1974). It is reasonable to speculate that cement line slippage or yielding accounts for much of this plastic ductility, but no one has compared its effects to those of porosity and mineralization.

These studies of the effects of osteonal remodeling on bone strength all used nonhuman bone because it is very difficult to obtain human specimens of primary bone large enough to test. (There are standards for specimen geometry which require that the minimum dimensions of the specimen be several times larger than those of its internal structural features.) In the long bones of adults, and even adolescents, almost the entire cortex is occupied by several generations of secondary Haversian systems. Vincentelli and Grigorov (1985) attempted to overcome this problem by very selectively taking specimens from primary bone near the periosteal and endosteal surfaces of the human tibia. These specimens were tested in tension and compared with similar specimens taken from Haversian regions of the same bones. (Such equal-sized controls are the recommended way of handling cases in which the standard geometry cannot be achieved.) While there was some mixing of the two types of bone in various specimens, it was possible to analyze samples containing either more than 75% or less than 25% Haversian tissue. It was found that both the strength and the stiffness of the predominantly Haversian bone were less than that in primary bone (Table 4.3).

Ascenzi and Bonucci showed that the mechanical properties of individual osteons are affected by collagen fiber orientation. Does osteon type— that is, different arrangements of collagen fibers inside osteons—affect whole bone strength as well? In 1969 Evans and Vincentelli reported that the **tensile** strength of tibial bone was positively correlated with the num-

TABLE 4.3. Tensile strength and stiffness of human primary and secondary bone

Tissue type	Ultimate stress, MPa	Elastic modulus, GPa
Primary bone	161 ± 11	19.4 ± 2.4
Haversian bone	130 ± 14	17.6 ± 2.0

From Vincentelli and Grigorov, 1985.

ber of type L osteons that it contained (Evans and Vincentelli, 1969). Because this work was done with embalmed material, they repeated the experiment with fresh human tibias (Vincentelli and Evans, 1971). They again found significantly greater tensile strengths associated with the presence of type L osteons, in concert with the greater tensile strength of single type L osteons. They also found, however, that bone with greater numbers of Type T osteons deformed more as it broke. This result may be related to the capacity of these osteons to exhibit longitudinal shear between their lamellae. Increased stiffness values were similarly correlated with Type L osteons, but not significantly so.

Later, this team reported the results of similar experiments on the *compressive* properties of osteonal bone (Evans and Vincentelli, 1974). They did not find that compressive strength was inversely correlated to increased numbers of type L osteons, as they had expected from Ascenzi and Bonucci's work. Instead, they found that the greatest compressive strength was associated with osteons having a mixed dark–light appearance; these osteons were apparently intermediate between type L and type A. They also found that deformation at failure was greatest when more type A osteons were present, in contrast to the relative brittleness of single type A osteons when compared to type L (see Table 4.2). These results show that bone's macroscopic mechanical properties are not entirely predictable simply on the basis of osteon fiber orientation. Nevertheless, this feature appears to play an important role in determining mechanical properties.

In addition to studying the mechanical effects of osteons having different collagen fiber orientations, one may examine the effects of the **average** collagen fiber orientation throughout the bone matrix on the mechanical properties of primary or secondary cortical bone. Using birefringence in polarized light to estimate average collagen fiber orientation in a whole specimen cross section, it has been found that collagen fiber orientation affects strength and stiffness at least as much as do density, porosity, and mineralization (Martin and Ishida, 1989; Martin and Boardman, 1993; Martin et al., 1996).

If osteonal collagen fiber orientation affects bone strength, is this aspect of bone structure governed by Wolff's law? Olivo and co-workers (Olivo, 1937; Olivo et al., 1937) were perhaps the first to postulate that osteonal collagen fiber orientation is determined by mechanical stress. If this is the case, one would expect that osteons with different fiber orientations would be preferentially found in regions of bone carrying particular kinds of stress. Pursuing this hypothesis, Portigliatti-Barbos et al. (1983, 1984) demonstrated that the orientation of collagen fibers in the shaft of the human femur is not random, but is correlated with tensile and compressive stresses determined by biomechanical analysis. These investigators quantified the brightness of 1-mm-square elemental areas within a complete femur cross section when viewed in polarized light. Thus, their analysis averaged the osteon fiber orientation of intact osteons and interstitial lamellae, rather than characterizing osteons as belonging to one of three types. They found that regions carrying tensile stress tend to have more longitudinally oriented fibers and those carrying compressive stress contain more transversely oriented fibers. This is shown in Fig. 4.10, which contains computer-generated diagrams of human femur cross sections located 1 cm apart (upper and lower panels, respectively), showing collagen fiber orientation (left panels) and radiographic density (right panels). The diagonal lines approximate the position of the neutral axis for bending because of the hip load (Pauwels, 1980). The darker pixels of the left panels represent more longitudinally oriented collagen fibers and are concentrated in the lateroanterior regions carrying tensile stresses. The lighter pixels of the right panels represent greater volumetric mineralization and are concentrated in the medioposterior regions carrying more compressive stresses.

These observations are consistent with reports that porosity (Martin et al., 1980) and bone stiffness (Ashman et al., 1983) also vary in a regular fashion with angular position in the femoral cortex. They support the compound hypothesis that (a) osteon collagen fiber orientation and mineralization help determine cortical bone's mechanical properties and (b) osteonal remodeling adapts internal structure to functional stresses. However, because in healthy bone osteonal mineralization is only supposed to be a function of time since formation, it is not clear how radiographic density could be reduced in the tensile cortex. Is mineralization of osteoid enhanced by compressive stress? Is the rate of remodeling greater on the tensile side, so that its osteons are, on average, younger? Perhaps a better hypothesis is that the radiographic density reflected decreased porosity rather than mineralization of the tissue. (See the discussion under Mineralization, next.)

Anisotropy of Cortical Bone Mechanical Properties

Cortical bone exhibits highly ordered internal arrangements of collagen fibers and mineral crystals; this is true for both the plexiform and circumferential lamellar forms of primary bone, as well as osteonal bone. X-Ray diffraction techniques show that the average orientation of the mineralized collagen fiber matrix is approximately parallel to the bone's long axis (Chen and Gundjian, 1974). The fact that this causes marked differences in the mechanical properties associated with the longitudinal, radial, and circumferential directions was recognized long ago (Hulsen, 1898). The radial and circumferential directions are visually distinguishable in both forms of primary bone (refer to Figs. 2.7

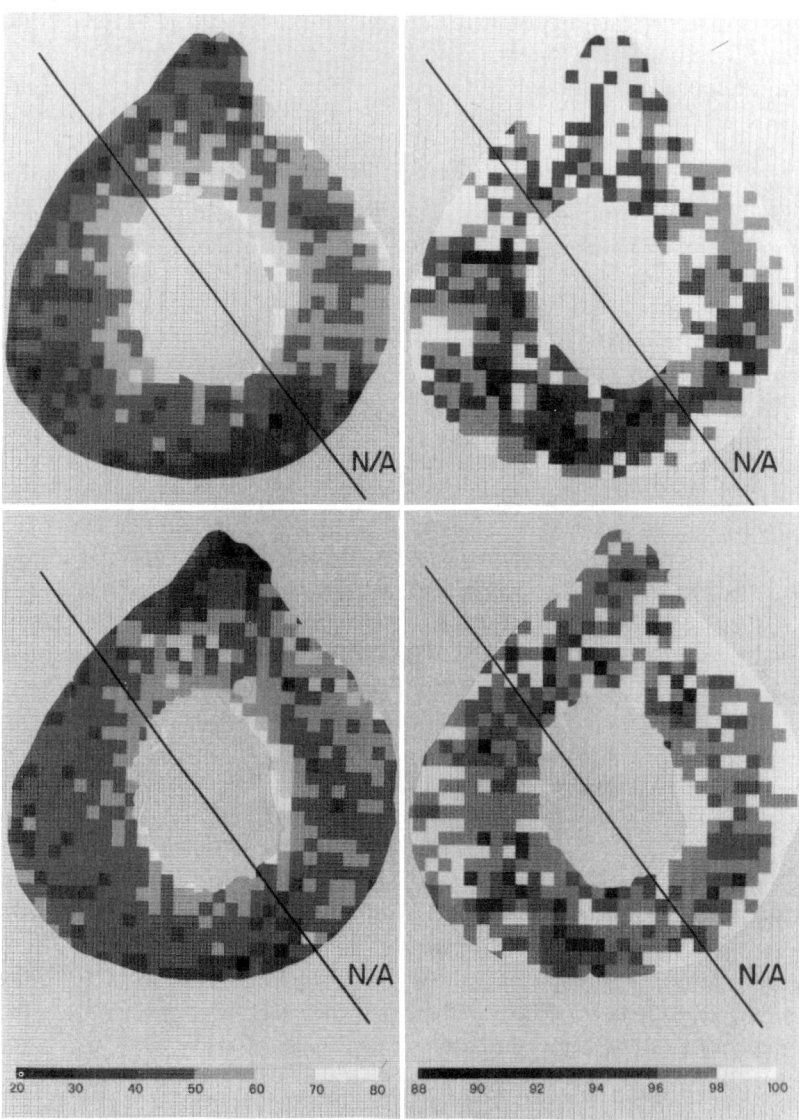

FIGURE 4.10. Plots of distribution of collagen fiber orientation *(left panels)* and radi-ographic density *(right panels)* in two cross sections from a human femur. These sec-tions were located 1 cm apart in the central diaphysis. Superimposed on the plots are the approximate locations of the neutral axis (N/A) for bending produced by loading at the hip (Pauwels, 1980). The darker pixels of the left panels, which have more lon-gitudinally oriented collagen fibers and greater tensile strength, are concentrated in the lateroanterior regions, carrying more tensile stresses. The lighter pixels of the *right panels*, representing greater volumetric mineralization, are more common in the medioposterior regions on the compression side of the neutral axis. Reproduced from Portigliatti-Barbos et al., 1983

and 2.8), but heavily remodeled bone appears to the eye to be transversely isotropic (e.g., the osteonal region in Fig. 2.3). We now examine the effects of osteonal remodeling on the mechanical anisotropy of cortical bone.

Anisotropy Effects Determined by Load-Deformation Testing

Classic contributions in this area are the work of Reilly, Burstein, and Frankel (1974; see also Reilly and Burstein, 1975). Using femurs, they measured the elastic moduli and strength of Haversian and primary bovine bone in different directions and compared them to those of human Haversian bone. Their data (summarized in Table 4.4) show that the strength and stiffness of bovine Haversian bone are similar to that of human bone when measured transverse to the long axis, but that cow bone seems to give consistently higher values in the longitudinal direction. In bovine bone, the ratios of elastic moduli in the longitudinal and transverse directions are similar in tension and compression and in primary and Haversian bone. These anisotropy ratios are less in human Haversian bone than in bovine Haversian bone (for both modes of loading) because the human bone is less stiff in the longitudinal direction. Both types of cow bone are stronger in compression than in tension, regardless of the loading direction. Human Haversian bone is stronger in compression when loaded transversely but not when loaded longitudinally. The strength anisotropy of Haversian bone from both species is more pronounced in tension than in compression, and

TABLE 4.4. Anisotropy of bovine and human bone

Elastic Modulus, GPa	Tension Longitudinal	Tension Transverse	Compression Longitudinal	Compression Transverse
Human				
Haversian	17.9 ± 0.9	10.1 ± 2.4	18.2 ± 0.9	11.7 ± 1.0
Bovine				
Haversian	23.1 ± 3.2	10.4 ± 1.6	22.3 ± 4.6	10.1 ± 1.8
Primary	26.5 ± 5.4	11.0 ± 0.2	—	—

Ultimate stress, MPa	Tension Longitudinal	Tension Transverse	Compression Longitudinal	Compression Transverse
Human				
Haversian	135 ± 16	53 ± 11	105 ± 17	131 ± 21
Bovine				
Haversian	150 ± 11[a]	49 ± 7[a]	272 ± 3	146 ± 32
Primary	167 ± 9	55 ± 9[a]	—	—

[a]Standard deviations approximate because groups were averaged.

From Reilly et al., 1974; Reilly and Burstein, 1975.

it appears that bovine bone anisotropy is somewhat greater than that of human bone. Thus, Haversian cow bone is a reasonable model for human cortical bone, but it can be expected to be more anisotropic because of its enhanced strength and stiffness in the longitudinal direction.

Reilly and Burstein's data suggest that osteonal remodeling has two important effects on mechanical anisotropy. First, secondary osteons reduce tensile strength and stiffness more in the longitudinal direction than transversely, suggesting that the differences between bovine and human Haversian bone result from the more extensive intracortical remodeling of human bone. Second, they serve to transform the bone from an orthotropic material into a more transversely isotropic one. The "transverse" direction referred to in Table 4.4 is the circumferential direction. Reilly and Burstein also tested bovine specimens in the radial direction. Recalling the structure of plexiform bone, one would expect to obtain different results in these two directions when this tissue is tested, and that is what was observed. Tensile strengths in the longitudinal, circumferential, and radial directions had proportionalities of about 3:1:0.4 in primary bone. In Haversian specimens the radial tensile strength was increased so that the ratios were about 3:1:0.7, and the compression strength ratios were equal in the two transverse directions: 3:1:1. Although distinct differences were found to persist in the deformability and energy to failure in the two transverse directions of bovine Haversian bone, these data show that osteons make whole bone transversely isotropic with respect to strength.

Knets (1978) determined the elastic constants of human tibial bone using conventional mechanical testing. He found that a statistically significant degree of orthotropy was present: C_{11} = 11.6 GPa, C_{22} = 14.4 GPa, C_{33} = 22.5 GPa. These values for human osteonal bone are in contrast to the ultrasonic data for bovine osteonal bone presented in the next section.

Anisotropy Effects Determined by Ultrasonic Measurements

Similar results have been obtained when ultrasonic techniques were used to measure stiffness values. In principle, the velocity of sound in a homogeneous continuum (i.e., a nonporous material) is proportional to the square root of its elastic modulus, so that measurements of sound velocity can be used to nondestructively determine material stiffness for very small deformations (Abendschein and Hyatt, 1970; Ashman et al., 1984). (In composite materials such as bone, this relationship becomes complex when the wavelength of the sound approaches the dimensions of the voids.) Yoon and Katz (1976a,b) have described in detail how the anisotropic elastic properties of bone can be nondestructively determined by measuring the velocity of sound waves propagated through a specimen in various directions.

Van Buskirk et al. (1981), Bonfield and Tully (1982), and Katz and co-workers (Lipson and Katz, 1984; Katz et al., 1984) have studied the effects of osteonal remodeling on the ultrasonic stiffness of bovine cortical bone. Their methods show that osteonal remodeling tends to convert plexiform bone from an orthotropic material to a transversely isotropic one, at least in the ultrasonic realm of very small strains (Katz et al., 1984). Table 4.5 shows that Haversian bone has very similar properties in the transverse (1 and 2) directions, but plexiform bone does not. Katz et al. (1984) have also demonstrated that osteo-porosis has no effect on anisotropy, but osteopetrosis greatly reduces it. The latter finding was confirmed by Ashman and co-workers (1985a). Because osteopetrotic bone is largely calcified cartilage and contains few osteons, it is easy to see why it would be less anisotropic than normal bone.

In the inferior cortex of the canine mandible, which is a site of heavy osteonal remodeling, Ashman et al. (1985b) initially obtained values for E, G, and Poisson's ratio that were similar to those for osteopetrotic bone; that is, isotropic and less stiff than transversely isotropic bone. Later, Ashman and Van Buskirk (1987) reported additional measurements indicating that mandibular bone is transversely isotropic after all. There is a need for the mechanical properties of the bone inside the mandible to be determined with certainty. This bone has a complicated geometry and is loaded by a complex system of muscle and dental forces. It is of considerable interest to physical anthropologists, because its structure may provide information about the diet of its owner (Hylander, 1981), and to dental clinicians for obvious reasons. Although long bones are usually used for testing theories of Wolff's law, the mandible may be a better choice in some respects.

Turner et al. (1995) used an acoustic microscope (which produces images based on local differences in sound velocity) to investigate the source of cortical bone anisotropy. They measured sound velocity in the longitudinal and transverse directions in whole canine bone, then repeated the measurements after the specimens had been either decalcified or reduced to the mineral phase by dissolving their collagen. Table 4.6 shows

TABLE 4.5. Elastic coefficients in GPa

	Plexiform	Haversian
C_{11}	22.4 ± 0.6	21.2 ± 0.5
C_{22}	25.0 ± 1.0	21.0 ± 1.4
C_{33}	35.0 ± 2.0	29.0 ± 1.0
C_{44}	8.2 ± 0.4	6.3 ± 0.4
C_{55}	7.1 ± 0.3	6.3 ± 0.2
C_{66}	6.1 ± 0.2	5.4 ± 0.2
C_{12}	14.8 ± 0.8	11.7 ± 0.7
C_{23}	13.6 ± 0.7	11.1 ± 0.8
C_{13}	15.8 ± 0.8	12.7 ± 0.8

The C_{ij} subscripts refer to the following directions: 1 = radial direction; 2 = cir-cumferential direction; 3 = longitudinal direction; 4 = circumferential-longitudi-nal shear; 5 = radial-longitudinal shear; 6 = radial-circumferential shear.
From Katz et al., 1984.

TABLE 4.6. Anisotropy ratios in osteonal bone

Material	E_L/E_T
Whole bone	1.50
Mineral phase	1.50
Organic phase	1.18

From Turner et al., 1995; Turner and Burr, 1997.

"anisotropy ratios," defined as the ratio of the longitudinal elastic modulus to the transverse modulus, for the three types of specimens. It is seen that the mineral phase is as anisotropic as the whole bone but the organic phase is more nearly isotropic. These data need somehow to be squared with the idea, held for some time, that it is the orientation of bone's collagen fibers which produce its anisotropic qualities (Turner and Burr, 1997).

In summary, there are three important points concerning cortical bone anisotropy. First, the predominantly longitudinal orientation of collagen fibers and mineral crystals makes bone stronger and stiffer in this direction. Second, osteonal remodeling tends to convert bone from an orthotropic to a transversely isotropic material. Finally, certain metabolic diseases may render bone virtually isotropic, but it is clear that this is not functionally advantageous.

Determinants of Osteonal Bone Mechanical Properties

In addition to the presence of osteons, a number of other factors affect the mechanical properties of cortical and osteonal bone. This section summarizes the more significant of these.

Porosity
Obviously, the introduction of holes in a structure should weaken it, and this is certainly what happens in bone. It is one thing to demonstrate this effect, however, and quite another to explain in some detail the nature of the relationship. Empirical formulae have been developed relating porosity to the mechanical properties of various porous materials, but a comprehensive theory useful over a broad range of porosity has not been developed. These problems are explored later in this chapter.

In Chapter 2, we defined porosity as the ratio of void volume to total volume, and noted that volumetric porosity is equivalent to that measured two dimensionally on a cross section (Underwood, 1970). Bone contains voids that range from a few to several hundred micrometers in diameter. Usually, however, the smallest of these (i.e., canaliculi and osteocyte lacunae) are not considered to affect its mechanical properties because, even though they probably do, they are rather uniformly distributed throughout all kinds of bone. They are therefore considered to be an intrinsic property of the bone material substance. In cortical bone, then, mechanical properties are considered to be affected by Haversian canals and related resorption cavities and vascular channels. (In trabecular bone they are affected by the marrow spaces between the trabeculae.) Several investigators have proposed empirical relationships between mechanical properties and porosity. For example, Schaffler and Burr (1988) found that the tensile elastic modulus (in GPa) of bovine cortical bone was related to porosity by the empirical equation

$$E = 33.9 \, (1 - p)^{10.9} \tag{4.8}$$

[Note that $(1 - p)$ = bone volume fraction.] Currey (1988) obtained, for cortical bone from a wide variety of species loaded in tension,

$$E = 23.4 \, (1 - p)^{5.74} \tag{4.9}$$

Comparing these results suggests that the modulus of bone becomes more sensitive to changes in porosity as its value decreases. Schaffler and Burr's maximum porosity value was 7.8%; Currey's was 31%. Pursuing this, McElhaney et al. (1970) found that the modulus of human skull bone (containing a very porous diploe layer) increased as the cube of $(1 - p)$. Similarly, modulus has been shown to increase as apparent density raised to the 7.4^{th} power for cortical bone, as the cube of density for cortical and cancellous bone samples pooled together, and as the square of density when a sample containing only cancellous bone was analyzed (Schaffler and Burr, 1988; Carter and Hayes, 1977a; Rice et al., 1988).

Figure 4.11 shows the relationship between porosity and strength as compiled from several experiments. Curves of this shape obtain in other

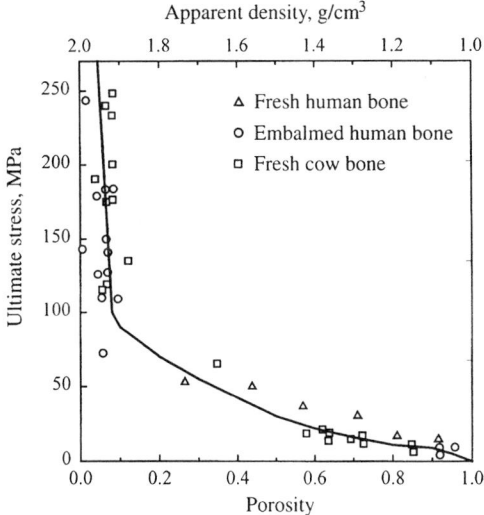

FIGURE 4.11. The relationships between ultimate compressive stress, porosity, and apparent density for fresh human bone (triangles) (Behrens et al., 1974), embalmed human bone (circles), and fresh bovine bone (squares) (Martin, 1984). Very small porosity increases greatly affect cortical bone strength. (Replotted from data in Martin, 1984.)

porous materials as well, including ceramics (Gannon et al., 1965). Clearly, the slope of the curve is quite different for cortical bone, in which most of the material is solid matrix, and trabecular bone, in which most of the material is void space (filled with soft tissues, but not rigid). Presumably, different mechanical principles determine the slopes in these two regions, but they have not been elucidated.

In Fig. 4.11, porosity is implicitly regarded as a **material** feature of bone. Alternatively, one may think of porosity as a **structural** feature. If the porosity is uniformly distributed, the bone is loaded in simple tension or compression, and one is concerned only with macroscopic phenomena, the former point of view is convenient and acceptable. On the other hand, in a bent or twisted bone, voids located in the outer cortex, where stresses are high, weaken the bone more than those near the medullary canal. If there is a porosity gradient in the cortical wall, it makes a substantial difference whether the more porous region is in the outer, more highly stressed material, or near the medullary canal. Bones such as the femur are frequently characterized by porosity gradients across the cortical wall. Smith and Walmsley (1959) showed that gradients typical of those in elderly people can reduce the bending strength of a rectangular bone test specimen machined from a cortical wall by 20% or so. However, the gradient is actually directed to enhance bone strength by placing most voids near the center of the bone. This is clear from Fig. 4.6, which compares the porosity distribution in the shaft of an elderly person's femur to that in a younger person. Not only are the void spaces concentrated near the medullary canal, but they are also preferentially located near the neutral axis.

Mineralization

As with porosity, it is intuitively clear that variations in the mineral content of bone should affect its mechanical properties, and experiments confirm this hypothesis. One may begin the analysis of this problem by breaking it into three parts according to the degree and cause of the mineralization difference. In the first place, because there is a delay in the mineralization of newly formed bone, bones that are being remodeled at a higher rate exhibit a reduced mineral content. This effect can be related to age, trauma, or metabolic disease. Reduced mineral content can also be caused by a lengthening of the osteonal refilling period because a delay in completion of new osteons means either that more osteoid or more void space is present for a given activation frequency. Much larger differences in mineralization may be produced by metabolic bone diseases, such as rickets or osteomalacia, in which the rate of turnover and refilling time are normal but the mineralization lag time is exaggerated.

It is very important to be clear about how "mineralization" is defined. One may define the amount of mineral per unit volume of whole bone as *volumetric mineralization*, and the amount per volume of bone matrix, exclusive of the volume of the void spaces in the structure, as *specific mineralization*. Clearly, volumetric mineralization is a function of both porosity and mineralization of the organic matrix; specific mineralization discounts the effects of porosity and is therefore more indicative of the mineralization process itself. Increased rate of bone turnover and increased refilling period, therefore, create both a volumetric and a specific mineralization deficit, while the inhibition of mineralization of osteoid creates only a specific mineralization deficit. In reading the literature one must be careful to understand which definition of "percent mineral" is being used. Similar comments could be applied to use of the term "density."

Most noninvasive determinations of mineral content (e.g., photon absorptiometry and computed tomography) measure volumetric mineralization in some fashion, and that is why these procedures are not generally capable of separating diseases with reduced volumetric mineralization (e.g., osteoporoses) from those that involve reduced specific mineralization (e.g., osteopenias such as osteomalacia). Specific mineralization is usually measured by ashing the bone and calculating its ash fraction (recall Eq. 2.9).

When Vose and Kubala (1959) correlated bending strength of human femurs with radiographic density, they were using a measure of specific mineralization because they pulverized and compressed the bone specimens before X-raying them. They found that small differences in matrix mineralization made very large differences in breaking stress: increasing the ash content from 63% to 71% caused the strength to rise 3.7 fold. Currey (1969a) became interested in this relationship, and wondered why animals did not take better advantage of the apparently disproportionate enhancement of bone strength afforded by adding a few grams of mineral to each bone. To explain this conundrum, he postulated that more highly mineralized bone is weaker at higher strain rates, and presented experimental data suggesting that the optimal specific mineralization for cortical bone is 66%–67%. The same year (Currey, 1969b), he further hypothesized that the slope of the mineralization-stiffness (and strength) curve was especially steep because as mineralization became complete, the ends of the apatite crystals came into contact and fused. In a later publication, Currey and co-workers (Wainwright et al., 1976) concluded that 66%–67% mineralization was the point of maximum toughness. A few years later, Bonfield and Clark (1973) reached a similar conclusion (see following).

Currey (1986) related the elastic modulus of cortical bone from 17 vertebrate species, in both bending and tension, to porosity and specific mineralization (expressed in the form of calcium content). He found that these two variables together accounted for about 84% of the stiffness variation, but he did not discuss their relative effects. He observed that such high correlations with porosity and mineralization left little room for dependency on such histologic features as collagen fiber orientation, but he also suggested that was probably not true for strength.

Finally, Schaffler and Burr (1988) found, for bovine cortical bone in tension,

$$E = 89.1 \, A^{3.91} \tag{4.10}$$

where A is specific mineralization expressed as percent ash by mass. However, percent ash was not a statistically significant predictor of E, perhaps because A varied so little, mineralization being tightly regulated.

Box 4.4 Technical Note
Noninvasive Measurement of Bone Mass

There are several ways of noninvasively measuring bone mass, that is, the amount or density of bone in a particular part of the skeleton. Most of these methods depend on the absorption of photons by the bone, and because the mineral absorbs photons much more readily than the collagen, water, and soft tissues, these methods essentially measure bone mineral. However, because the variability of mineralization in adult bones is much less than that of porosity, these methods provide information about bone size and porosity as well.

   The oldest and simplest of these methods is a simple radiographic film or "X-ray." However, from variability in exposure and processing, a 30%–50% reduction in bone density may occur without being evident on a standard clinical radiograph. Calibrated *computed tomography* (CT) images (called quantitative CT, or QCT) are considerably more accurate, but both these methods involve significant radiation

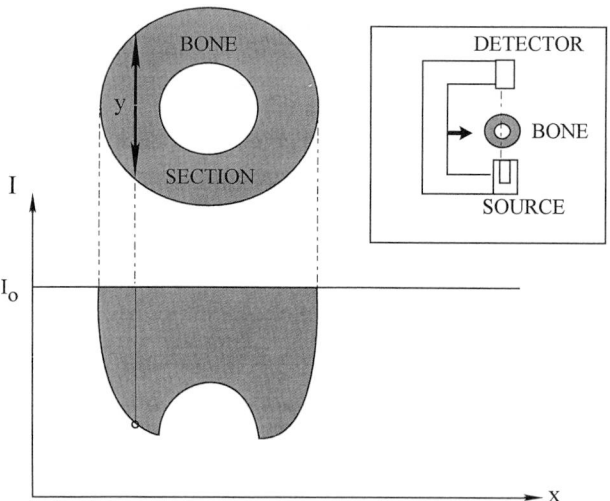

An absorption curve (I vs. x) in relationship to the cross section of the bone being scanned. The distance from the baseline (I_0) to the absorption curve is proportional to the distance y across the cortex and the volumetric density of the mineral along this path. Inset: schematic diagram of the scanner's source and detector passing across the bone.

doses. In the 1970s, this problem was significantly reduced by the development of *photon absorptiometry* by Mazess and Cameron (1972). This method involves scanning across a bone with a beam of photons emitted from a low-level radionuclide source such as ^{125}I. The source is fixed opposite a detector, the beam is scanned across the bone, and the intensity of the radiation received by the detector, I, is plotted as a function of scan position, x. The intensity of the radiation reaching the detector is

$$I = I_0 \, e^{-ky}$$

where I_0 is the intensity of the radiation from the source, y is the total thickness of the mineral in the beam's path through the bone, and k is an absorption coefficient. The area within the resulting absorption curve is proportional to the mineral content of the bone in the path of the beam. The width of the curve gives the width of the bone, and the dip in the middle is caused by the medullary canal. The most common site for such measurements has been the bones in the distal forearm. Results are reported as bone mineral content (BMC, g mineral/cm bone length) and bone mineral density (BMC/bone width, g mineral/cm^2). Later, it was realized that by scanning with two radionuclides emitting photons with different absorption coefficients, correction could be made for the fact that soft tissues absorb some radiation; this is known as *dual energy photon absorptiometry*.

In the 1990s, an alternative version of this known as *dual energy x-ray absorptiometry* (DEXA or DXA) was developed. This technique substitutes low-energy X-rays for the radionuclide source, and involves scanning back and forth along some length of the bone to produce a two-dimensional image. Thus, the hip, the spine, and even the whole body can be scanned for bone mineral content. These widely used absorptiometry methods are accurate to within 1%–2% and expose the patient to a fraction of the radiation dose of a standard chest X-ray.

Density

The material properties of bone are frequently discussed in terms of its "density." Used alone, this word would be expected to refer to the specific gravity of the solid matrix. The term *apparent density*, on the other hand, refers to the mass per unit volume of a region of bulk bone, which includes Haversian canals, marrow spaces, and other soft tissue spaces. Therefore, apparent density is a function of both the porosity and mineralization of the bone material. Carter and Hayes (1977a) showed that the compressive strength (σ_{cf}) and elastic modulus of bone specimens having a wide range of apparent densities[1] (d = 0.5–1.9 g/ml) were correlated with the square and cube, respectively, of density in g/ml:

$$\sigma_{cf} = 68 \, \dot{\varepsilon}^{0.06} d^2 \qquad \text{(MPa)} \qquad (4.11)$$

---

[1]In this experiment, apparent density was measured after the void spaces had been emptied of soft tissues, so they contained only air. That is why the symbol d is used here instead of the ρ that was used for apparent density in Section 2.4. How would you change Eq. 2.6 to obtain an expression for d?

$$E = 3790 \; \dot{\varepsilon}^{0.06} d^3 \qquad \text{(GPa)} \qquad (4.12)$$

In these equations, $\dot{\varepsilon}$ is strain rate in s^{-1}. The latter relationship has been disputed by Rice et al. (1988). Through extensive multiple regression analysis of a large collection of data for cancellous bone, in which they attempted to control for differences in species, type of load, and load direction, Rice et al. found that both E and σ_{cf} correlate better with d^2 than d^3. Other studies of cancellous bone have also obtained exponents closer to 2 than 3 (Bensusan et al., 1983; Ashman and Rho, 1988). It appears that the reason for this discrepancy is that the data of Carter and Hayes included cortical as well as cancellous bone specimens, and these points increased the exponent from 2 to 3.

Histologic Architecture

Two regions of bone having similar porosity and mineralization may still have quite different material properties because the organization of the solid matrix is highly variable. As noted in Chapter 2, compact bone may have a great variety of internal architectures. One may first distinguish between primary lamellar bone (mostly formed by modeling on periosteal or endosteal surfaces during growth) and lamellar bone composed of secondary osteons (formed by remodeling). We have seen that secondary bone is weaker than primary bone (Reilly and Burstein, 1975). Primary cortical bone itself may also consist of two nonlamellar types of tissue. Plexiform bone is similar to lamellar bone but is laid down more rapidly in growing individuals (Enlow and Brown, 1956, 1957, 1958). Much of the data on primary bone's mechanical properties is based on this material because it occurs in cows, a ready source of material for testing. How these properties compare to the "pure" lamellar bone usually found in humans is unclear because remodeling occurs so early in humans that it is difficult to obtain sufficient material to test.

Collagen Fiber Organization

Given a particular porosity and mineralization, and a particular trabecular or osteonal architecture, it remains possible for two regions of bone to have different material properties. There could be several reasons for this, but the next variable to be considered here is the orientation of constituent collagen fibers. This aspect of structure is apparently controlled both by cell functions and by extracellular physical processes. Frost (1986) has called these processes "micromodeling."

Lamellar bone, with lamellae of parallel collagen fibers, is stronger than woven bone in which the collagen fibers are more or less randomly oriented. Furthermore, the predominant orientation of the collagen fibers affects tissue strength in relation to the mode of loading: longitudinal fibers promote strength in tension, while transverse fibers are associated with strength in compression. This has been shown for individual osteons

(Ascenzi and Bonucci, 1967, 1968) as well as bone in bulk, and it was seen in Fig. 4.10 that fiber orientations vary with position relative to the neutral axis of the femur, so that regions in habitual tension have more longitudinal fibers and those in compression have fewer (Portigliatti-Barbos et al., 1983). In cortical bone, which generally exhibits little variability in porosity or mineralization, collagen fiber orientation may be the predominant factor in determining tensile strength (Martin and Ishida, 1989). No studies have been found of collagen fiber orientation effects in cancellous bone, but it seem likely, for example, that trabeculae composed of woven bone must behave differently from those containing lamellar bone.

Fatigue Damage
Another factor that may affect the material properties of bone is microdamage. It is known that fatigue microdamage reduces the elastic modulus (Carter and Hayes, 1977b; Schaffler et al., 1989), and as techniques for measuring such damage improve, this may be added to the list of those things studied as determinants of bone's mechanical properties. When loaded repetitively in the laboratory, bone experiences fatigue damage and may fail from this like any other structural material (Gray and Korbacher, 1974; Carter et al., 1976; Carter and Hayes, 1976). The in vivo fatigue life of bone depends on the rate at which remodeling repairs fatigue damage. In Chapter 5 we consider the relationship between the creation of fatigue damage and its repair by remodeling and describe a model for stress fracture, which is what clinicians call the fatigue fracture that occurs when bone's repair process fails to keep up with damage formation.

Rate of Deformation
Cortical bone is a viscoelastic material; its mechanical properties depend considerably on the rate at which it is deformed. One may speculate that features of internal structure associated with remodeling and osteons could contribute to such behavior in several ways, including the presence of osteoid in varying amounts, the flow of fluid through Haversian canals, canaliculi, or the interstitial spaces in bone matrix, and the presence of cement lines. There is, however, very little information available regarding the effects of osteons on bone's viscoelastic behavior.

The classic paper on bone's viscoelasticity is that of McElhaney (1966). He studied bovine femoral bone over a wide range of deformation speeds, building a pneumatic cannon to achieve an upper strain rate of 1500/s. In general, as shown in Fig. 4.12, cortical bone becomes stronger, stiffer, and more brittle as strain rate increases. Although he did not quantify the histologic structure of his specimens, McElhaney remarked that those loaded at high strain rates tended to fail along cement lines, while those loaded at low strain rates failed along shear planes passing through osteons. It is interesting, and perhaps functionally significant, that the energy absorbed to failure is greatest not at the

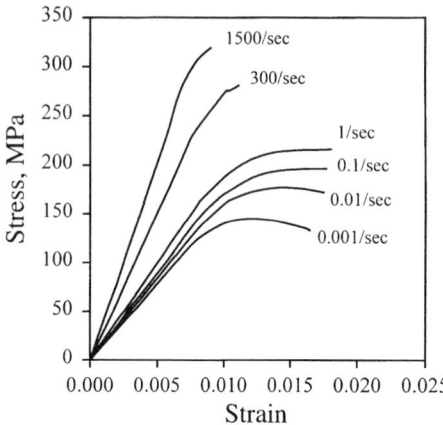

FIGURE 4.12. Stress-strain curves for cow bone loaded in compression at various strain rates. (Redrawn with permission from McElhaney, 1966.)

lowest or highest strain rates, but in the range 0.01–0.10 s^{-1}. Assuming a bone the size of an adult femur is loaded in bending, this corresponds to a mean impact velocity much less than 1 m/s or 2 miles/h, well below that expected in trauma, even before the advent of modern transportation and weapons. Margel-Robertson and Smith (1978) tested bovine femoral bone in compression at strain rates below this range and found a second local maximum in energy absorption at 0.00001–0.0001 s^{-1}. Others who have experimentally studied the viscoelastic properties of cortical bone include Smith and Keiper (1965), Sedlin (1965), Burstein and Frankel (1968), Panjabi et al. (1973), Crowninshield and Pope (1974), and Bargren et al. (1974). Sedlin proposed a rheological model for cortical bone consisting of a three parameter solid with a frictional element to account for plastic deformation at higher strains, but the data of Bargren and co-workers suggest that a two-element Kelvin model satisfactorily describes bone in the physiologic strain range. On the other hand, Laird and Kingsbury's (1973) tests on cow bone indicated that a three parameter model was inadequate to describe vibration phenomena in bone (frequencies of the order of kilohertz).

Ramaekers (1977, 1978, 1979a,b,c) published a series of papers on the rheological properties of bone. In one of these (1978), he made the interesting suggestion that one of the viscoelastic functions of bone is to damp out the individual twitches of muscles. This hypothesis was prompted by the fact that one of the peaks in the shear damping frequency response corresponds to the twitch frequency of mammalian muscle.

Lakes, Katz, and Yoon have contributed a number of experiments and mathematical analyses of the effects of viscoelasticity on cortical bone mechanical properties (Lakes and Katz, 1974, 1979a,b; Lakes and Saha,

1980; Lakes et al., 1979, 1986). This work considers primarily the deformation of bone at sonic or ultrasonic frequencies and is related to Katz's ultrasonic studies of stiffness anisotropy. The work included torsional and biaxial loading modes, and the suggestion was made that peaks in the frequency response of the material properties could have histologic and physiologic significance. Johnson and Katz (1984) pointed out erroneous assumptions in these studies and attempted to correct them. They concluded that fluid flow through vascular channels is probably not important in cortical bone viscoelastic behavior; rather, it is the interactions of the fluids with the calcified matrix (which is not strictly solid, but contains minute water-filled spaces) that are important. Much remains to be done in extending these viscoelasticity studies to provide insights and predictive capabilities more relevant to bone fracture and remodeling.

In summary, three of the fundamental determinants of the mechanical properties of cortical bone's material include porosity and mineralization (which together determine apparent density), and collagen fiber orientation. These variables are strongly affected by histologic structure, starting with distinctions between primary and osteonal bone structure. Clearly, these factors are largely controlled by remodeling, particularly in adults. Fatigue damage may also intrinsically affect mechanical properties, and finally, strain rate is an important extrinsic factor.

4.5 Material Properties of Cancellous Bone

We now turn our attention to cancellous bone. As one would expect from looking at its porous structure (recall Fig. 2.2), cancellous bone is substantially weaker and more compliant than cortical bone. Table 4.7 gives some typical values of cancellous bone's compressive strength and elastic modulus. Note the considerable variability in these values. Although bone porosity is bimodally distributed (i.e., either cortical or cancellous, with few regions between), it varies continuously, and there are not sharp geometric, densitometric, or mechanical distinctions among various examples of cancellous bone or between dense cancellous bone and porous cortical bone.

Stress-Strain Curves for Cancellous Bone

Figure 4.13 shows typical load-deformation curves for cubes or cylinders of cancellous bone loaded in compression. As the specimen compresses, it yields and reaches an ultimate stress, as does a cortical bone specimen. Thereafter, however, the load does not fall to zero because the specimen does not fracture into pieces. Instead, its void spaces collapse and its trabeculae compact against one another. As this process continues, the load is typically maintained, or may even rise far above the previous ultimate load. In addition to these behaviors, mechanical testing of cancellous bone

TABLE 4.7. Compressive strength and stiffness of cancellous bone

	Ultimate Stress, MPa	Modulus, MPa
Human		
Yamada, 1973[a]	1.86–1.37	90–70
Neil et al., 1983[b]	2.54 ± 0.62	272 ± 195
Kuhn et al., 1989a[c]	5.6 ± 3.8	424 ± 208
Rohl et al., 1991[d]	2.22 ± 1.42	489 ± 331
Canine		
Vahey et al., 1987[e]	12.1 ± 5.7	434 ± 174
Kuhn et al., 1989a[c]	7.12 ± 4.6	264 ± 132
Norrdin et al., 1990[f]	9.60 ± 0.80	231 ± 22

[a]Means for people in their forties and sixties, respectively; lumbar vertebrae.
[b]Lumbar vertebrae from men aged 54–90 years, 12-mm-diameter cylinders, 25–30 mm long.
[c]8-mm cubes from distal femur.
[d]Proximal tibia, 4 men and 3 women, aged 42–76 years.
[e]5-mm cubes from head and neck of the femur in 2 dogs.
[f]21 vertebral specimens from 7 dogs.

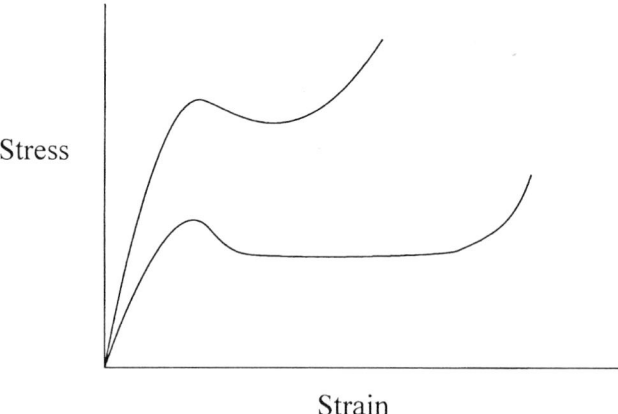

FIGURE 4.13. Load-deformation curves for two specimens of cancellous bone having different porosities. Yield and ultimate loads are apparent in the early portion of each curve. Subsequently, reduction and plateauing of the load occur. As void spaces collapse, the specimen densifies and the load rises steeply.

is much more problematic than testing cortical bone. When a compression test is done between platens and strains are based on platen displacement, errors can occur because artifactual damage in the regions adjacent to the platens—damage caused by interruption of the trabecular network when the specimen is machined, friction against the platens, and stress concentrations. Other problems arises because supporting trabeculae have been removed along the sides of the specimen and the marrow in the void spaces is not confined as it is in the intact bone.

Keaveny and co-workers (1994) have developed testing methods designed to alleviate the problems associated with end effects at the platen interfaces. Cylindrical (40 mm long, 8.3 mm in diameter) specimens are cored from trabecular bone and their ends are cemented into brass endcaps. The specimen is then frozen and the central 8-mm region turned to a diameter of 6 mm on a lathe. Strain is measured using a device called an extensometer, which fastens to and measures the distance between two points on the central region of the specimen. These precautions eliminate or reduce the problems with end effects.

Strength in Tension Vs. Compression
Carter et al. (1980), Bensusan et al. (1983), and Rice et al. (1988) all found that the failure stress of cancellous bone was the same in tension as compression. Kaplan et al. (1985), on the other hand, found that cancellous bone was stronger in compression for any given apparent density. Cowin (1989, p. 138) suggests that this may be because the specimens in the Kaplan experiment came from a region particularly well adapted to compressive loading in the tested direction.

Caution: Continuum Limits Nearby
As with cortical bone, we treat cancellous bone here as a porous material used to construct whole bones and try to identify and quantify the features (e.g., apparent density) that govern its mechanical behavior. However, in cancellous bone the situation is different and a bit tricky because of the relative sizes of the porosities in the bone.

The thickness of trabecular struts and plates is typically in the range of 100 to 300 μm, with the spacing between adjacent trabeculae of the order of 300 to 1500 μm. This is true regardless of the size of the animal being considered. Therefore, a line passing through the femoral head of an elephant would encounter hundreds of trabeculae, and a similar line through the femoral head of a mouse would encounter only a few. Although we can imagine taking 1 cm^3 of cancellous bone from the elephant and treating it as a material of which the strength and modulus can be measured, in the case of the mouse the cancellous bone of the femoral head must be modeled as a structure rather than as a material, because slight geometric variations in shape and orientation of those few trabeculae could greatly alter the mechanical properties. One of the requirements of continuum mechanics is that the minimum dimension of a sample that is treated as a material must be significantly larger than the dimension of its structural subunits (Fung, 1969). In the case of trabecular bone, the minimum continuum length has been estimated to be about five trabecular spacings (i.e., 5–10 mm, depending on the sample; Brown and Furguson, 1980; Harrigan et al., 1988). This is an order of magnitude greater than the continuum length for compact bone, where the porosity is much less and the void spaces are much smaller.

A practical implication of this is that in speaking of cancellous bone as a **porous material**, we need to be thinking in terms of samples of about 1 cm³ (Linde et al., 1992). This leads to a conundrum. Cancellous bone can be quite heterogeneous, with its properties varying millimeter by millimeter, as shown, for example, by Fig. 2.2 and in penetration test[2] data (Hvid and Hansen, 1985; Hvid, 1988). A sample that is big enough to satisfy the requirements of the continuum length assumption, therefore, is quite likely to contain local heterogeneities in mechanical properties. When strength is measured on such a sample, the measurement often reflects the weakest portion of the sample. Consequently, local variations in mechanical properties tend to be overlooked in the results. Therefore, attempts to treat cancellous bone as a porous, continuum-level material must be evaluated very cautiously. With this reservation in mind, we can proceed to examine cancellous bone as a porous material.

The Three Determinants of Cancellous Bone Mechanical Properties

Insofar as we can treat cancellous bone porosity as a material property, we shall find that its mechanical properties are governed by three things: the magnitude of the porosity (or apparent density), the anisotropy of the trabecular architecture, and the material properties of the tissue in the individual trabeculae. We now discuss each of these, beginning with the latter.

Cancellous Bone Tissue Properties

The typical range of apparent density in cancellous bone is 1.0 to 1.4 g/cm³, compared to about 1.8–2.0 g/cm³ for compact bone (see Fig. 4.11). This decrease in apparent density primarily results from the presence of marrow-filled spaces; the density of trabecular bone's calcified matrix (or *tissue*) itself is similar to that of compact bone. Wolff (1892), in formulating his ideas about bone's mechanically optimized structure, assumed that cor-

---

[2]A penetration test involves pushing a needle about 2.5 mm in diameter into the flat, machined surface of a region of cancellous bone. A graph of the resistive force vs. depth of penetration looks somewhat like a compression load-deformation plot. The correlation between the peak force seen in the penetration test and compressive strength is about $R = 0.9$ (Hvid and Hansen, 1985).

tical and trabecular bone were both made of the same kind of tissue, so that the differences in their mechanical properties only resulted from the varying sizes and shapes of the soft tissue spaces. This concept has received more critical attention in recent years, however. There are apparently some important differences in the microstructure of the calcified bone matrix in cancellous and in cortical bone, and these evidently are responsible for measured differences in material properties at the tissue level. Let us explore these differences.

The tissue in cancellous bone is usually made in the same way, and of the same lamellar material, as cortical bone. In both cases, osteoblasts sit on a bone surface and emit an organic matrix that calcifies to produce an anisotropic, lamellar composite. There is little evidence that, at the lamellar level, this basic structure is fundamentally different in cortical and in cancellous bone. Studies of osteocyte lacunar sizes and numbers in cancellous and cortical bone have not been able to establish any clear differences. However, the mineral content and tissue density of cancellous bone have been found to be less than in cortical bone (Gong et al., 1964; Dyson and Whitehouse, 1968; Norrdin et al., 1977). Consistent with this, the water content of cancellous tissue has been reported to be greater than that of cortical bone. These results are consistent with the higher rate of turnover in trabecular bone; the average age of the bone matrix is simply less, so it is less mineralized.

Once one gets beyond the lamellar level and examines the geometry of the bone structural units created by the remodeling BMUs, another difference between cortical and cancellous bone tissues is apparent. Cortical bone osteons are parallel cylindrical structures that extend for several millimeters through the bone tissue. Their cement lines are closed surfaces. The trabecular equivalents, on the other hand, are dish-shaped "patches" having cement lines that intersect the surface of the trabecula (Fig. 4.14). Therefore, cortical osteons act like reinforcing fibers running through the tissue, but trabeculae are composed of multiple dish-shaped patches. If a cement line is disrupted by stress in cortical bone, the disruption is contained within the bone, but in cancellous bone it may propagate to the surface, disconnecting the structural unit from the parent trabecula. Thus, remodeled cortical bone structure may be inherently more damage resistant than remodeled cancellous bone structure.

These mineralization and structural issues lead to the question of whether cancellous tissue has the same mechanical properties as cortical tissue. There have been two ways of approaching this question. The first is to measure the property (e.g., elastic modulus) in about 1-cm^3 machined specimens of cortical and cancellous specimens and plot the results against the specimens' apparent density. If both kinds of specimen have the same tissue-level value of the property, they should fall on a single, smooth curve. Rho and co-workers (1993) did this with human bone, measuring the modulus ultrasonically (Fig. 4.15); they found that the modulus of the cor-

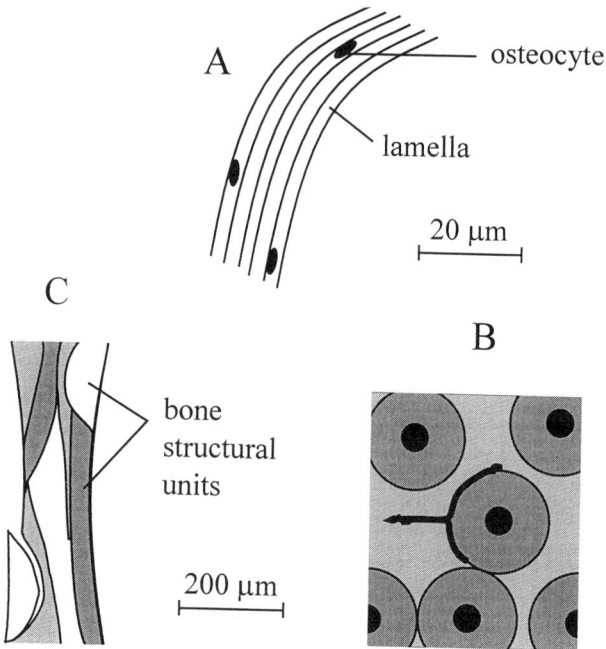

FIGURE 4.14. **A** At the level of lamellae, there is little to distinguish between osteonal and trabecular bone tissue. **B** In osteonal bone, the structural units produced by remodeling are largely intact, relatively long, fiberlike osteons. Their cement lines are internal to the bone; cement line disruptions are self-contained and do not compromise longitudinal loadbearing. **C** In trabeculae, the structural units produced by remodeling are much smaller and dish shaped with cement lines that, if disrupted, exfoliate the structural unit.

tical bone specimens fell outside, and above, the confidence limits of the extrapolated regression line for the cancellous specimens. The triangular data points on the curve represent values ultrasonically measured in just the cancellous bone material. (This was done by increasing the frequency of the sound so it only propagated through the trabecular tissue.) These points fall within the extrapolated cancellous values. This graph provides strong evidence that cancellous bone has a reduced tissue-level modulus.

The other method for approaching the question is to test either individual trabeculae, or very small specimens machined from individual trabeculae, and compare the results with those from similarly sized specimens machined from cortical bone. Rho et al. (1993) also did this, measuring the modulus both ultrasonically and in tension tests in whole trabeculae and microspecimens of cortical bone. The tension tests showed the modulus of cancellous material to be 56% of that of cortical bone (10.4 ± 3.5 vs. 18.6 ± 3.5 GPa). The ultrasonically measured values also showed the cancellous material was less stiff (14.8 ± 1.4 vs. 20.7 ± 1.9 GPa). Others have obtained different results when making the comparison between machined

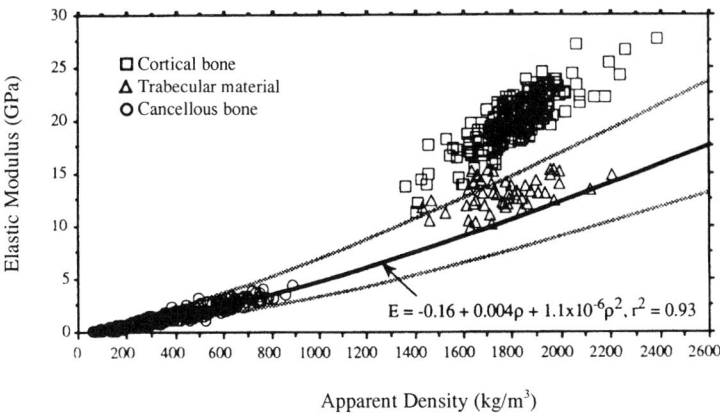

$$E = -0.16 + 0.004\rho + 1.1 \times 10^{-6}\rho^2, \ r^2 = 0.93$$

Apparent Density (kg/m^3)

FIGURE 4.15. Ultrasonically determined elastic modulus vs. apparent density for continuum-level cancellous *(circles)* and cortical *(squares)* bone specimens. A linear regression line and its 95% confidence limits are shown for the cancellous specimens. The fact that the cortical specimens have modulus values above the upper confidence limit for the cancellous data indicates that the cortical bone tissue has a greater modulus than that of cancellous bone. When the tissue modulus of cancellous bone was measured and plotted against the apparent density of the tissue, the data *(triangles)* fell inside the confidence limits. (Reproduced from *Journal of Biomechanics*, Vol. 26, Rho et al., Young's modulus of trabecular and cortical bone material: ultrasonic and microtensile specimens, 111–119, 1993, with kind permission from Elsevier Science Ltd., The Boulevard, Langford Lane, Kidlington OX5 1GB, UK.)

microspecimens of cancellous as well as cortical bone. In three-point bending, Kuhn et al. (1989b) found a larger modulus for cortical than cancellous bone (4.89 vs. 3.81 GPa), but it was substantially less than typical values for large cortical specimens (18–20 GPa). Choi et al. (1990) obtained similar bending results, and furthermore they found that the modulus of cortical specimens decreased steadily with the dimensions of the specimens, from 14.9 GPa for a beam with a square cross section 1000 µm on each edge to 5.5 GPa for a similar beam 100 µm on each side.

Taken together, these and other studies suggest that the mechanical properties of cancellous bone tissue are different from those of cortical bone tissue. The difference between the tension test results of Rho and the bending results of Kuhn and Choi, and the beam size results of the latter group, indicate that the elimination of intact osteons may reduce the modulus, and much more so in bending than in uniaxial tension. When cortical specimens as small as trabeculae are made, their osteons must be cut into fragments not unlike those of trabecular remodeling packets, and this probably reduces the modulus, especially in bending. The reduced mineralization of cancellous bone tissue, at least in part caused by the increased rate of remodeling, may also explain some of the modulus reduction.

There may be a hierarchy of average, measured tissue moduli in descending order from primary cortical bone, to osteonal cortical bone, to cancellous bone tissue. However, it is also important to bear in mind that there is no "correct" value of elastic modulus for any of these tissues because of normal variations in mineralization, collagen fiber orientation, and other factors. It is important to understand the differences between cortical and trabecular bone tissue properties, but it is probably pointless to pursue increasingly "accurate" values for these properties.

Now let us move back up in scale once more, and consider how the architecture of a substantial group of trabeculae determine the mechanical properties of a region of cancellous bone a few millimeters across. There are two key aspects of this architecture: the **apparent density** and the **orientation** of the trabeculae.

Apparent Density

As we noted earlier, Carter and Hayes (1977a) found that, when elastic modulus is plotted against apparent density for cortical and cancellous bone specimens combined, E varies with d^3 (eq. 4.12). However, Rice and co-workers (1988), in an extensive analysis of data from several experiments, found that the exponent was 2 when only cancellous bone was included. Typically, such regression analyses show that apparent density or porosity account for 70%–80% of the variability in elastic modulus if one is considering a particular species, type of stress (tension or compression), and direction of loading relative to anatomical axes. The percentage goes down if these factors vary.

Rice and co-workers also confirmed that the compressive strength of cancellous bone varies with d^2, and this correlation accounts for about 73% of the variability. Restrictions regarding species and directionality applied to this case as well. Thus, about three-fourths of the variability in the strength and stiffness of cancellous bone can be accounted for by apparent density. Presumably, the balance depends on other factors, one of which is thought to be the directionality of the trabeculae relative to the loading direction.

Anisotropy of Cancellous Bone Architecture: Mean Intercept Length

It is intuitively clear, and abundant experimental data support the notion, that the orientation and spacing of trabeculae influence the elastic modulus and strength of cancellous bone as well as the anisotropy of these variables (Raux et al., 1975; Townsend et al., 1975; Goldstein, 1987). Both trabecular and cortical bone are anisotropic. In cortical bone the solid matrix dominates the space, and anisotropy is primarily determined by lamellar and osteonal orientation. In cancellous bone, soft tissues dominate the space, and anisotropy is governed by trabecular orientation. The relative numbers and sizes of trabeculae oriented in different directions control the anisotropy of the macroscopically measured material properties. However, the nature of this relationship was slow to emerge. The development of this understanding began with the work of Whitehouse and Dyson (1974;

Whitehouse, 1974a,b), who developed a method for measuring the microscopic structural anisotropy of trabecular bone. They measured the average distance, along test lines laid out in various directions, from one trabecular surface to another through the marrow; this distance is called *mean intercept length*, or MIL (Fig. 4.16). Whitehouse and Dyson found that, for a two-dimensional section, a graph of MIL vs. the measurement direction produced an ellipse whose eccentricity is a measure of anisotropy. Harrigan and Mann (1984) pointed out that a three-dimensional MIL ellipse can be expressed by a second-rank tensor (the MIL tensor). Soon after, Cowin (1985) put the major piece of the puzzle into place when he showed how the elastic constants of an anisotropic material can be calculated from apparent density (or porosity) and a *fabric tensor* that is equal to the

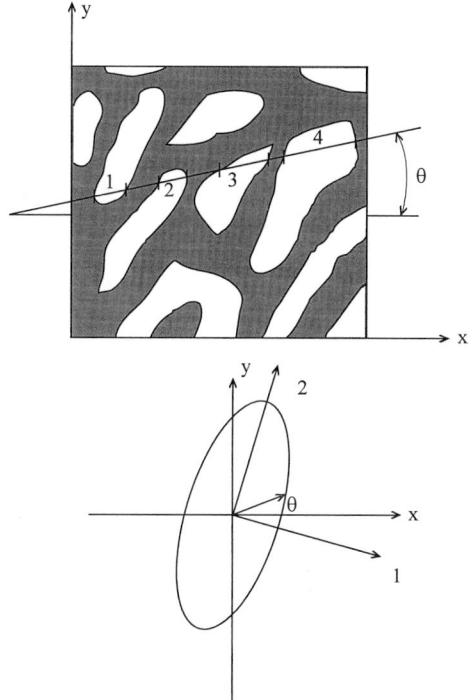

FIGURE 4.16. *Upper:* Schematic diagram of anisotropic cancellous bone structure as seen in a two-dimensional section also depicts an example of a test line used to measure mean intercept length. This line makes an angle Θ with the x-axis and the intercept lengths measured along it are shown as line segments *1–4. Lower:* Plot of mean intercept length (MIL) vs. Θ is an ellipse. The vector for test lines having orientation Θ is shown inside the ellipse. The ellipse has minor and major axes (labeled *1* and *2*) aligned with the principal directions of the two-dimensional version of the MIL tensor. Note that the maximum principal axis of the MIL ellipse is aligned with the general orientation of the trabeculae in the upper diagram.

inverse square root of the MIL tensor. Thus, Cowin was able to mathematically formulate what other investigators suspected but could not fathom: that the stiffness of cancellous bone depends on both its porosity and trabecular orientations.

The equations relating the elastic constants to porosity and the fabric tensor components are complicated and we shall not pursue them here. These relationships were subsequently further developed (Turner and Cowin, 1987; Cowin and Mehrabadi, 1989), and Turner et al. (1990) provided experimental evidence that strongly supports the theory. These data indicate that Cowin's theory accounts for 72%–94% of the variability in the elastic constants for cancellous bone.

Because the theory assumes that the solid matrix of the bone is homogeneous and isotropic, some of the unexplained variability probably results from organizational factors within individual trabeculae. For example, individual trabeculae more closely aligned with average principal **tensile** stresses may have collagen-mineral structures (and thus modulus values) that are different from those trabeculae closely aligned with average principal **compressive** stress directions.

Modeling Cancellous Bone as a Cellular System of Plates or Struts

Gibson (1985) explored the relationships between mechanical properties and apparent density of cancellous bone using open- and closed-cell porous structures like those shown in Fig. 4.17. She found that for a closed-cell structure (more representative of cortical bone or dense cancellous

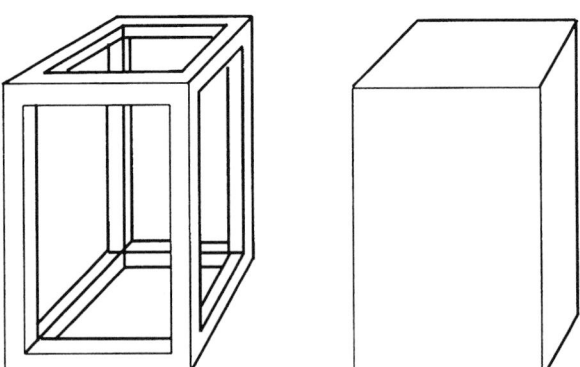

Figure 4.17. Types of open (left) and closed (right) cell structures used by Gibson (1985) to model cancellous bone behavior. (The "box" at right is hollow.) It is understood that in bone the solid plates forming the closed cell walls are actually perforated so that the marrow spaces communicate.

bone with unperforated plates), stiffness varied as d^3, but for an open-cell structure (like most cancellous bone), stiffness varied as d^2. This result is consistent with the different apparent density relationships for cortical and cancellous bone found by Carter and Hayes (1977a) and Rice et al. (1988). Gibson also found that for bone arranged in columnar cells, both strength and stiffness should be linearly related to apparent density. This finding agreed with data from Williams and Lewis (1982) for such bone.

Invariance of Yield Strain

We have noted that both the strength and modulus of cancellous bone depend on apparent density squared. It follows that strength is linearly proportional to modulus. This concept has been confirmed in several studies (Currey, 1969a; Brown and Ferguson, 1980; Goldstein et al., 1983; Bensusan et al., 1983; Vahey et al., 1987) and has an important implication for defining a failure criterion for cancellous bone. If yield stress, σ_y, and modulus, E, each vary with apparent density but are related by a constant, k, then

$$\sigma_y = \mathrm{k}\, E \qquad (4.13)$$

One has from Hooke's law (which applies virtually up to the yield point),

$$\sigma_y = E\, \varepsilon_y \qquad (4.14)$$

where ε_y is the yield strain. Eliminating σ_y/E from these equations yields

$$\varepsilon_y = \mathrm{k} \qquad (4.15)$$

That is, the yield strain is a constant, equal to the slope of the strength-modulus graph, and immune to the large variations in strength and stiffness caused by differences in apparent density.

Turner (1989) tested this concept using bovine bone. He found that compressive yield stress varied from 1.4 to 20 MPa, but yield strain only varied from 0.52% to 1.21%. The value of k was 0.88%, very close to the mean yield strain of 0.92%. Furthermore, ε_y was independent of trabecular orientation as measured by MIL. Later, Keaveny et al. (1994) confirmed the isotropy and relative constancy of ε_y. However, they found that yield strain was about 28% higher in compression than tension. Shear yield strain is also isotropic and independent of apparent density, particularly when the specimen is twisted about its own longitudinal axis (Ford and Keaveny, 1996).

Finally, there is a connection between Turner's (1989) work and Gibson's (1985) models. Gibson assumed that individual trabeculae could deform (and fail) either by bending of struts oriented perpendicular to the load direction or by axial compression (and buckling) of struts oriented parallel to the load direction. Cowin (1989) has pointed out that the structure would be much stiffer and stronger in the latter case, but would be capable of absorbing more energy if bending of trabeculae occurred. Therefore, this is an interesting issue for several reasons. Turner had found that there was a slight but statistically significant increase in yield strain with bone volume fraction (and, presumably, apparent density). He performed an analysis of Gibson's model to show that this would be consistent with failure by buckling but not with failure by bending of trabeculae. More experiments are needed to learn how trabeculae fail individually, and what determines whether or not this leads to collapse of a whole region of cancellous bone.

4.6 Predicting Material Properties: Bone as a Composite Material

A number of studies have attempted to mathematically model bone as a composite material. The most general approach has been to assume that bone is composed of two or more phases (for example, collagen and mineral, or soft and hard tissue), and that these phases contribute to a mechanical property P in direct proportion to the volume fraction F that they occupy. (Most commonly, P is the elastic modulus.) One may simplify the structure of such a model to consist of alternating slabs of the two phases; if two such slabs are arranged so that they have equal strains when they are loaded, one has the Voigt model (at left in Fig. 4.18), in which the composite property is given by

$$P_c = P_1 F_1 + P_2 F_2 \qquad (4.16)$$

Here the subscripts 1 and 2 refer to the two component phases and P_c designates the composite property. If the slabs have equal stresses, the result is known as the Reuss model (at right in Fig. 4.18), described by

$$P_c = \frac{P_1 P_2}{P_1 (1 - F_1) + P_2 (1 - F_2)} \qquad (4.17)$$

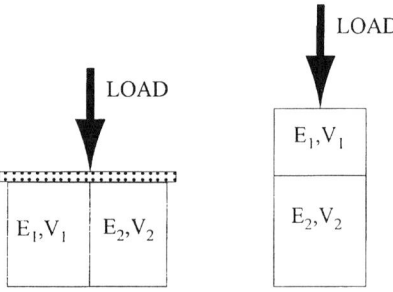

FIGURE 4.18. Voigt *(left)* and Reuss *(right)* models for composite material formed from a mixture of two materials. In the Viogt model, each material strains the same amount; in the Reuss model, each material carries the same stress.

It is known that the Voigt model provides an upper bound, and the Reuss model a lower bound, for the actual value of the composite property (Hill, 1952, 1963). Unfortunately, these bounds are typically so far apart as to be virtually useless. The Hashin–Shtrickman model (Hashin and Shtrikman, 1963) gives somewhat narrower bounds. Several authors (Currey, 1964; Katz, 1971; Bonfield and Li, 1967; Piekarski, 1973a,b) have assumed that compact bone consists of collagen (organic phase) and mineral (inorganic phase) and used these models to derive certain mechanical properties. These efforts have met with limited success, partly because the properties of the component phases are not well known, partly because their interactions are not well understood (see Burstein et al., 1975), and perhaps because the void space was ignored. However, Martin (1984) had no better success when he assumed that bone consists of calcified matrix and voids and used the same methods to compute the elastic modulus. Others who have analyzed solid-void two-phase materials include Hirsch (1962) and Brown et al. (1964). Stech (1967) analyzed the anisotropy of compact bone using a model containing parallel cylindrical voids representing Haversian canals. Bonfield and Clark (1973) had some success in combining mixture theories to explain an age-related increase in the elastic modulus of rabbit bone that seemed to be caused by changes in both porosity and specific mineralization. They used a formulation by MacKenzie (1950),

$$E = E_o \,(1 - 1.9\,p + 0.9\,p^2) \tag{4.18}$$

where p is porosity and E_o is the modulus of the calcified matrix, to account for porosity variation, and an equation similar to Eq. 4.10 to account for changes in mineralization. However, the analysis did not account for the entire increase in stiffness, supporting Currey's (1969b) hypothesis that bone apatite experiences age-related structural changes that increase its elastic modulus.

Many of these models contain constants that are more or less arbitrary and can be manipulated to fit various data. So far, however, no one has developed a theory of bone structure that can satisfactorily explain its mechanical behavior on the basis of its internal structure and composition.

4.7 Summary and Additional Reading

In this chapter, we have seen how the mechanical properties of cortical and cancellous bone are influenced by the histologic structures described and analyzed in Chapters 2 and 3. Bone behaves as an elastic (and also viscoelastic), anisotropic, composite material. The feature of its histologic structure that dominates its mechanical properties is porosity, also expressed as bone volume fraction (and closely related to apparent density). Mineralization and collagen fiber orientation are also important determinants of mechanical properties. In cortical bone, secondary osteons significantly alter anisotropy and strength, and in cancellous bone the orientations of trabeculae play a similar role. The elastic modulus and strength of cancellous bone are both heavily dependent on apparent density squared, and this results in yield strain being nearly isotropic and constant. Mixture theories may be used to formulate models for the mechanical behavior of bone, but the results have not been very satisfactory.

Recommendations for additional reading include the excellent book, *Bone Mechanics*, edited by Cowin (1989). This work contains chapters that provide a detailed mathematical presentation of elasticity and strength of materials theory, as well as compilations of data on the mechanical properties of cortical and cancellous bone. The chapter on bone's physical properties in Albright and Brand (1987) and the tutorial paper by Turner and Burr (1993) are excellent introductions to this subject for nonengineers. Another resource intended for clinicians is the textbook *Fundamentals of Biomechanics* by Ozkaya and Nordin (1991). Alexander's monograph, *Optima for Animals* (1996), contains a section on optimizing the dimensions of tubular bones with respect to strength and weight (pp. 13–23), and presents a succinct overview of optimization theory from the biological perspective.

4.8 Exercises

4.1. Show by analysis of the units of stress and strain that the area under a stress-strain curve has units of work/volume.

4.2. Suppose a human bone has a cylindrical shaft whose periosteal and endosteal diameters are 2 and 1 cm, respectively. Using the data in Table 4.1, calculate the yield and ultimate load for this bone when loaded in simple compression. (Ignore buckling as a possibility.) Then calculate the torques (twisting moments) required to yield and break the bone by twisting it. When the outer fibers of the bone have reached the ultimate shear stress, are the fibers on the endosteal surface above or below the shear yield stress?

4.3. Suppose the bone described in Exercise 4.2 has its shear strength diminished by 20% by the accumulation of Haversian canals with age. Assume also that 1 mm of bone is lost from the endosteal surface during this period of time. How much must the periosteal diameter be increased to compensate for these changes? If all these changes occurred simultaneously over a 30-year period, what would the mean periosteal apposition rate be, and how would it compare to typical values during the refilling of an osteonal BMU?

4.4. Suppose a long bone diaphysis has cleanly fractured straight across and is healing through the development of a fracture callus as described in Chapter 2. Sketch a cross section through the callus assuming circular geometry for it and the original bone. The outer diameter of the callus is 3 cm. Sketch in the original bone cross section, which had periosteal and endosteal diameters of 2 cm and 1 cm, respectively. Assume that no strength has developed between the cortices of the original bone, so that the strength of the healing fracture is entirely provided the periosteal and medullary calluses. Calculate the bending moment required to fracture the original, intact cortex, assuming that its ultimate bending stress was 200 MPa. Now calculate the cross-sectional moment of inertia of the periosteal and endosteal calluses and compare them, individually and together, to that of the original bone. Assume the material in these calluses is homogeneous. What would the ultimate stress of the callus material have to be for it to provide the same strength as the original bone? What percentage of this strength is provided by the periosteal callus?

4.5. Derive the Voigt and Reuss mixture theory equations (Eqs. 4.16 and 4.17) from the concepts shown in Fig. 4.18.

4.6. Assume that the two phases denoted by subscripts 1 and 2 in Eqs. 4.16 and 4.17 represent bone voids (soft tissue spaces) and calcified matrix, respectively. In that case, volume fraction F_1 becomes porosity (p) and F_2 = 1 − p. Let the material property P be elastic modulus, so that P_2 = E_o, the modulus of the calcified matrix. Try to plot Eqs. 4.8, 4.9, 4.16, and 4.17 on the same graph. You will have a problem with Eq. 4.17 if you assume the modulus of soft tissues to be 0, so choose a value small in proportion to E_o. What sort of insight into mixture theory and the effects of porosity on bone's mechanical properties can you obtain from your graph?

4.7. Assume that osteonal bone consists of parallel, longitudinally arranged secondary osteons in an interstitial matrix of primary bone and osteon remnants. Suppose that this composite is loaded in longitudinal tension so that the simple rule of mixtures (Eq. 4.16) applies to the elastic modulus. Schaffler (1985) has measured the elastic modulus of cow bone as a function of the volume fraction of secondary osteons, obtaining the following results:

Elastic modulus, GPa	Osteon volume fraction	Elastic modulus, GPa	Osteon volume fraction
19.90	0.00	20.70	0.21
26.20	0.00	18.10	0.41
24.00	0.00	21.50	0.21
24.90	0.00	14.90	0.37
24.50	0.00	15.70	0.40
26.00	0.08	21.00	0.27
21.80	0.00	19.10	0.32
22.10	0.00	18.10	0.47
23.30	0.00	25.20	0.05
22.80	0.03	25.20	0.03

Use Eq. 4.16 to derive a relationship between the elastic modulus of the composite bone (E_b), the osteon volume fraction (F_o), and the moduli of osteons (E_o) and the interstitial matrix (E_m). Based on this relationship, how are E_o and E_m related to the slope and intercept of a plot of E_b vs. F_o? Obtain estimates of E_o and E_m from a plot of Schaffler's data. Which is greater? Why?

4.8. As described in Box 4.4, bone mass may be noninvasively measured by photon absorptiometry. The absorption curve shown in the box is a typical result for a long bone diaphysis. The area above the curve is proportional to the area occupied by mineral in the scanned section, and the distance across the curve is proportional to the periosteal diameter of the bone. It might be of mechanical interest to noninvasively determine the bending resistance of the bone's cross section. Derive a method for determining (a) the centroid of the cross-section's mineral in the scan direction and (b) the cross sectional moment of inertia of the mineral about an axis through the centroid and perpendicular to the scan direction.

4.9. Box 2.1 described the apparatus developed by Burstein and Frankel for testing the torsional strength of a long bone diaphysis under conditions of rapid loading. This machine uses the energy of a swinging pendulum to apply an impulsive torque to the specimen. The mass and length of the pendulum are 14 kg and 23 cm, respectively. To ensure that the impact will be sufficiently forceful, the energy in the swinging pendulum must be much greater than that required to break the bone. Assume each "arbitrary torque unit" in Fig. 2.23 is equal to 10 N-m. Estimate the energy to failure of the strongest of the specimens in this figure, and compare this energy with the potential energy of the pendulum at the start of the experiment. What percent of the pendulum's kinetic energy is required to break the healed rabbit bone, assuming no other energy losses?

4.10. Consider the stress in the neck of the femur of a 70-kg person during the stance phase of gait. Use what you learned in Chapter 1 about the magnitude of the hip joint force, and assume that it is directed at an angle of 20° to the neck axis. Calculate the compressive and bending stresses produced by the joint force components parallel and perpendicular to the neck axis, respectively. Choose reasonable values for the bony dimensions. Then estimate the total axial stress on the superior and inferior surfaces of the neck. How do these stresses compare to the compressive strengths of cancellous bone shown in Table 4.7? Discuss how you might account for the effect of a shell of cortical bone around the neck of the femur.

5

Fatigue and Fracture Resistance of Bone

Textbooks and Heaven only are ideal;
Solidity is an imperfect state.
Within the cracked and dislocated Real
Nonstochiometric Crystals dominate.
John Updike in "The Dance of the Solids"

5.1 Introduction

In Chapter 4, we examined the mechanical properties of bone when it is fractured monotonically by a load sufficient to exceed the failure stress of the material. Structures may also fail more gradually. There are two principal ways in which this can happen: creep and fatigue. In both, a stress less than the ultimate stress is applied, and damage from this stress grows and accumulates until failure occurs. In creep, the stress is applied continuously; in fatigue, the stress is applied cyclically. Obviously, these phenomena may occur simultaneously in the components of an engineering structure like a bridge, where the weight of the structure provides creep loading and the periodic passage of vehicles produces fatigue. Similarly, in the skeleton, bones often support more or less constant loads for prolonged periods of time (such as the vertebral bodies in your spine as you sit reading this book), and equally often carry cyclic loads (such as when you walk to class). In this chapter we shall see that creep and fatigue are closely related and explore the ways in which bone limits and repairs the damage that they produce. We also discuss what seems to happen when the limits of this capacity for damage control are exceeded and the bone fails by what is called a *stress fracture*.

At least some of the damage caused by creep and fatigue can be seen in histologic sections of normal, healthy bones (Fig. 5.1). These *microcracks* can, under continued loading, multiply and grow into "macrocracks," eventually resulting in complete fracture. Ordinarily, microcracks (or other kinds of microdamage) are removed by remodeling. If fatigue damage in bone accumulates more rapidly than remodeling can remove it, fatigue failure or a "stress fracture" can occur. These fractures can be partial or complete, and tend to occur in soldiers and joggers, as well as ballet dancers and other elite athletes, including racehorses. In principle, inhi-

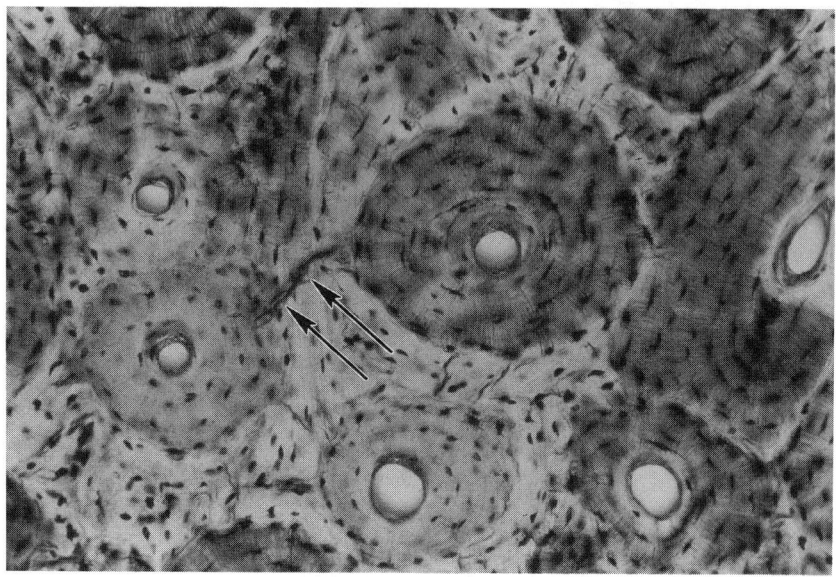

FIGURE 5.1. A typical microcrack *(arrows)* as seen in a 100-μm-thick histologic cross section of human cortical bone. The specimen was stained in basic fuchsin en bloc to mark the surfaces of cracks (such as this one, which would be red in a color photograph) existing before the section was cut. Artifactual cracks produced by the sectioning process are not stained (Burr and Stafford, 1990). this crack is about 80 μm long in the plane of the section, and its ends abut the cement lines of two secondary osteons. Most microcracks in cortical bone are of this type.

bition of normal bone remodeling could cause an accumulation of unrepaired fatigue damage that, over time, could lead to fracture.

It is important to understand, however, that the prevention of fatigue failure is not entirely dependent on repair by remodeling. The growth of microdamage is also controlled by the lamellar structure of bone. In ways completely analogous to the crack deflection and blunting effects achieved in man-made composite materials, bone's internal structure limits the development of fatigue damage. Much of the information contained in this chapter has been extrapolated from the extensive experimental data on the fatigue behavior of nonbiologic composite laminates (e.g., graphite-reinforced epoxies, or polyester laminates). Knowledge of their behavior allows us to understand how various features of lamellar bone structure function to improve its fatigue properties. Before considering such crack-resistant materials, however, we should look at what is known about cracking in simpler materials.

5.2 Basic Fracture Mechanics

Fracture mechanics considers the conditions under which a flaw or crack in a load-bearing structure grows and causes catastrophic failure. In conventional engineering applications, the flaws of concern may be produced as the structure is manufactured or by overloading or repetitive loading of the structure thereafter. In the case of skeletal structures, "manufacturing flaws" are usually (i.e., in healthy people) of less concern than those arising from heavy usage. However, genetic errors can result in abnormal bone tissue (usually because of a collagen defect) that becomes damaged under normal loads. Sometimes damage is introduced by surgeons as they try to correct other difficulties. Whatever the cause of the cracking, bones have an exceptional ability to control it. The purpose of this section is to provide a brief summary of fracture mechanics to help us understand how cracks are propagated generally and how they are controlled in bone.

Linear Elastic Fracture Mechanics

Perhaps the simplest way to approach this topic is to consider it in the context of stress concentrations. Consider the stress concentration on the edge of an elliptical hole in an infinite plate (Figure 5.2A). The plate bears a tensile stress, s, normal to the long axis of the ellipse. The stress at each end of this axis is

$$\sigma_A = s \, (1 + 2a/b) \tag{5.1}$$

where b and a are the minor and major radii of the ellipse. (For a circular hole, this reduces to 3s.) However, this equation loses its validity if $a \gg b$ and the ellipse becomes cracklike (Figure 5.2B). In that case, the stresses at a point given by polar coordinates (r, Θ) relative to the end of the b-axis are

$$\sigma_x = s \, (a/2r)^{1/2} \cos \Theta/2 \, (1 - \sin \Theta/2 \sin 3\Theta/2) \tag{5.2a}$$

$$\sigma_y = s \, (a/2r)^{1/2} \cos \Theta/2 \, (1 + \sin \Theta/2 \sin 3\Theta/2) \tag{5.2b}$$

$$\sigma_{xy} = s \, (a/2r)^{1/2} \cos \Theta/2 \sin \Theta/2 \cos 3\Theta/2 \tag{5.2c}$$

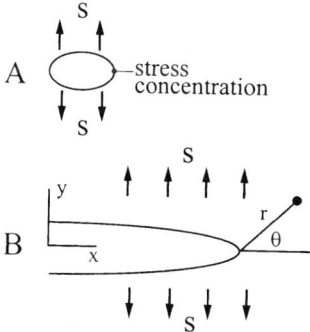

FIGURE 5.2. **A** An elliptical hole in a plate or solid structure bearing a tension stress s has a stress concentration several times s at the point shown. **B** A crack may be thought of as a very eccentric elliptical hole. The stress at a point near the crack tip (dot) is a function of its distance r and angular position θ.

The stress in the y direction along the $\Theta = 0$ line is

$$\sigma_y = s\,(a/2r)^{1/2} \qquad (5.3)$$

As one approaches the crack tip, r becomes very small relative to the crack length, a, and the stress concentration very large. At the crack tip itself, $r = 0$ and σ_y is singular (i.e., it is infinite). To avoid this difficulty, one may move the $2r$ term to the other side of the equation, obtaining $\sigma_y(2r)^{1/2}$, which is not singular at the crack tip but is still proportional to the "intensity" of the stress at the crack tip. One can then define a

$$\textit{Stress intensity} = \sigma_y\,(2r)^{1/2} = s\,a^{1/2} \qquad (5.4)$$

That is, the stress intensity in the vicinity of the crack tip is proportional to the nominal stress, s, and the square root of the crack length, a. This stress intensity is what drives the crack forward.

G.R. Irwin (1957) used linear elasticity theory to obtain equations for the stresses near the tips of cracks with several different geometries and loading situations. He found that in general

$$\sigma = k(2\pi r)^{-1/2} + \text{other terms} \qquad (5.5)$$

where k is a coefficient peculiar to the particular problem. As one approaches the crack tip, the "other terms" became less important and can be ignored. Because of the π in these equations, a formal *stress intensity factor*, K, was defined so as to include π as well:

$$K = C\,s\,(\pi a)^{1/2} \qquad (5.6)$$

K has units of Pa-m$^{1/2}$ (compare Eq. 5.6 with Eq. 5.4), C is a dimensionless constant depending on the size and shape of the flaw and the size, shape, and mode of loading of the specimen, and s is a pertinent stress component in the specimen beyond the influence of the crack, for example, tensile stress perpendicular to the plane of the crack. K characterizes the state of stress near the crack tip; therefore it is known as a *characterizing parameter*. Its utility stems from the fact that **if K exceeds a critical value, the crack will propagate; otherwise, it will not.** There is no need to make calculations of the local stresses to gain this information.

To reiterate, the fundamental concept in linear elastic fracture mechanics is that K determines whether a crack will propagate or not, and if so, how fast. In practice, K bears a subscript denoting the type of cracking that is being considered.

Kinds of Cracks

Of the various ways in which a specimen may be loaded to extend cracks, three are commonly discussed in fracture mechanics (Fig. 5.3). Mode I cracks are propagated perpendicular to the line of action of a tensile load, as when a wedge splits a log lengthwise. Mode II cracks result when shearing forces are applied parallel to the plane of the crack and parallel to the direction of its propagation. Mode III cracks are a "tearing" phenomenon;

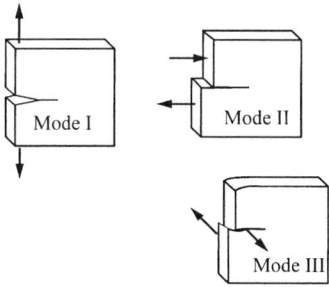

FIGURE 5.3. Modes of cracking.

in this case, the shearing is perpendicular to the crack propagation direction. Mode I is the most commonly discussed type of cracking.

As an example, consider the Mode I cracking that occurs when a plate of width b, containing a crack of length a on one edge, is loaded in tension. In this case, the C in Eq. 5.6 turns out to be

$$C = [(\tan q)/q]^{1/2} \quad \frac{0.752 + 2.02(a/b) + 0.37(1 - \sin q)}{\cos q} \qquad (5.7)$$

where $q = \pi a/2b$. As shown in Eq. 5.6, when multiplied by $(\pi a)^{1/2}$ and the tensile stress in the plate, this C gives the stress intensity factor, K_I, for this situation. If this K_I is greater than a critical stress intensity factor, K_{Ic}, the crack propagates rapidly clear through the structure. K_{Ic} is also called the *fracture toughness*; it is a material property, like yield stress. Table 5.1 gives some typical values of K_{Ic} for metals, ceramics, polymers, and cortical bone (Pilkey, 1994). By conducting many experiments and tabulating the results, the means for calculating K_I in various situations and comparing them to the critical values have been amassed for many conventional engineering applications. Table 5.2 gives formulae for C in a few simple cases.

TABLE 5.1. Fracture toughness for several materials

Material	K_{Ic}, MPa-m$^{1/2}$
2024 Aluminum	20–40
4330V Steel	86–110
Ti-6Al-4V	106–123
Concrete	0.23–1.43
Al_2O_3 ceramic	3–5.3
SiC ceramic	3.4
PMMA polymer	0.8–1.75
Polycarbonate polymer	2.75–3.3
Cortical bone	2.2–6.3

TABLE 5.2. Equations for the constant C in three simple cases

Crack geometry	C

Plate of width W in tension, Mode I crack of length $2a$ in center.

$C = [1 - 0.1(a/W)^2 + 0.96(a/W)^4]\,[\sec(\pi a/W)]^{1/2}$

Plate of width W in tension, Mode I crack of length a on one edge.

$C = [(sW/\pi a)\tan(\pi a/2W)]^{1/2}\bullet$

$$\frac{0.752 + 2.02(a/W) + 0.37[1 - \sin(\pi a/2W)]}{\cos(\pi a/2W)}$$

Plate of width W in tension, Mode I crack of length a on each edge.

$C = [1 + 0.122\cos^4(\pi a/W)]\,[(W/\pi a)\tan(\pi a/W)]^{1/2}$

The measurement of K is beyond the scope of this book because it is complicated by the need to ensure that all the assumptions underlying its calculation are fulfilled. The basic process is simple, however, as illustrated in Fig. 5.4 for mode I cracking. A specimen with a standardized crack is manufactured, and the crack is propagated across the specimen by loading it in tension. The applied load and crack opening displacement are measured and plotted. A fail-

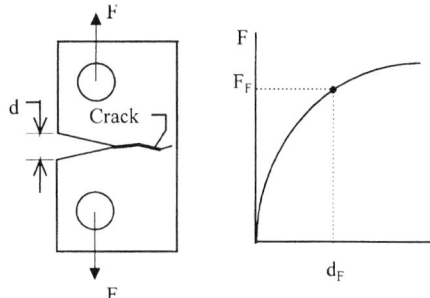

FIGURE 5.4. The basics of measuring of K_{Ic}. A specimen with a notch is loaded in tension to drive a crack to the *right* as shown. The tensile force F is plotted against the crack opening displacement, d. A failure point, $P(F_F, d_F)$, on the curve is defined by certain rules, and K_{Ic} is calculated from F_F and d_F using a formula.

ure load is identified on the load-displacement curve using several rules. K_{Ic} is then calculated by multiplying the failure load by a formula derived from the length of the crack and the shape and dimensions of the specimen.

Several investigators have measured K_{Ic} for cortical bone (Table 5.3). Values for bone are an order of magnitude less than those for structural metals but similar to a tough ceramic like Al_2O_3. In making these comparisons, it needs to be emphasized that the values for bone are for crack propagation parallel to the long axis of the bone, which is to say "with the grain." If one tries to propagate a crack transversely, or perpendicular to the primary and secondary lamellar structure, the crack immediately turns and goes "with the grain." Thus, the fracture toughness of bone is decidedly anisotropic, and the values shown are for the direction offering the least resistance to cracking.

TABLE 5.3. Experimental values of K_c and G_c for bone

Bone Type	K_c, MPa-m$^{1/2}$	G_c, Jm^{-2}	Source
Mode I, transverse fracture			
Bovine femur	5.49	3100–5500	Melvin and Evans, 1973
Bovine tibia	2.2–4.6	780–1120	Bonfield and Datta, 1976
Equine metacarpus	7.5	2340–2680	Alto and Pope, 1979
Human tibia	2.2–5.7	350–900	Norman et al., 1992
Mode I, longitudinal fracture			
Bovine femur	3.21	1388–2557	Melvin and Evans, 1973
Bovine tibia	2.8–6.3	630–2238	Behiri and Bonfield, 1984
Human femur	2.2–5.7	350–900	Norman et al., 1992
Mode II			
Human tibia	2.2–2.7	365	Norman et al., 1992

Maximum Tolerable Flaw Size

By rearranging Eq. 5.6 and setting the stress intensity factor equal to the critical value, one may solve for the flaw size that will precipitate unstable crack propagation. For human tibial compact bone, Norman et al. (1995) found that $K_{Ic} \approx 4$ MPa-m$^{1/2}$. For a semicircular crack of radius a at the surface of a semi-infinite body loaded in tension (e.g., on the tension side of a large bone loaded in bending,) the stress intensity factor is (see Pilkey, 1994, p. 343)

$$K_I = 2 \, (a/\pi)^{1/2} \, s \, [1.211 - 0.186 \, (\cos \theta)^{1/2}] \qquad (5.8)$$

Here, θ is as defined in Fig. 5.2 and s is the stress elsewhere in the body, far from the crack. Setting $\theta = 0$, the maximum tolerable crack radius would be

$$a_c = 0.238 \, \pi \, K_{Ic}^{2} \, / \, s^2 \qquad (5.9)$$

Assuming that $E = 20$ GPa and the strain (ε) is a physiologic 2000 microstrain, one has $s = E\varepsilon = (20 \times 10^9) \, (2 \times 10^{-3}) = 40$ MPa and

$$a_c = 0.238 \, \pi \, (4^2/40^2) = 0.0075 \text{ m} = 7.5 \text{ mm} \qquad (5.10)$$

which is a rather large flaw. Remember also that the K_{Ic} used here is probably considerably smaller than the one for a crack propagating **across** a bone.

It Takes Energy to Propagate a Crack

Before Irwin and others developed the concept of a stress intensity factor, A.A. Griffith (1920) formulated the problem of crack propagation in brittle solids in terms of energy. He reasoned that energy was required to create the surfaces of the extended crack because all surfaces have "surface energy" associated with them. He conducted many experiments with glass rods to develop his idea into a theory.

Suppose it takes energy or work, dW_c, to propagate a crack for a distance, da. This energy usually comes from the work done in deforming the specimen via the strain energy, U, stored throughout its interior. Conservation of energy requires that

$$dW_c/da = dU/da \qquad (5.11)$$

The right side of this equation is called the *strain energy release rate* and is denoted by G, after Griffith, who showed that for mode I propagation of a thin ellipsoidal crack the strain energy release rate is

$$G = dU/da = \pi \, a \, s^2/E \qquad (5.12)$$

where E is the elastic modulus. By setting up an experiment in which a crack can be slowly propagated in a material of modulus E, and its length (a) measured as a function of the applied stress (s), the strain energy release rate can be determined from Eq. 5.12.

FIGURE 5.5. A caricature of a crack feeding from a pot of strain energy. (Reproduced with permission from Finkel, 1985.)

The left side of Eq. 5.11 is called the *crack growth resistance, R*:

$$R = dW_c/da \tag{5.13}$$

For a crack to grow, strain energy must be produced and released at least as fast as it is consumed by breaking molecular bonds in the material to extend the crack. For the crack to grow precipitously, G must be greater than a critical value, which is R; that is, $G \geq R$.

By comparing Eq. 5.6 and Eq. 5.12, it is apparent that G is proportional to K^2. For mode I cracking in a plane stress situation,

$$G_I = K_I^2/E \tag{5.14}$$

It is important to understand that a propagating crack is "nourished" by the elastic energy of the mass under stress. Figure 5.5 was used by Finkel (1985) to illustrate this point. Finkel also pointed out that although cracks are driven by the strain energy stored in the mass under stress, they are themselves without inertia, and thus capable of turning instantaneously in any direction that promises more elastic energy.

Beyond the Linear Theory: Real Cracks Have Ears

In most materials, and certainly in bone, linear elasticity theory does not hold when stresses become high, and that includes the situation at the tip of a crack. There is a zone of plastic deformation around the crack tip that

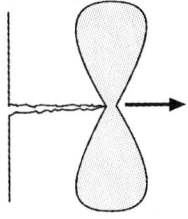

FIGURE 5.6. The plastic zone at the leading edge of a propagating crack has an ear-like characteristic shape.

is generally shaped as depicted in Fig. 5.6. Note the "Mickey Mouse ears" shape of the plastic zone. Finkel has called the plastic zone the "ears" of the crack, and refers to computations showing that these ears may "wiggle" as the crack propagates.

For mode I cracking, the distance directly ahead of the crack that is plastically deformed is given by

$$r_p = K_I^2 / k \, \pi \, \sigma_y^2 \qquad (5.15)$$

where σ_y is the yield stress; $k = 2$ for plane stress or 18 for plane strain.[1] One way to accommodate the presence of a plastic zone in linear elastic fracture mechanics theory is to add this r_p (or some portion of it) to the crack length in the equations for K and G. Equivalently, when using the energy approach, another term must be added to Eq. 5.14: the rate at which energy is consumed by plastic deformation per unit length of crack extension.

Box 5.1 Technical Note
Dynamic Fracture Mechanics

Things break differently when they break extremely rapidly. Dynamic fracture mechanics is the study of cracking under these conditions (Freund, 1990). When a bone is broken very rapidly in a situation involving a great deal of kinetic energy, the fracture is highly comminuted (i.e., the bone breaks into many pieces). The reasons for this are complex, but stem from two important phenomena. When a crack propagates slowly through a structure, it causes the adjacent material to be unloaded. This load change prevents nucleation of, or arrests, other cracks in the vicinity. When a structure is impulsively loaded, a stress wave passes through it at the speed of sound (about 3000 m/s for bone). This results in crack initiation and propagation at many sites simultaneously, before any one crack can arrest others. When these many cracks grow to intersect one another, many fracture fragments are produced. Emergency room physicians see this result in victims of high energy trauma.

The second phenomenon promoting fragmentation at high fracture rates is bone's viscoelastic nature. The more rapidly bone is loaded, the stiffer it is (see Fig. 4.12). The velocity of the stress wave increases as the square root of the elastic modulus, so that the stress field more rapidly fills the bone, accentuating the effects just described. On the other hand, bone strength increases with loading rate. This effect is seen in other materials, too. The reasons for it are not clear.

In summary, when a bone is rapidly loaded, it is more fracture resistant, but if it breaks, the fracture is more fragmented. Also, logically enough, the severity of damage to the soft tissues tends to parallel the degree of damage to the bone; thus, the seriousness of severe fractures is compounded by impairment of the associated soft tissues, on which fracture healing depends.

---

[1] In a plane strain situation, the strains are all within a plane, i.e., deformations perpendicular to the plane may be assumed zero. Plane stress is similarly defined in terms of stress components perpendicular to the plane. Thanks to Poisson's ratio, the two situations are not the same.

Crack Growth and Fatigue

Paris (1964) proposed that the number of cycles required for fatigue failure of a specimen with a crack could be related to the crack's stress intensity factor. His idea was that the crack length increase per load cycle, da/dN, would be proportional to the range of K values experienced during the loading: $\Delta K = K_{max} - K_{min}$. A commonly suggested form for such a relationship is

$$da/dN = \xi (\Delta K)^{\zeta} \tag{5.16}$$

where ξ and ζ are constants. Unfortunately, the Paris equation has not proven to be a consistently valid relationship. It describes the results at intermediate crack growth rates better than at low or high values of da/dN. Within that range, if crack propagation data are presented as a log-log plot of da/dN vs. ΔK, the slope and intercept should yield the coefficients ζ and ξ, respectively.

5.3 Fatigue Behavior of Bone

Depending on the fracture toughness of a material, and the magnitude of the applied loads, cracks start and grow more or less rapidly under repetitive loading. In this section we examine the basic fatigue and fatigue damage behavior of bone.

The S-N Curve

The basic means of presenting fatigue data is an S-N curve: a plot of the applied stress or strain vs. the number of cycles required for failure at that load level. A fatigue S-N curve was first measured for bone by King and Evans in 1967 (Fig. 5.7). Subsequently, in the 1970s and 1980s, several other

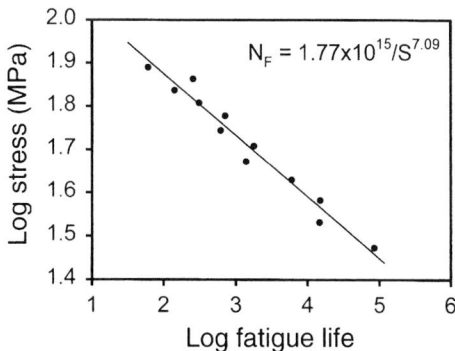

FIGURE 5.7. An early S-N curve for bone with its regression equation, for which $R^2 = .96$. (Replotted from data by King and Evans, 1967.)

investigators (notably Carter and co-workers) reported similar experiments, establishing that bone behaves similar to engineering materials in fatigue. That is, its elastic modulus declines as fatigue progresses, the log of N is linearly proportional to the log of S, and cracks can be seen within fatigued bone which are apparently fatigue damage. The concept that the fatigue behavior of bone is similar to that of composite materials was also established in this period.

Fatigue life, which we now represent by N_F, is related to the applied stress in a very nonlinear way. When one regresses log N_F vs. log S for bone specimens fatigued in various modes of loading, one finds that

$$N_F = c/S^q \tag{5.17}$$

where c is a coefficient and the exponent q is a number between 5 and 15, depending on the type of bone and the mode of loading. Clearly, small changes in load magnitude (S), or the kind of bone and mode of loading (q), can result in very large changes in fatigue life. Caler and Carter (1989) and Pattin et al. (1996) have shown that the fatigue life of human femoral cortex is longer in compression than in tension. Testing in uniaxial tensile and compressive fatigue at 2 Hz, Pattin et al. produced data that, when fitted to Eq. 5.17, had the following coefficients:

$$N_F = 1.445 \times 10^{53} / S^{14.1} \text{ (tension)} \tag{5.18}$$

$$N_F = 9.333 \times 10^{40} / S^{10.3} \text{ (compression)} \tag{5.19}$$

where the stress range, S, is from zero to a peak stress normalized by the initial elastic modulus. (Therefore, S is also the initial strain, and has units of microstrain.) These data were obtained at superphysiologic strains (2600–6600 $\mu\varepsilon$). If one may extrapolate the results to a physiologic peak strain of 2000 $\mu\varepsilon$, the fatigue life would be 4.1 million cycles in tension and 9.3 million cycles in compression.

Fatigue Damage in Bone

Figure 5.1 illustrates a typical microcrack caused by fatigue in bone, but fatigue damage is not a narrowly defined quantity in bone or other materials. In the engineering literature, it has usually been defined theoretically as a parameter that varies from zero in virgin structures to unity in failed structures, based on changes in mechanical properties, but without any specific reference to cracks or other evidence of physical damage, which may be difficult to detect. For example, damage is often defined as the fractional reduction in elastic modulus. Alternatively, when cracks are apparent, as in certain kinds of composite materials, damage may be defined in terms of the numbers or lengths of microscopic cracks that appear after loading.

Cracks are relatively easy to find in normal bone as it comes from the body. Frost (1960a) first described such damage, and linked it to fatigue. Burr and Stafford (1990) counted 0.014 such cracks/mm^2 in the ribs of a 60-year-old man. The number of such cracks increases with age in the human femoral diaphysis and neck cortex (Schaffler et al., 1995), and in the trabecular bone of the femoral head (Mori et al., 1997). Damage accumulates exponentially, and more rapidly in women than men, after age 40 (Schaffler et al., 1995). Although the data suggest increased skeletal fragility caused by microdamage in aging women, they do not show why some older women fracture and others do not. Mori et al. (1997) found that older women have significantly more microcracks than younger women, but older women who fracture did not have more microcracks than older women who did not fracture.

Microdamage burdens may be higher in trabecular than cortical bone; Wenzel et al. (1996) found approximately 5 microcracks/mm^2 in trabecular bone from the spine, without a significant increase with age. In the spine, disk pathology may contribute to microdamage (Hasegawa et al., 1995), either because loads are improperly attenuated or because stress concentrations are created in the vertebral body.

About 80%–90% of all microcracks in cortical bone are found in the interstitial matrix between osteons (Schaffler et al., 1995; Norman and Wang, 1997); see Fig. 5.1, for example. Such cracks are commonly 50–100 μm long. Ten thousand cycles of bending to physiologic strain levels have been shown to increase interstitial crack counts in canine bone (Burr et al., 1985). Debonding of osteonal cement lines has also been observed following in vitro loading (Carter and Hayes, 1977c). This phenomenon is apparently related to "osteon pullout" seen on fracture surfaces (Piekarski, 1970; Stover et al., 1995), in which osteons or their lamellae have debonded and broken transversely, either slightly above or below the main fracture surface.

5.4 Creep Behavior of Bone and Its Relationship to Fatigue

As we noted in Chapter 4, bone is viscoelastic. It therefore creeps, that is, it deforms for a prolonged period of time under a constant load. If a bone specimen is loaded to a constant stress well below the ultimate stress, it gradually deforms and eventually fractures. Figure 5.8A shows this behavior for cortical bone from a human femur. The initial strain is immediate and therefore elastic; then, as time goes by, the material deforms nonelastically as the result of damage processes. The strain-time curve is similar to the damage-cycle number curve for fatigue (Fig. 5.10). After a period of rapid creep there is a plateau, and the rate of damage then increases again just before failure. If groups of specimens are loaded to various stress levels, σ, and the time to failure, t_F, is recorded, the log-log graph of σ vs. t_F is linear (Fig. 5.8B) and described by an equation similar to Eq. 5.17:

$$t_F = A/\sigma^B \qquad (5.20)$$

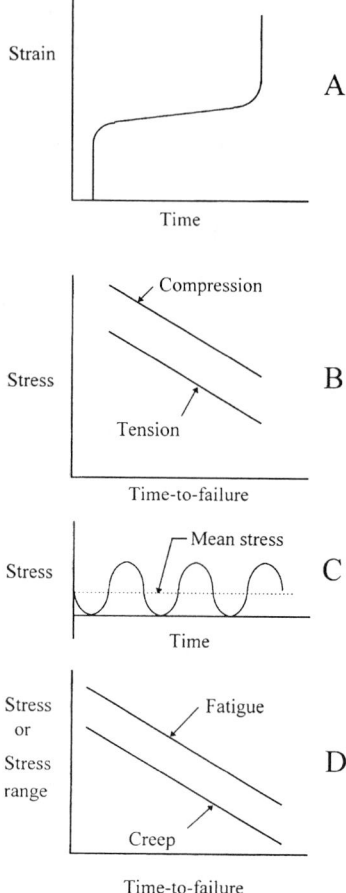

FIGURE 5.8A–D. **A** Typical graph of strain vs. time for cortical bone creep test to failure. The initial elastic response is followed by relatively constant strain during slow damage accumulation, eventually leading to rapid damage accumulation and failure. **B** Log-log plot of applied stress vs. creep failure time is linear with different behaviors for compression and tension. **C** Cyclic loading from zero to a tensile or compressive stress produces a nonzero mean stress that superimposes creep effects on fatigue behavior. **D** Comparison of creep and fatigue responses for human femoral bone.

where A and B are coefficients peculiar to the kind of bone and the mode of loading. Caler and Carter (1989) found that the exponent B was about 17 and similar for tensile and compressive creep. On the other hand, the coefficient A was greater for compression that for tension.

If the creep damage (D_c) occurs at a constant rate, its formation rate would be

$$dD_c/dt = \sigma^B/A \tag{5.21}$$

If the stress varies with time, the creep damage at any time, t, is

$$D_c(t) = \int_0^t [\sigma(t)^B/A \ ]dt \qquad (5.22)$$

In this formulation, failure occurs when $D_c = 1$, and the time required can be obtained by solving Eq. 5.22 for this condition.

Carter and Caler pointed out that creep and fatigue often occur simultaneously. If a specimen is loaded in tensile fatigue, for example, the stress varies from zero to a peak tensile value. The loading is cyclic and therefore it is "fatigue loading," but it also has a creep component because there is an average load value that is not zero (see Fig. 5.8C). The creep component is only zero if the fatigue loading is "fully reversed," that is, it cycles between peak tensile and compressive stresses that are equal in magnitude.

If fatigue loading occurs at a constant stress range, S, then the fatigue damage may be written as

$$D_f(t) = R_L \ t \ S^\beta/\alpha \qquad (5.23)$$

where R_L is the loading rate (number of load cycles/day) and α and β are coefficients. If fracture occurs from fatigue damage alone, then the time to fracture is obtained by setting $D_F = 1$ and solving for $t = t_F$. When damage is accumulating in response to simultaneous creep and fatigue effects, it might be possible to formulate the result as a simple summation:

$$D_{total} = D_c + D_f \qquad (5.24)$$

with $D_{total} = 1$ at failure. For cyclic loading with a non-zero mean stress, creep and fatigue damage combine to produce failure after N_F cycles and at time t_F, and

$$D_{total} = \int_0^{t_F} [\sigma(t)^B/A] \ dt + R_L \ t_F \ S^\beta/\alpha = 1 \qquad (5.25)$$

For a zero-tension fatigue test in which stress is a haversine function of time and the peak stress is S, the stress history may be written as

$$\sigma(t) = (1 + \sin 2\pi R_L t) \ S/2 \qquad (5.26)$$

Substituting this function into Eq. 5.22 gives, at failure,

$$1 = (1/A) \ (S/2)^B \int_0^{t_F} (1 + \sin 2\pi R_L t)^B \ dt \qquad (5.27)$$

which with two changes of variables can be solved to give

$$t_F = KA/S^B \qquad (5.28)$$

for the time to failure in the zero-tension fatigue test. Here,

$$K = \frac{2^B}{\displaystyle\int_0^1 (1 + \sin 2\pi t^*)^B \, dt^*} \; , \quad t^* = t/t_F \qquad (5.29)$$

Now compare Eq. 5.20, for time-to-failure in a creep test, to Eq. 5.28, for time-to-failure in a zero-tension fatigue test. The only difference is the coefficient K in Eq. 5.28. If we rewrite these equations in log-log form so as to place stress on the y-axis and time on the x-axis, as is customary, we have

$$\log S = B \log A - B \log t_F \quad \text{(creep)} \qquad (5.30)$$

and

$$\log S = B \log KA - B \log t_F \quad \text{(zero-tension fatigue)} \qquad (5.31)$$

The slopes of the two equations are the same but their intercepts are different. Figure 5.8D shows the results of creep and zero-tension fatigue experiments on human femoral bone performed by Carter and Caler (1983). Note that the fatigue data fall along a line with the same slope as the creep data, but displaced up and to the right. The specimens loaded in fatigue required longer to fail at a given stress level because the stress noted on the y-axis was only applied part of the time; in the case of the creep specimens, it was applied continuously.

In later papers, Carter and Caler (1985; Caler and Carter, 1989) showed that when fatigue stresses are high, so that N_F and t_F are low, creep damage dominates, and vice versa. This is significant because most fatigue experiments are done with superphysiologic stresses to keep the time necessary to bring the specimen to failure reasonable (see Exercise 5.3); damage in these experimental specimens is mostly caused by creep. Conversely, bones are fatigued in vivo at relatively low stresses, so N_F and t_F are large; damage then is mostly caused by fatigue. Insofar as the mechanisms of damage caused by creep and by fatigue are different, the results of laboratory experiments cannot be extrapolated to the in vivo situation. On the other hand, differences between creep and fatigue damage have not been identified. These are important caveats to bear in mind as you read the remainder of this chapter and work through the exercises at the end.

5.5 Fatigue Behavior of Fiber-Reinforced Composite Laminates

Engineers have learned to manufacture materials with exceptional creep and fatigue resistance. These materials limit damage through the presence of lamellae or fibers in their structure. Because bone is a lamellar structure in which both ultrastructural (i.e., collagen fibrils) and microstructural (i.e., osteons) fibers figure prominently, we can gain some insight about bone's damage resistance by examining the behaviors of these man-made materials.

The Birth and Growth of Cracks

The resistance of any material to fatigue failure is a function of its resistance to both the *initiation* and *propagation* of cracks. These two factors are in turn highly dependent on the microstructural characteristics of the material. Materials that resist the initiation of microcracks often cannot effectively resist their propagation; on the other hand, those materials in which cracks are easily started are often difficult to propagate a crack through. Because the fatigue life of a material is a function of both crack nucleation and growth, a material in which cracks are easily started, but find it difficult to grow, can often provide greater resistance to fatigue failure than one in which cracks are difficult to initiate. In fact, many highly fatigue-resistant materials derive this property primarily from their resistance to crack growth rather than initiation. Bone is such a material.

Material Strength and Fiber Diameter

There are two primary microstructural factors that affect crack initiation and growth: *material strength* and *fiber diameter*. In general, decreased yield strength *increases* the potential for crack initiation, but it also *increases* the resistance to crack growth (Fig. 5.9). This is because plastic flow at the leading edge of the advancing crack blunts the tip, reducing the stress concentration and the strain energy to be found at this critical location. As discussed more fully later in this chapter, crack initiation

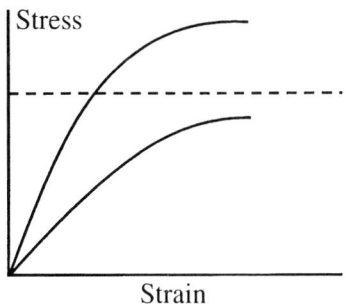

FIGURE 5.9. Stress-strain curves for materials having high and low yield stress. If the latter material's yield strength is less than the stress required to propagate a crack (*dashed line*), it may suffer from plastic deformation, but cracks cannot grow because the stresses never become high enough.

and growth also depend on the size and number of fibers running through the material. (In cortical bone, osteons serve as large fibers and collagen fibrils as small fibers.) The interface between a fiber and the surrounding matrix is a source of weakness that can initiate cracks. On the other hand, the cracks tend to remain small and follow the fibers instead of propagating across the structure. Therefore, it is important that the fibers be oriented so as to lead the cracks in harmless directions; for example, along rather than across a beam.

The Road to Failure

The ability of fibrous, lamellar composites to keep cracks small and running in harmless directions gives them a characteristic, three-phase damage accumulation history as they fail. Consequently, the long-term fatigue behavior of fibrous composite laminates is often divided into three regions, as shown in Fig. 5.10 (Reifsnider, 1990). We may think of the path to failure as a road running through these three regions.

Region I: Initiation
The first stage of fatigue is characterized by a relatively rapid but limited loss of stiffness coincident with the initiation of microdamage. The kind and amount of microdamage is determined by the properties of the composite structure. As illustrated in Fig. 5.11, in a cross-ply lamellar structure, cracks begin at structural discontinuities within a lamella and propagate to the interface with an adjacent lamella, where they are apt to stop, particularly if the fibers in the adjacent laminae are aligned in a different direction. These self-limiting cracks redistribute the stresses in the material, reducing the stress in the material alongside the cracks and increasing the stresses near the crack tips, at lamellar interfaces.

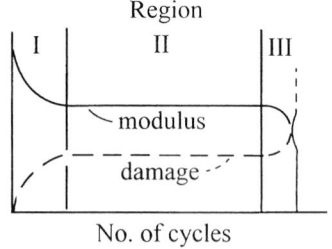

FIGURE 5.10. The road to fatigue failure for a typical composite material. In *Region I*, elastic modulus declines as damage begins under the initial loading. In *Region II*, the rates of change of damage and modulus are much less. In *Region III*, damage builds up rapidly to failure, and the modulus plummets.

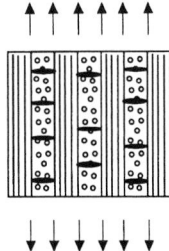

FIGURE 5.11. Typical cracking in a cross-ply laminate loaded in tension. Cracks open in the lamina having fibers perpendicular to the load direction, and propagate to the lamellar interface, where they stop. The material above and below each crack is unloaded because of the crack's free surfaces. When the cracks become so numerous that the peak stress between them is sufficiently low, no more cracks can form; a saturation damage state has been reached.

Region II: Benign Accumulation

The rate of damage accumulation and the stiffness loss quickly stabilize, generally within the first 25% of the fatigue life, and this marks the transition to a period of benign damage accumulation. Stabilization occurs for two reasons. First, the crack tips are stopped at lamellar interfaces, so their progress across the specimen is halted. Second, as the cracks become more numerous, and the distance between their sides diminishes, the peak stresses between them become so low that additional cracks can no longer be initiated in the intact material. The level of damage at which this stabilization occurs is called the *characteristic damage state*. Because this intralamellar damage is now shut off, damage accumulation in region II shifts to interlamellar debonding or delamination. These cracks travel parallel to the lamellar interfaces which, if the design is right, are aligned with the principal stress directions. Consequently, these cracks cannot directly cause catastrophic failure and do not even change the elastic modulus very much. The surface area of the resulting secondary cracks is very large, serving as an energy sink and preventing the growth of a single, large, and more dangerous crack. Region I damage tends to be the same regardless of whether the material is loaded cyclically or monotonically; however, the kind of lamellar debonding damage typical of region II is characteristic of cyclic rather than monotonic loading.

Region III: Failure

Near the end of fibrous laminate fatigue life, stiffness declines rapidly. Eventually, debonding of lamellar interfaces, combined with matrix degeneration, increases the stress in the load-aligned fibers or lamellae, and they begin to fracture. Although the damage formed in region I is self-limiting because cracks stop at fiber or lamellar interfaces, the damage predominating in region III is self-feeding because as fibers or load-aligned lamellae break, they increase the stress in those elements that remain intact. This process usually occurs rapidly, constituting the final 10% or so of the total fatigue life, and ends in ultimate failure of the structure.

5.6 Osteonal Bone as a Fibrous Lamellar Composite Material

We saw in Chapter 4 that Haversian bone is mechanically weaker than primary bone in monotonic bending and tension, in part because of its greater porosity, and probably because of the presence of cement lines as well. However, there is evidence to suggest that the unique microstructural arrangement of the various components of Haversian bone is adapted to control crack growth and thereby extend fatigue life. Bone tissue contains collagen fibers organized into lamellae; this is a fibrous laminate on a microscopic scale. In secondary bone, these fibrous lamellae are in turn organized into larger fibers (osteons) running through a matrix (primary bone or osteon fragments). An important hypothesis of bone mechanics is that this structure is responsible for the fact that osteonal bone subjected to cyclic loading behaves much like an engineered fibrous-lamellar composite.

Osteonal Bone's Road to Failure

In the upper part of Fig. 5.12, the tensile elastic modulus of human femoral bone declines over its fatigue life (Pattin et al., 1996). There is an early, rapid diminishment of stiffness, followed by a prolonged, much more gradual loss until, very suddenly, the modulus plummets, as failure occurs. This sort of modulus degradation curve is characteristic of osteonal bone loaded in tension or bending. The lower part of Fig. 5.12 shows that this behavior is substantially different for compression: the modulus falls slowly at first, but at an ever increasing rate until failure. Later in this chapter we attempt to explain this difference based on the kind of damage associated with each mode of loading.

There is evidence that if the imposed loads are physiologic in magnitude, region II of the road to failure may persist for many millions of cycles. Schaffler (1985) tested secondary and primary bovine cortical bone in tension at a strain range of 0–1200 microstrain and a frequency of 4 Hz. In both cases, there was an initial rapid stiffness loss over the first 1–3 million loading cycles (region I), after which no further stiffness loss occurred (region II), up to 45 million loading cycles. Thus, the results looked just like Fig. 5.10, but region III was not reached. It is believed that bone's region I is associated with multiple crack initiation events, just as in other composite materials. Much of this damage is apparently in the interstitial bone outside of the cement lines of *intact* osteons (i.e., those osteons not overlapped by more recent osteons). However, at physiologic strains and strain rates, region III and failure were beyond the capacity of Schaffler's experi-

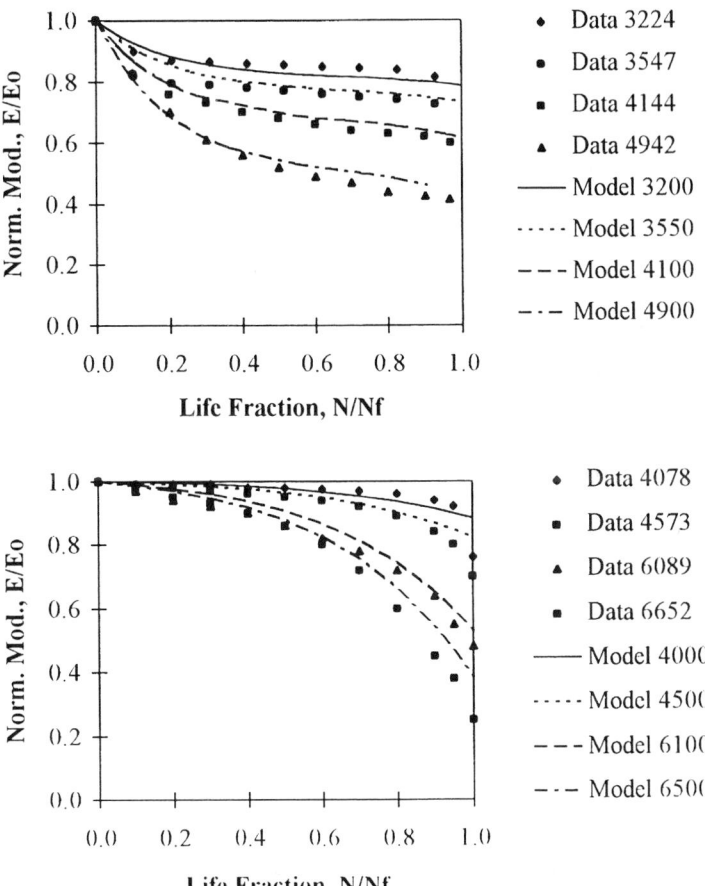

FIGURE 5.12. Degradation of elastic modulus during the fatigue life for human femoral bone specimens loaded in tension *(upper graph)* and compression *(lower graph)*. Symbols are experimental data from Pattin et al. (1996); curves are from the theory described in Section 5.7. In both cases, numbers shown in the legend represent initial strain values in $\mu\varepsilon$. (Reprinted from *J Biomechanics*, Vol. 29, Pattin, CA, Caler, WE, Carter, DR, Cyclic mechanical property degradation during fatigue loading of cortical bone, 69–79, 1996, with kind permission from Elsevier Science, Ltd., The Boulevard, Langford Lane, Kidlington 0X5 1GB, UK.)

mental apparatus to keep cycling. Bone has excellent fatigue properties **at physiologic strains and strain rates.**

This fatigue behavior may be contrasted with that seen by Carter et al. (1976) when bovine bone was fatigued at superphysiologic strains and strain rates (Fig. 5.13). Under these conditions remodeled bone had less fatigue resistance than primary bone, and both had fatigue lives of fewer than a million cycles at 3300–5500 microstrain.

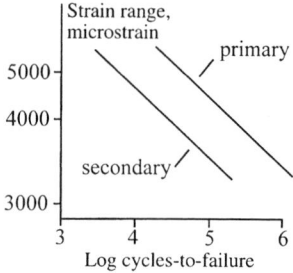

FIGURE 5.13. Effects of remodeling on fatigue life of bovine bone. For a given strain range, primary bone specimens had a longer fatigue life than those composed of secondary bone. (Reprinted from *J Biomechanics*, Vol. 9, Fatigue life of compact bone—II. Effects of microstructure and density, 211–218, 1976, with kind permission from Elsevier Science, Ltd., The Boulevard, Langford Lane, Kidlington OX5 1GB, UK.)

Bone Toughness

We have defined toughness in two ways: as the resistance to crack growth and as the amount of energy required to produce complete fracture. Clearly, if more effort is required to propagate cracks through a specimen, more energy is required to produce complete fracture, too. Moyle and Bowden (1984) found that bone's work-to-failure was positively correlated ($r^2 = .31$) to the fractional area occupied by osteons. Haversian bone is very similar in its osteonal arrangement to very tough composite materials reinforced with discontinuous fibers. Fiber-reinforced composites are often highly anisotropic and have reduced transverse strength because of the directionality of the fibers. Aligned fibers provide high strength and stiffness in the predominant load direction at the expense of properties in the circumferential and radial directions. Secondary osteons, each only 5–10 mm long (Cohen and Harris, 1958) and oriented nearly in the direction of principal strain (Lanyon and Bourn, 1979), behave like oriented, discontinuous fibers within a matrix of more brittle material. The osteonal "fibers" each have a cylindrically lamellar structure inside and a relatively weak osteon–matrix interface. Failure at these weak interfaces by interlamellar debonding inside the osteon, or delamination at the cement lines, stops or deflects crack propagation. Crack deflection serves the dual purposes of energy absorption and reorientation so that the applied stresses no longer spread apart the crack tip. Both these mechanisms increase fracture toughness while at the same time minimizing damage to vascular and neural structures within the Haversian canal.

Controlled Crack Propagation Studies

Controlled crack propagation may be accomplished by machining a wedge-shaped specimen (Fig. 5.14A). This specimen is loaded very slowly in

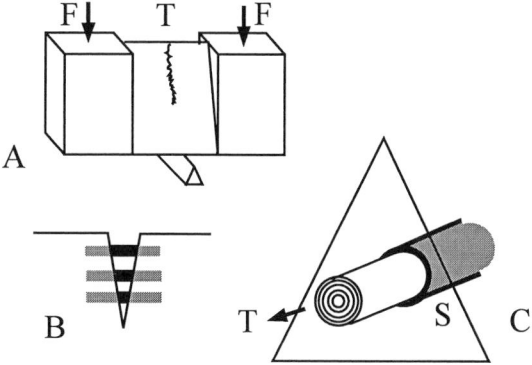

FIGURE 5.14. **A** Specimen geometry for controlled crack propagation studies. The central region is wedge shaped and loaded in bending. As the crack propagates from the tensile edge *(T)*, it encounters more bone, and this helps control the crack velocity. **B** As the crack propagates through the bone, it tends to leave osteons in its path intact, bridging the opening. **C** An osteon bridging the crack experiences a tensile load *(T)* in the gap as well as a shear load *(S)* on the embedded portion.

three-point bending so that the apex of the central wedge is in tension, and the area under the load-deformation curve is taken as the work-to-failure, or toughness, of the specimen. The fact that the width of the crack front increases as it moves across the wedge helps keep the crack from accelerating. The first such experiment on bone was reported by Piekarski (1970), using bovine femurs. He showed that the microscopic appearance of the fracture surface depended a great deal on the rate at which the crack propagated across the wedge. Typically, cracks traveled slowly at first, and then suddenly accelerated to produce a catastrophic failure. The initial portion of the fracture surface was always rough and exhibited many pulled-out osteons; the surface produced by the subsequent rapid failure was consistently smooth, with osteons breaking across rather than pulling out. When the crack was advancing slowly it tended to enter and follow cement lines, thereby avoiding Haversian canals. Work-to-failure was very dependent upon the rate of crack propagation. The slow, controlled cracks required almost 60 kilojoules/m^2 of crack surface to propel them, while the catastrophic cracks required less than 1 kJ/m^2.

Mathematical Analysis of Osteonal Pullout

Applying techniques used to study reinforced plastics, Pope and Murphy (1974) mathematically analyzed the phenomenon of osteon pullout. Their analysis was directly applicable to the sort of experiment done by Piekarski and raised important questions about the mechanisms by which histologic structure may affect crack propagation. Consider a region of osteonal bone loaded in longitudinal tension as shown in Fig. 5.14B. A crack is propagating transversely through the cortex. As the leading edge of the crack envelops individual osteons, they experience two forces. First, they continue to be stretched by the tensile stress in the cortex. The tensile force from the left side of the crack attempts to pull the osteon out of the bone matrix on the right side of the crack, and vice versa. The tensile pull-out forces are resisted by the shear forces that the surrounding bone exerts on the walls of the embedded portions of the osteon. If the tensile strength of the portion of the osteon bridging the crack is great enough, these shear forces may exceed the shear strength of the cement line and the osteon will begin to pull out of the bone matrix. Subsequently, one would expect the osteon to fail not within the primary, transverse crack, but away from the primary crack, at the stress concentration at the end of the debonded region. If, on the other hand, the osteon is relatively weak in tension, it will break within the primary crack, without pulling out. Let us consider a simplified version of Pope and Murphy's mathematical formulation of this problem.

The tensile force acting on the osteon in the primary crack is

$$T = \pi R_o^2 \, \sigma_t \qquad (5.32)$$

where σ_t is the tensile stress and R_o is the radius of the osteon. The shear force on the walls of the embedded portion of the osteon on, say, the right side is given by

$$S = 2 \, \pi R_o \, L \, \sigma_s \qquad (5.33)$$

where σ_s is the shear stress in the cement line and L is the distance over which the surrounding bone grips the osteon. The osteon will pull out when the tensile force exceeds the resisting shear force; that is, when $T > S$. By substituting the foregoing equations into this inequality, one finds that pullout occurs when

$$\sigma_{tf} > 2L/R_o \, \sigma_{sf} \qquad (5.34)$$

where σ_{sf} is the shear failure stress for the cement line and σ_{tf} is the tensile failure stress of the osteon. Note that it is easier for pullout to occur if R_o is relatively large. Conversely, smaller osteons are more likely to break off when forced to bridge a transverse crack.

Comparing Theory with Data

Moyle and co-workers (1978) performed an experiment similar to Piekarski's on canine femoral bone. They confirmed the correlation between slow cracking and osteon pullout, and they also found that the work-to-failure for slow cracking was significantly higher in the lateral aspect of the dog's femur than the medial side (Table 5.4). Because the bending caused by eccentric loading at the hip joint produces tensile stress

TABLE 5.4. Work-to-failure in different regions of canine femur

Lateral side ($N = 10$)	$11,800 \pm 3,000$ J/m^2
Medial side ($N = 13$)	$6,960 \pm 1,350$ J/m^2

From Moyle et al., 1978.

TABLE 5.5. Effects of osteon diameter and density on crack propagation

Type failure	Mean osteon diameter, μm	Osteon density, mm^{-2}
Slow cracking ($N = 23$)	148 ± 22	20 ± 7
Fast cracking ($N = 45$)	202 ± 29	14 ± 5

From Moyle et al., 1978.

on the lateral side of femurs and compression on the medial side, this difference in resistance to tensile cracking may be a manifestation of mechanical adaptation. Like Piekarski, Moyle's group found that most specimens failed catastrophically, but some cracked slowly all the way across the wedge. These specimens turned out to have significantly *smaller* osteons than did the catastrophic failures (Table 5.5), in direct contradiction to the analytical result obtained above.

Osteon size and control of crack propagation are also important in fatigue resistance. In flexural fatigue experiments, Gibson et al. (1996) found that the dorsal region of the racehorse "cannon" (third metacarpal) bone had a longer fatigue life than the lateral region, which was monotonically stronger and stiffer. Fatigue resistance was strongly associated with signs of osteon pullout on the fracture surfaces; there were no signs of pullout in the lateral specimens. The dorsal region had smaller osteons than the lateral region: 156 ± 19 vs. 179 ± 13 μm. The reason for this repeated disagreement with theory may be that osteon pullout depends as much on the number of osteons as it does on their diameter. In Chapter 3 it was shown that as secondary osteons accumulate in a section, their population asymptotically approaches a limit that is the reciprocal of their mean individual area. Thus, if the osteons are smaller, more of them can be packed into a section, and there are more to pull out.

Thus, the energy dissipated by osteon pullout should depend on both the likelihood of pullout (which increases with osteon diameter) and the number of osteons available to be pulled out (which decreases with diameter). Figure 5.15 shows data supporting such an argument (Moyle et al., 1978; Moyle and Bowden, 1984). The left side of the graph shows that greater energy absorption can be achieved when there are many small osteons in the area of a developing crack. As osteon diameter increases, energy absorption declines. However, as shown on the right side of the graph, work of fracture again increases when osteon diameters are greater

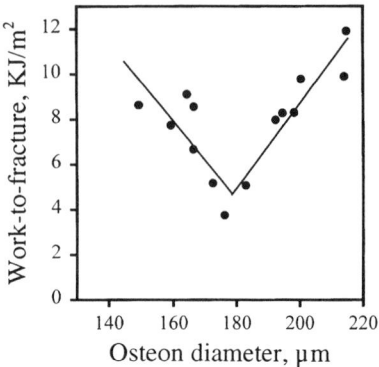

FIGURE 5.15. Relationship between work-to-failure and osteon diameter. (Reprinted from *J Biomechanics*, Vol. 17, Moyle, OD, Bowden, RW, Fracture of human femoral bone, 203–213, 1984, with kind permission from Elsevier Science, Ltd., The Boulevard, Langford Lane, Kidlington 0X5 1GB, UK.)

than 200 μm. The reason might be that these larger osteons are more likely to pull out rather than break off, as predicted by Eq. 5.34. Corondan and Haworth (1986) presented additional data which further demonstrate that fracture is inhibited by increased numbers of osteons as well as by larger osteons. Comparing the morphology of fractured surfaces of bone with that of sections adjacent to the fracture, they found that bone broke where there are fewer *and* smaller osteons. Bone *did not* break in adjacent regions where osteon packing *and* size were greater.

Crack Initiation

Persistent slip bands are "pre-failure" or plastic slip lines representing partial failure within a material, generally along maximum shear planes. These bands may represent the starting point for microcracks within a material. There are features in cyclically loaded bone that look like irreversible microscopic slip bands which occur in metals from plastic deformation (Fig. 5.16). Frost first demonstrated these in 1960. As in metals, slip bands in bone may form a focus for subsequent crack formation.

The relatively weak cement line interface between osteons and interstitial bone matrix may reduce the shear strength of osteonal bone (Saha, 1977; Frasca, 1981). At the same time, however, this interface may, by slipping, relax shear stresses, reduce strain energy, and slow crack propagation. Lakes and Saha (1979) suggested that cement line displacement may be a manifestation of shear failure in bone. Their observations of displacement at lamellar interfaces in bovine plexiform bone supported the notion that lamellar interfaces in bone reduce stress concentrations by reducing shear stress. In studying the viscoelastic creep behavior of human bone, Park and Lakes (1986) observed displacement at the cement line, but it was

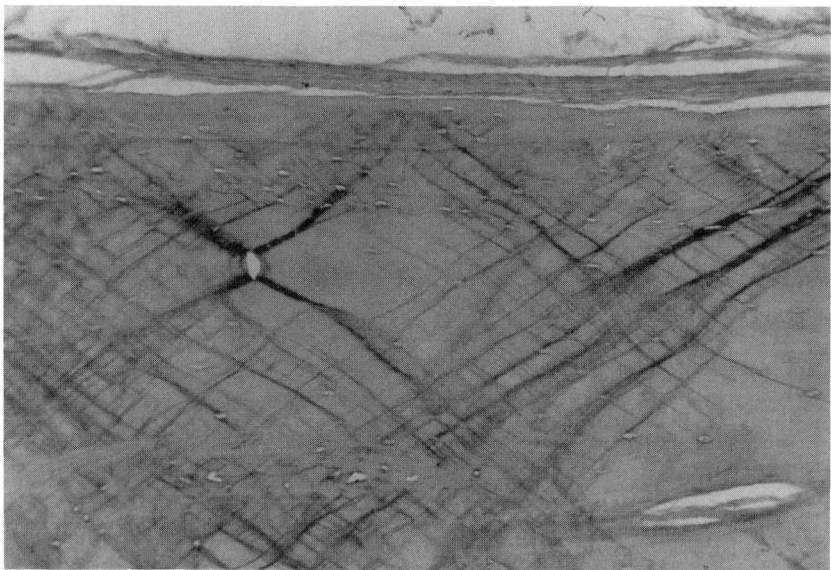

FIGURE 5.16. Slip bands in cortical bone. (Reproduced with permission from Prof. A. Chamay.)

one-third of what would be expected if it were the only source of the creep. This observation suggests that the osteonal cement line fails only after other damage (e.g., slip bands) has begun to grow in the matrix.

Crack Stopping

The fatigue life of Haversian bone is governed both by crack initiation and propagation. Because cracks can be removed through bone remodeling, it can be postulated that crack initiation is not so deleterious as the rapid propagation of cracks across the bone material. From this point of view, the most useful microstructural characteristics of bone tissue would be those able to slow, or alter the direction of, crack propagation, rather than those that inhibit crack nucleation. In fact, Haversian bone appears to be constructed in just this manner. As we have seen, the cement lines between osteons and the interstitial matrix are less mineralized than the bone matrix, and therefore probably more compliant. On reaching a more ductile region, a crack will accelerate and then decelerate, and finally be arrested when it reaches another stiff region (Kendall, 1975). Even if crack initiation occurs repeatedly during cyclic loading events, Haversian bone is postulated to stop the transverse growth of these cracks, diverting them along the bone axis and allowing them to expend the strain energy at the crack tip without seriously reducing structural integrity.

Strain Rate

The fracture toughness of bone may be reduced at high strain rates because crack velocity is generally increased under these conditions. When crack velocities exceed approximately 1 mm/s, the likelihood of osteon pullout is reduced and brittle failure is more common (Behiri and Bonfield, 1984). Increased crack velocity requires a higher stiffness ratio between bone matrix and the cement line to halt crack propagation. A change in mechanical properties of the cement line material from viscous to elastic at high strain rates would be expected to compound this problem (Balazs and Gibbs, 1970; Gottesman and Hashin, 1980). Even within the relatively narrow physiologic range of strain, Schaffler (1985) found that a strain rate of 0.03/s caused significantly more microscopic damage and greater loss of stiffness than 0.01/s.

Simkin and Robin (1974) found, in static tests at tensile strain rates of only 0.004/s, that nearly two-thirds (59%) of the microcracks were trapped either at cement lines or circumferential lamellae within an osteon; only 41% cut across these boundaries. At higher strain rates (0.4/s), osteons were capable of stopping fewer than 45% of the microcracks. Maximum strain rates during moderate activity range between 0.0034 and 0.015/s (Lanyon and Baggott, 1976; Rubin and Lanyon, 1984). Thus, at strain rates experienced by sedentary individuals, osteons might be expected to stop close to two-thirds of the propagating microcracks coming their way. These observations are in concert with those of Norman and Wang (1997), who found that 67% of all microcracks in cortical bone from human femurs were interstitial rather than intraosteonal, and most interstitial cracks abutted a cement line. The bones of more active individuals would be expected to experience higher strain rates and thus more intraosteonal cracking.

One must be careful to distinguish between strain rate and frequency effects. Ordinarily, strain rate increases as the frequency of loading increases, so that one might expect fatigue life to diminish as frequency increases. However, Caler and Carter (1989) found that the fatigue life of human femoral bone increased rather than decreased as the loading frequency increased from 0.02 to 2 Hz. This change is apparently related to creep effects. Even though the bone is more resistant to crack propagation at the low strain rate associated with a lower frequency, the amount of time that the bone is loaded for each cycle is longer at a low frequency. Thus, the number of cycles required for failure (i.e., the fatigue life) is less at a low frequency. Schaffler (1985) was careful to keep frequency constant (at 4 Hz) in his tests of the effects of strain rate; he did this by using a triangular waveform and varying the "off period" between "spikes" to vary frequency.

Effect of Remodeling

The effects of remodeling on fatigue life are unclear. Carter et al. (1976, 1981; Carter and Hayes, 1977b) tested bovine cortical bone in rotating bending fatigue at stress amplitudes of 65–108 MPa. Specimens with the most remodeling were significantly less fatigue resistant than primary bone specimens of equal density. Evans and Riolo (1970) found the opposite effect in human tibial specimens loaded in bending to 34 MPa. The source of this discrepancy could be the species tested, or it could be the different stress amplitudes. Assuming a modulus of 20 GPa, the Carter experiments produced superphysiologic strains (3250–5400 µε); the Evans–Riolo experiment did not (1700 µε). We have seen that much of the damage to bone at high strain magnitudes is caused by creep rather than fatigue, and osteons may affect fatigue and creep damage mechanisms differently. Strain rate may also have entered the picture: although Evans and Riolo did not specify their frequency, Carter et al. tested at 125 Hz. At the high strain rate associated with such a high frequency (and strain magnitude), the crack-stopping effects of osteons may disappear.

Schaffler's (1985) tensile fatigue tests at physiologic strains and strain rates revealed no differences between primary and secondary bone for either percent stiffness loss or microdamage accumulation. However, the modulus of the primary bone was greater than that of remodeled bone, both initially and just before failure, and the primary bone's modulus held up better over the fatigue life. These experiments suggest that the remodeling does not substantially reduce the fatigue life of bone *at physiologic strain magnitudes*. There may be a strain rate at which crack resistance and fatigue life are optimal. At very low strain rates, creeplike conditions occur and damage is more severe. At very high strain rates, the effectiveness of cement lines and other tissue interfaces as crack stoppers is diminished. Perhaps the bone structure is adapted to crack stopping best at physiologic strain rates.

5.7 Modeling Fatigue Damage Effects in Osteonal Bone

Krajcinovic et al. (1987) developed a continuum damage theory for osteonal bone, which assumed that when a crack propagates transversely through the cortex, osteons become debonded, break, and pull out as described previously. He calculated the modulus degradation and residual strength of the specimen as this process continued. He was able to predict both bending and tensile elastic moduli and the relationships between elastic modulus and tensile failure properties.

Griffin and co-workers (1997) recently proposed a more detailed model for fatigue of osteonal bone, based on the observation that cortical bone fatigue behaviors are different in tension and compression. Fatigue lives are longer in compression than in tension, and the shapes of the modulus degradation curves are different (see Fig. 5.12). In tension there are generally three distinct regimes of behavior with increasing numbers of cycles: the first is an initial reduction in modulus, the second is a plateau, and the third is a rapid modulus reduction before failure. In compression, on the other hand, the modulus diminishes very slowly at first, but at an ever increasing rate until failure.

Carter and Hayes (1977c) also observed differences in the kind of microdamage associated with fatigue in tension and compression. In tension,
most cracks were found in the interstitial matrix or in osteonal cement lines.
In compression, the cracks tended to run from one vascular channel to
another. Figure 5.17 shows similar behavior in the tensile and compressive

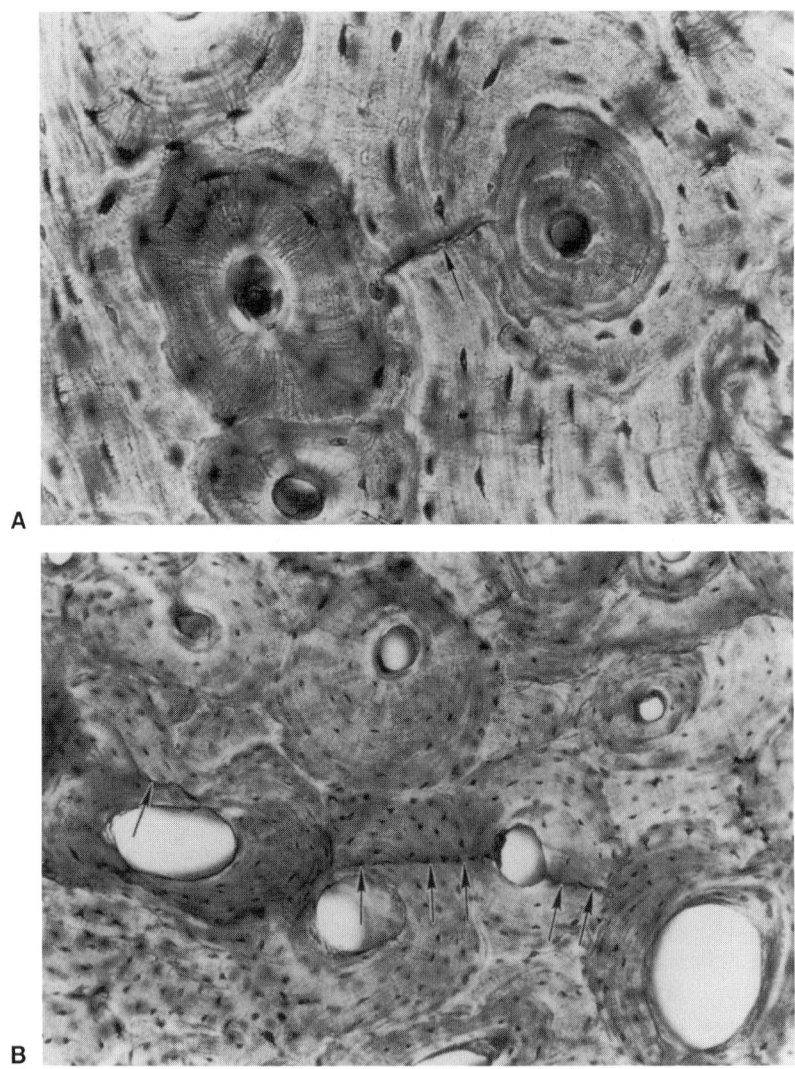

FIGURE 5.17A,B. Microdamage (indicated by *arrows*) typical of (**A**) tensile and (**B**)
compressive regions in a beam loaded in four-point bending. Human femoral cortex, 100-μm-thick section stained with basic fuchsin. Field width, ~500 μm.
(Reproduced with permission from Griffin et al., 1997.)

regions of human femur specimens fatigued in four-point bending. Is it possible to mechanistically account for the differences in modulus degradation in tension and compression on the basis of the differences in the nature of the damage?

The theory of Griffin et al. (1997) models cortical bone as a fiber-matrix composite, the fibers being osteons and the matrix consisting of primary bone and remodeled osteon fragments. Damage is partitioned into components that independently affect the fibers and the matrix, and the "damaged modulus" is obtained from the isostrain rule of mixtures given in Eq. 4.16. Here, we write this equation in the form

$$E = (1 - D_f) E_f V_f + (1 - D_m) E_m V_m \tag{5.35}$$

where E is the elastic modulus of the bone, V is volume fraction, and D represents damage. The subscripts m and f refer to matrix and fibers, respectively. D_f and D_m have values that range from zero, in an undamaged state, to one, in a failed state. When there is no damage, this equation is reduced to Eq. 4.16.

For ceramic matrix composites loaded in tension, the initial damage is caused by microcracks in the matrix. The number of these cracks does not increase indefinitely, but approaches a limit known as the saturation or characteristic damage state. It is postulated that for tensile fatigue of cortical bone, the stiffness loss is caused primarily by cracks that form in the interstitial matrix and are self-limiting in length because they consistently appear as seen in Figs. 5.1 or 5.17A. Therefore, damage to the matrix evolves as

$$dD_m/dN = K_1(D_s - D_m) \tag{5.36}$$

where D_m represents the matrix damage, N is the current cycle number, K_1 is a matrix damage rate parameter, and D_s represents the saturation level of matrix damage, a function of the applied stress.

Most matrix cracks intercept osteon cement lines, and can be expected to produce stress concentrations at the point of impingement. Matrix damage also results in load transfer from the matrix to the osteons. These factors should eventually lead to the tensile failure of osteons. As osteons break, the load is further redistributed to neighboring osteons, producing a positive feedback and resulting in osteon failure at an ever increasing rate. Therefore, the fiber damage, D_f, is assumed to accumulate as

$$dD_f/dN = K_2 D_s + K_3 D_m \tag{5.37}$$

where K_2 is an osteonal damage rate coefficient and K_3 is an osteon–matrix damage interaction parameter. Equations 5.36 and 5.37 can be solved simultaneously to obtain

$$D_m = D_s (1 - e^{-K_1 N'}) \tag{5.38}$$

$$D_f = D_s K_3 [K_2 \exp(-K_1 N') + K_1 \exp(K_2 N') - (K_1 + K_2)]/(K_1 + K_2) \tag{5.39}$$

where $N' = N/N_F$ is the normalized fatigue life in tension. Substitution of these two equations in Eq. 5.35 provides an expression for the reduction in modulus as a function of the normalized fatigue life.

In compression, microcracks appear less likely to stop at the osteon–matrix interfaces than they do in tensile loading. If so, then in bone that is fatigued in compression cracks may tend to grow in length and number without bound, degrading the material in a positive feedback manner, and damage is assumed to evolve as

$$dD_c/dN = K_4 (D_I + D_c) \tag{5.40}$$

where K_4 is a compression damage rate parameter and D_I represents incipient damage in the form of stress concentrators, such as Haversian or Volkmann's canals, which initiate or extend D_c. Equation 5.40 is readily solved to obtain

$$D_c = D_I [\exp(K_4 N'') - 1] \tag{5.41}$$

where N'' is the normalized fatigue life in compression. For the compression case, D_c is used for both damage terms in Eq. 5.35 because there do not appear to be distinct damage modes in osteons and interstitial matrix in this mode of loading.

By fitting the resulting modulus degradation equations to Pattin's (1996) data, the constants K_1, K_2, K_3, and K_4 were determined. Return to Fig. 5.12 to see examples of how well these results fit data for typical compression and tension specimens. These results support the hypothesis that the different shapes of these curves are caused by different damage modes in tension and compression. In particular, it appears that matrix damage under tensile loading is self-limiting, with cracks being stopped by osteonal cement lines, but this may not be the case in compressive loading.

Griffin et al. (1997) went on to demonstrate that a mathematical model for fatigue of a beam loaded in four-point bending which incorporated the foregoing equations to describe the development of damage on the tensile and compressive sides was able to predict modulus degradation and fatigue life reasonably well.

Why Would Cracks Be Stopped in Tension but Not in Compression?

What could account for the observation that osteons stop cracks better in uniaxial tensile fatigue than in compression? Figure 5.18 offers a possible explanation. When a specimen is loaded in uniaxial tension, the Poisson stress acting across the fibers is compressive and tends to close a matrix crack of the kind shown in Fig. 5.1. Thus, it is more difficult to propagate the crack and easier for a cement line to stop it. On the other hand, when uniaxial compression is applied, the Poisson stress is tensile and tends to open the crack. This should make it more difficult for cement lines to stop

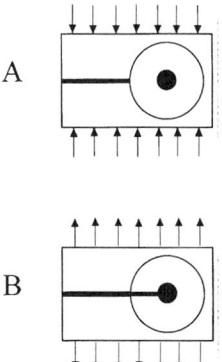

FIGURE 5.18A,B. Hypothetical basis for cracking differences in tensile and compressive fatigue. **A** In tensile loading, Poisson stress is compressive, closing cracks and making them more likely to stop at cement lines. **B** In compressive loading, Poisson stress is tensile, opening cracks and making them less likely to stop at cement lines.

the crack, consistent with the observation that such cracks travel from Haversian canal to Haversian canal, passing through cement lines.

If Cracks Are Self-Limiting in Tensile Fatigue but Not in Compressive Fatigue, Why Is Fatigue Life Longer in Compression Than in Tension?

Although we have argued that the toughness and fatigue resistance of bone are based on resistance to crack propagation rather than crack initiation, perhaps this is an oversimplification. Consider the following possibility. When cortical bone is loaded in uniaxial tension, cracks are easily initiated but difficult to propagate substantial distances. When loaded in uniaxial compression, on the other hand, the opposite situation may obtain: difficult initiation but then easier propagation. This hypothesis is consistent with the possibility that the Poisson effect drives crack propagation in compressive loading. It is also in concert with the modulus degradation curves of Pattin and co-workers. In tensile fatigue, the modulus drops rapidly at first, then plateaus until just before failure. This finding is consistent with a form of damage that is easily initiated but self-limiting, followed eventually by accelerating osteon breakage. In compression, on the other hand, the early, self-limiting form of damage may not occur because crack initiation is more difficult in this mode of loading. However, if continued loading eventually breaks down the tissue's resistance to crack initiation, the cracks may be of a kind that grow relentlessly, and the modulus may then diminish rapidly (see Fig. 5.12B). To test this hypothesis, more work is needed to elucidate the relative ease of crack initiation and crack propagation in different modes of loading oriented in various ways to the lamellar structure.

5.8 The Role of Fatigue in Activating Bone Remodeling

We have asserted that one of the primary functions of remodeling is to repair fatigue damage. In this section we discuss the evidence for that. To begin, it is useful to compare the fatigue life of bone to the rate at which bone is renewed or turned over by remodeling. We have seen Schaffler's data indicating that bovine cortical bone can endure 30–45 million load cycles at 1200 microstrain without fatigue failure.[2] Pattin's data for the fatigue life of human femoral bone suggest that the fatigue life at a peak strain range of 2000 $\mu\varepsilon$ is 4–10 million cycles. At 1200 $\mu\varepsilon$, this might be substantially extended. How long would it take an average person to accumulate 30 million load cycles in, say, the femur? If a person's daily activities were the equivalent of walking 5 miles/day, each femur would receive about 5 miles × 5280 feet/mile ÷ 6 feet/stride = 4400 load cycles/day, assuming each step covers 3 feet. This works out to be roughly 1.5 million load cycles/year, so that 20 years would be required to accumulate 30 million cycles. The turnover rate for the cortex of a bone like the femur is about 3% per year in adults (Parfitt, 1983), but this rate should be higher in children and the elderly. Assuming an average value of 5%, 20+ years is just about what would be required to replace all the bone in any given region. Thus, there is a rough correspondence between the fatigue life of cortical bone and the rate at which it is replaced.

Random or Directed Repair

Osteonal remodeling could replace microdamage in one of two ways: by directing the BMUs to the location of damage, or by simply randomly remodeling the cortex at a rate designed to keep up with the overall rate of damage accumulation. Random remodeling would not be ideal, however. Some regions can be expected to habitually have higher stresses than others (e.g., near points of muscle attachment). The process might be more efficient if osteonal activation occurred at sites where microdamage has actually developed.

It has been proposed that osteonal cement lines not only stop cracks, but that this event initiates the reparative remodeling process (Martin and Burr, 1982). This could happen because debonding of the osteon during crack trapping produces a local disuse state, which activates a new BMU. This new remodeling unit could then resorb the damaged region and replace it with new bone (Fig. 5.19A). Alternatively, Frost (1973b) postu-

---

[2]Unpublished data by Schaffler and co-workers extend this to 45 million cycles.

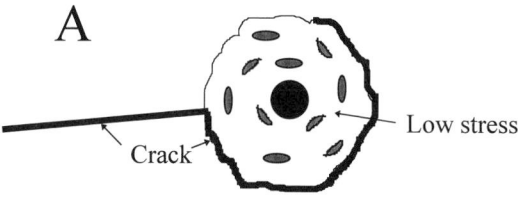

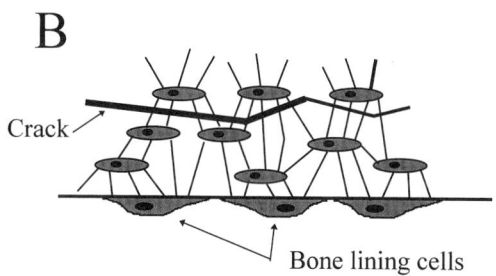

FIGURE 5.19A,B. Alternative hypotheses regarding the mechanism for activation of remodeling by fatigue damage. **A** debonding crack in cement line isolates osteon from stress, creating local disuse state that produces activation signal. **B** Damage to osteocytic network produces activation signal.

lated that damage triggers a response through disruption of the canalicular network. There is evidence that stress or strain provoke responses in osteocytes that are transmitted from cell to cell through their processes (Weinbuam et al., 1994; Cowin et al., 1995; Owan et al., 1997). The interruption of normal osteocyte signaling to bone lining cells could activate remodeling if these signals are inhibitory (Fig. 5.19B). These two theories provide very different mechanisms for activation of the repair of microdamage in bone, and a contrast that should allow for experimental testing of their validity. They are similar, however, in that they both use the same pathway; in one case the signal is not generated because cement line debonding relieves stress, and in the other the signal is blocked by intraosteonal damage.

The Historical Perspective

This concept is only a variation on an old idea. Tschantz and Rutishauser (1967) suggested that microcracks in the compressed ulnar cortex of dogs stimulated remodeling. Chamay and Tschantz (1972) proposed that loading stimulates bone remodeling. Rahn et al. (1972) showed that overload zones adjacent to fractures in the process of repair are also remodeled, resulting in the resorption of damage and the formation of more secondary osteons.

Comparative primate morphology studies (Schaffler and Burr, 1984) show that osteonal remodeling is more intense when bones are impulsively and repetitively loaded, that is, in very active animals, suggesting that remodeling may be initiated by such loads. Frost's (1960a) earlier data on human ribs support the same concept: the rib is loaded many times each day by respiration. Typically, ribs have more microcracks and also a greater rate of remodeling than any other cortical bone site in the body. These observations are consistent with the hypothesis that bone remodeling is activated by microdamage, but they do not implicate microdamage itself as the direct cause of the remodeling process.

Burr and co-workers (1985; Mori and Burr, 1993) performed in vivo fatigue experiments to verify a cause-and-effect relationship between microcracks and BMU activation in cortical bone. The results showed that microcracks were associated with bone resorption spaces several times more often than expected by chance alone, even when the magnitude of the strain and the duration of loading were within reasonable physiological limits. They also demonstrated that cracks caused the resorption spaces, and not the other way around. However, the mechanisms behind this association have not been elucidated, and it remains to be shown how long remodeling takes to repair a "batch" of fatigue damage.

More recently, Bentolila and co-workers (1997) demonstrated that remodeling can be activated in association with fatigue damage in a bone that ordinarily lacks osteons and therefore has no cement lines. By cyclically loading the ulnas of rats, which do not ordinarily remodel their cortical bone, they were able to demonstrate microcracks in association with BMU resorption cavities. This experiment is important for two reasons. First, it suggests that cement line debonding, as proposed by Martin and Burr (1982), need not play a role in activation of BMUs by fatigue damage because the rat's ulnar cortex initially has no osteons. Instead, it supports Frost's proposal that damage acts directly on osteocytes. Second, it supports the hypothesis that the control of fatigue damage is an important function of remodeling, because in the normal rat ulna there is neither damage nor remodeling to be seen, but when damage is introduced, remodeling is activated within days. (Remodeling may also be activated in these animals by calcium deficiency, however.)

Microdamage and Bone Fragility in the Elderly

Recently, the idea that fatigue damage may be an important component of bone fragility in the elderly has gained credence among bone researchers (Heaney, 1992). This idea is supported by data showing that microdamage increases exponentially with age in human femoral bone, especially in women (Fig. 5.20) (Schaffler et al., 1995; Mori et al., 1997). This increase could occur because reduced bone mass increases bone strains, and fatigue damage is probably very sensitive to strain magnitude. It follows from this

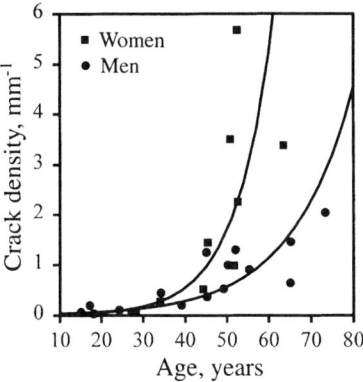

FIGURE 5.20. Microcrack density vs. age for men *(circles)* and women *(squares).* (Reproduced from Schaffler et al., 1995.)

view than many nontraumatic fractures in the elderly (i.e., osteoporotic fractures) are essentially stress fractures. This hypothesis is supported by the fact that callus-like woven bone formations have been observed on periosteal surfaces adjacent to femoral neck fractures (Rutishauser and Defour, as cited in Rutishauser and Majno, 1950). If this concept is correct, then one may ask whether giving drugs like bisphosphonates, which conserve bone mass by reducing postmenopausal bone remodeling rates, will in the long run reduce or increase fracture rates. Can a remodeling rate be achieved that will conserve bone mass and repair bone microdamage too?

5.9 Modeling Stress Fractures

Stress fractures are the clinical manifestation of the accumulation of fatigue damage in bone. They are an important medical problem among military recruits; ballet dancers, race horses, and other high performance athletes; and in recreational athletes. Their diagnosis and treatment is complicated by their capricious nature and the difficulty in predicting when bone pain is indicative of an incipient fracture. We have argued for the importance of remodeling in repairing fatigue damage in bone to prevent stress fracture. We now attempt to learn by analysis more about how this repair system may work.

Nash (1966) made an early attempt to mathematically analyze the accumulation of fatigue damage in a self-repairing structure. Prendergast and Taylor (1994) used damage as a stimulus for adaptive remodeling on bone's external surfaces, but did not consider the problem of damage repair itself. Later, Martin (1995) studied the consequences of assuming that damage itself activates remodeling, and found that there could be negative as well as positive repercussions. When a bone experiences a "fatigue challenge," its mechanical properties may be changed not only by damage, but also by increased remodeling space (resorption cavities and refilling osteons) produced by the induced remodeling. It has been postulated that this may actually accelerate progression to a stress fracture (Scully and Besterman, 1982). That is, increased remodeling introduces more porosity, which decreases the elastic modulus, elevates the strains produced by the applied loads, and further increases damage formation. Thus, internal remodeling is required to repair fatigue damage, but in the process it may make the bone more susceptible to damage. Modeling responses on the periosteal surfaces of a fatigued bone can offset the negative effects of internal remodeling. Sufficient fatigue damage can stimulate the proliferation of woven bone from the periosteal surface (Johnson et al., 1963; Uhthoff and Jaworski, 1985). Presumably, this external modeling response reduces strains in the weakened structure by increasing its cross-sectional area. It is of great interest to know what determines whether the remodeling and modeling responses to fatigue ultimately have a positive or negative effect.

Let us use some of the methods discussed in Chapter 3 and the approach of Martin (1995) to analyze this problem. Consider the fatigue damage and remodeling in a representative cross section of a long bone diaphysis uniformly loaded in simple axial compression. A sequence of calculations simulates daily mechanical and remodeling changes in the bone. For convenience, fatigue damage is defined as millimeters of observable crack length per square millimeter of bone cross section. The damage formation rate is assumed to be a function of the loading rate (R_L in cycles/day) and the resulting strain range (S). Based on Eqs. 5.16 and 5.23, the damage formation rate may be assumed to be proportional to the strain range raised to a power, q:

$$dD_F/dt = k_D \, S^q \, R_L \tag{5.42}$$

where k_D is a damage rate coefficient that depends on loading conditions and bone structure.

The BMU activation frequency within the section is f_a. As each BMU passes through the section, its resorption cavity expands to the diameter of an osteonal cement line, removing a portion of the cortical area (which is subsequently refilled) and any damage it contains. Assuming that fatigue damage can only be removed by this process, and that the rate of removal is proportional to the amount of existing damage (because the more there is, the greater the probability that a BMU will remove it), the rate of damage removal (or repair) is

$$dD_R/dt = D f_a \, \pi \, R_c^2 \, F_S \tag{5.43}$$

where D is the existing damage and R_c is the osteonal cement line radius. (Note that $f_a \, \pi \, R_c^2$ is the amount of cross section removed by all the BMUs created in 1 day.) F_S is a *damage repair specificity factor*, which accounts for any mechanisms that serve to direct remodeling to fatigue-damaged sites, as opposed to strictly random remodeling. The net rate of accretion of fatigue damage is $dD/dt = dD_F/dt - dD_R/dt$. In the equilibrium state, $dD_R dt = dD_F/dt$ and the corresponding burden of damage waiting to be repaired is

$$D_0 = (k_D \, S^q \, R_L)/(\pi \, R_c^2 \, f_{a0} \, F_S) \tag{5.44}$$

where the "0" subscripts indicate equilibrium values.

The "dose–response" relationship between damage and activation frequency is unknown, but many such curves in biology are sigmoidal, so that is a reasonable assumption. Mathematically, this may be expressed as

$$f_a = \frac{f_{a0} \, f_{amax}}{f_{a0} + (f_{amax} - f_{a0}) \, \exp[k_R f_{amax}(D-D_0)/D_0]} \tag{5.45}$$

where f_{a0} and f_{amax} are the minimum and maximum allowable activation frequencies and k_R is a coefficient that determines the steepness of the response curve. The next part of the analysis is a bit involved, but you have already explored similar problems in the Chapter 3 exercises. How does the bone's porosity change as a result of the changes in f_a? This depends on the current numbers of resorbing (N_R) and refilling (N_F) BMUs in the section. In Exercises 3.2 through 3.4 you saw that the activation frequency could be integrated over past time intervals to obtain N_R and N_F. Then, in Exercises 3.5 and 3.6 you saw that the net change in bone volume fraction could be obtained by calculating

$$Q = Q_B N_F - Q_C N_R \tag{5.46}$$

where Q_B and Q_C are the rates (mm^2/day) at which bone is added and removed by refilling and resorbing BMUs, respectively. Because the time rate of change of porosity is just $-Q$, porosity can be tracked by constantly recalculating N_R, N_F, and Q as f_a changes.

The elastic modulus, E, is computed from porosity, P, using a relationship similar to those seen in Chapter 4:

$$E = E_o(1 - P)^3 \tag{5.47}$$

where E_o is the modulus of the bone tissue. It is recognized that E may also be diminished by fatigue damage, but this effect is ignored for the moment.

Damage is also assumed to stimulate periosteal bone formation, increasing the area of the cross section. No dose–response data are available for this effect, but Turner et al. (1993a) found that when rat tibias were cyclically loaded, surface bone formation was lamellar at lower loads and converted to woven bone at higher loads. Lamellar bone apposition rates were linearly proportional to load or strain magnitude, but woven bone production was an "all or nothing" response, that is, the rate of bone formation was not correlated with load. Mathematically, this may be represented by

$$M_P = k_P (D - D_0)/D_0 \quad \text{for} \quad D < D_C \tag{5.48}$$

$$M_P = M_W \quad \text{for} \quad D > D_C \tag{5.49}$$

where k_P is a coefficient and M_W is the apposition rate for woven bone formation. If M_P^* is the maximum apposition rate for lamellar bone formation, achieved when the damage reaches a critical value, D_C, then

$$k_P = M_P^*/[(D_C/D_0) - 1] \tag{5.50}$$

The daily change in periosteal radius is calculated as $\Delta r_P = M_P \Delta t$, and cortical area is calculated assuming the endosteal radius is constant. Strain is computed by dividing the applied load by the model's cross-sectional area and elastic modulus.

All these calculations may be incorporated in a computer program and the behavior of the system under various conditions explored. The model's parameters can be partially calibrated against the results of the Burr experiments (Burr et al., 1985; Mori and Burr, 1993), in which remodeling was activated by fatiguing canine bones in vivo. The model can then be used to examine the general behavior of the system and its sensitivity to its various parameters. Suppose that R_{LE} cycles per day (CPD) of additional loading, initially producing 2500 $\mu\epsilon$, is superimposed on the equilibrium loading. When the periosteal response is disabled and R_{LE} is 90 CPD, the additional damage increases the remodeling rate and porosity, and the bone becomes less stiff, but a new equilibrium is reached. However, when R_{LE} surpasses a critical value, the remodeling never catches up with the increased damage formation rate. Porosity, strain, and damage increase at an ever increasing rate, without limit. It is reasonable to consider this instability to be fatigue failure in the model (i.e., a complete "stress fracture"). If the loading is only slightly supercritical, this instability can be quite insidious. Damage levels off at what seems to be a new equilibrium level. Then, suddenly, after what may be a long time, the damage "explodes," rising rapidly at a very high rate.

Thus, without a periosteal response, the model can withstand only relatively small fatigue challenges because of the increased strain caused by the remodeling space. When a periosteal callus is included in the model, the results are quite different (Fig. 5.21). Now the porosity increases from remodeling are offset by periosteal diameter changes, and the porosity comes back down as the bone's cross-sectional area increases. The addition of external bone acts to reduce strain, and the system returns toward a new equilibrium level after a transient response lasting about a year Increasing the loading causes larger transients, which end more quickly. Eventually, however, another critical load level is reached, and instability implying failure ensues once more. Again, the inherent instability of the system is remarkable. As loading increases, remodeling responds ever more vigorously, and the challenge is overcome more and more quickly. Then, when just one

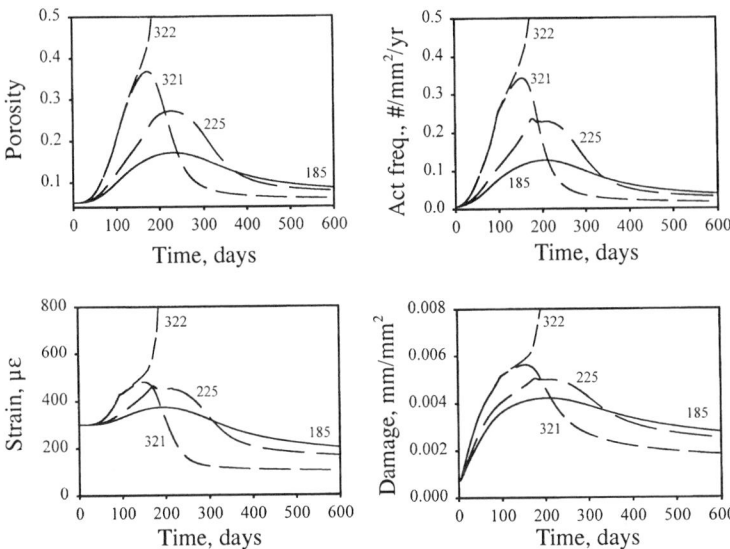

FIGURE 5.21. Stress fracture responses for model with periosteal modeling are shown for loading at four different loading levels: R_{LE} = 185, 225, 321, and 322 cycles per day (CPD). The first of these rapidly produced a *stress fracture* in the model lacking a periosteal response. Now the model survives this loading easily, showing only transient responses in damage, activation frequency, porosity, and strain. As the loading increases, the responses increase in magnitude but recovery is quicker, until suddenly the system "blows up." (Reproduced with permission from Martin, 1995.)

more cycle per day of loading is added, instead of damage simply not decreasing, the system precipitously fails.

Such instability is typical of feedback control systems that are nonlinear or contain delays in their responses. Bone remodeling in response to fatigue would have both these characteristics. It is clear that damage is a very nonlinear function of strain, and modulus is a nonlinear function of porosity. The activation of remodeling may be a fairly linear function of damage, but there are delays in the development of porosity changes after f_a changes. These traits are fully consistent with the response of the model, particularly the fact that small changes in the loading conditions or the system parameters can produce large changes in the outcome if the system is near a critical state. This kind of behavior in the model, and the difficulty in predicting when someone will suffer a stress fracture and when they will not, are in concert with one another. However, much remains to be learned about the intriguing interactions between remodeling and mechanical behavior in the skeleton. We explore these interactions further in the next chapter.

5.11 Summary and Additional Reading

Bone is a natural composite material that shares with engineered composite materials many mechanisms for resisting crack propagation. Fracture mechanics teaches us that cracks are propagated through materials when the stress concentration at the crack tip becomes high enough to exceed a critical stress intensity factor, K_c. Because of the lamellar structure of bone, K_c is lower for longitudinal than transverse crack propagation, so cracks find it difficult to travel catastrophically across the diaphysis of a long bone. Cracks are driven by strain energy, and the cement lines and lamellar interfaces of bone serve to consume energy in the propagation of relatively harmless longitudinal cracks, preventing the propagation of dangerous transverse cracks. Both the numbers and the sizes of osteons appear to be important in maximizing fatigue resistance and the amount of energy required to fracture a bone.

Creep and fatigue damage of bone are closely related phenomena, and both kinds of loading are common in daily life. Experimentally, it is difficult to study fatigue damage mechanisms in bone because experiments at physiologic strain levels require very prolonged loading times. It appears that most of the fatigue failure data on bone involve primarily creep damage caused by superphysiologic stresses. Consequently, we still do not know very much about the differences between creep and fatigue damage in bone. At relatively high stress levels, crack damage appears to be different for cyclic tensile and compressive loading. During most of the tensile fatigue life cracks seem to be self-limiting, occurring primarily in interstitial regions between osteons. In compression, cracks seem to form later, running between Haversian canals and crossing cement lines. The differences in the shapes of the modulus degradation curves seen in tensile and compressive fatigue seem to be attributable to these differences in crack propagation characteristics.

Cyclic loading at physiologic stress levels produces microcracks in osteonal bone and activates increased intracortical remodeling; the mechanism behind this is unclear. If one assumes that bone remodeling is activated in proportion to fatigue damage, and that the increased porosity produced by this remodeling reduces the elastic modulus of the bone, increasing strains and therefore damage, it becomes clear that limits exist on the ability of remodeling to control fatigue damage under increased loading. Mathematically modeling this system suggests that periosteal callus formation is essential to maintaining its stability. Even so, the nonlinearity of the system seems to make it critically stable, so that small differences in skeletal physiology may greatly alter the ability of individuals to avoid a stress fracture under conditions of increased physical activity.

Additional reading suggestions include the early clinical perspective of Devas (1975) on stress fractures and the more recent review by Jones et al. (1989). A survey of the work of D.R. Carter between 1975 and 1990 will produce much of the literature on the mechanics of fatigue and creep in bone. Recent reviews by Burr (1997) and Burr et al. (1997) address the relationships between fatigue damage, bone remodeling, and the fragility of the aging skeleton.

5.12 Exercises

5.1. Equations 5.6 and 5.7 may be used together to calculate the stress intensity factor for a transverse crack in the edge of a plate loaded in tension. Plot a graph showing how both C and K/s change as the crack length increases from 1% to 90% of the plate width $b = 10$ cm.

5.2. Cut a strip of paper 10 cm wide (x-dimension) and approximately 28 cm (11 in.) long (y-dimension). Mark the middle of its length. About 4 cm above the middle, cut a center slit (crack) 5 cm long parallel to the x-axis. About 4 cm below the middle, cut two edge slits, one from each side, each 1 cm long, again parallel to the x-axis. Do a calculation to determine where the strip of paper will fail if it is loaded in tension in the y-direction (see Table 5.2). Then wrap the ends of the strip around pencils and do the experiment to check your results.

5.3. Equations 5.18 and 5.19 give the number of cycles to failure for fatigue of human cortical bone in longitudinal tension and compression, respectively. How many days would it take to achieve fatigue failure in each of these modes of loading if the normalized applied stress range, S, is selected to produce 2000 $\mu\varepsilon$ in the specimen? Assume the loading rate is 2 Hz. What experimental problems would arise from these long fatigue times? How long would the experiment take if S produced a superphysiologic 5000 $\mu\varepsilon$? Why would you not want to run the experiments at 20 Hz to reduce their duration?

5.4. Equation 5.44 states a relationship between remodeling variables and the damage burden (D_0) waiting to be repaired in an equilibrium state: $D_0 = (k_D \, S^q \, R_L) / (\pi \, R_c^2 \, f_{a0} \, F_S)$. If the loading increases, an increase in the damage burden can be avoided by increasing either the activation frequency or the diameter of the osteons, or both. Does one of these options have any advantages over the other? Think about the differences in the porosity increase that would occur during and after the remodeling, the metabolic energy required to support the cells doing the remodeling, and the effects of the altered osteonal structure on fatigue damage resistance. What are the limitations on increasing R_c? How about f_{a0}?

5.5. Muscle fatigue may be an important cause of stress fractures (Devas, 1975; Yoshikawa et al., 1994) We saw in Chapter 1 that there are more muscles to control our movements than we actually need. The muscle fatigue theory of stress fracture holds that the additional muscles not only make refinements of movement possible, but also reduce the stresses in the bones when the most common movements are performed. For example, there are six principal muscles that control the foot during the toe-off phase of walking. Sharkey et al. (1995) found that the compressive strain on the dorsal surface of the human second metatarsal bone at toe-off increases from about 1700 $\mu\varepsilon$ to about 2190 $\mu\varepsilon$ when just one of these (flexor hallucis longus) stops functioning. Using Eq. 5.17, find the percent decrease in fatigue life to be expected from this increase in strain.

5.6. The Pacific Crest Trail runs from Mexico to Canada through the Sierra Nevada and Cascade mountain ranges. Each year a few people succeed in walking the entire trail, starting in early spring and finishing in the fall. Many fail because of physical problems, including stress fractures. Using Eq. 5.42, consider the relative effects of the distance walked and the weight of a backpack on the rate of fatigue damage formation in one of your foot or leg bones if you undertook this walk. Suppose that you ordinarily walk 5 miles/day. Plot a graph of the percent increase in damage formation rate as a function of the weight of your pack, expressed as a percent of your body weight, assuming you continued walking just 5 miles/day. Try relatively high and low values of q. Then convert this graph to a three-dimensional plot of percent change in damage formation rate vs. pack weight and distance walked per day.

5.7. Show that Eq. 5.41 is the solution to Eq. 5.40.

5.8. Show that Eqs. 5.38 and 5.39 are the solutions to Eqs. 5.36 and 5.37.

5.9. Write a computer program that simulates the generation and repair of fatigue damage in a segment of cortical bone using Eqs. 5.42–5.47. See if you can reproduce the responses described for the model **without** a periosteal response.

6

Mechanical Adaptability of the Skeleton

> Every change in the form and function of ... bone[s] or of their function alone is followed by certain definite changes in their internal architecture, and equally definite secondary alterations in their external conformation, in accordance with mathematical laws.
>
> J. Wolff as quoted by Keith (1918).

6.1 Introduction

In Chapter 4 we learned that the mechanical properties of bone are strongly affected by porosity, mineralization, collagen fiber orientation, and other aspects of histologic structure. We also saw evidence that the variations in structure from region to region within a bone are correlated with mechanical factors—for example, greater mineral content and more transversely oriented collagen fibers on the more compressed side of the human femur. Then, in Chapter 5, we learned that the histologic structure of bone is also important in providing bone's exceptional capacity for resisting catastrophic crack propagation and fatigue failure. Furthermore, we saw that in addition to providing the structures (e.g., cement lines) that control crack formation, remodeling also rids the bone of microdamage by replacing old bone with new. Finally, there is strong evidence that microdamage itself initiates remodeling. In reading about these features of bone, it is difficult to avoid the impression that local mechanical factors are influencing the bone cells as each portion of the skeleton is formed and remodeled. If this thought captured your imagination, you are not alone. Many others have had similar notions. Indeed, this idea that bone structure is somehow controlled locally to suit its current mechanical function has become a central tenet of orthopaedic medicine and science. The goal of this chapter is to come to grips with this idea, which we briefly introduced in Chapter 2 as *Wolff's law*. As a scientific hypothesis, the putative ability of bone not only to repair itself, but to adapt to variations in imposed stresses, is an extremely appealing idea to engineer and biologist alike. However, developing this idea into a scientific theory supported by appropriate experimental results has proven to be a hefty scientific problem. In the following pages we take up this problem, trying to organize into a coherent theory the principal hypotheses and experiments found in the literature on this subject. First, however, it is helpful to review the origins of Wolff's law and the concepts that it represents.

6.2 The Historical Context

What is today called Wolff's law incorporates several different concepts, most of which Wolff himself said nothing about. Following the lead of Roesler (1981, 1987), we focus on three key concepts that arose in the nineteenth century:

1. Optimization of strength with respect to weight
2. Alignment of trabeculae with principal stress directions
3. Self-regulation of bone structure by cells responding to a mechanical stimulus

The idea that bone senses and adapts to its mechanical environment is an old one. Julius Wolff is routinely given credit for this idea, but he was certainly not the first to make the observation. Galileo (1638) pointed out the mechanical implications of the shapes of bones, as did other early writers (e.g., Monro, 1776). Bell (1827) observed that cancellous bone structure had "reference to the forces acting on the bone." Both he and Bourgery (1832) apparently recognized that trabecular bone architecture is influenced by mechanical forces, and thought that it maximized strength relative to the amount of material used. This concept is the first important concept that we want to recognize: optimization of bone strength (or stiffness) with respect to its weight. This is vitally important because the density of bone tissue (about 2 g/cm^3) is twice that of other tissues in the body. The metabolic cost of carrying about extra bone is very high. It is perhaps unlikely that Bell and Bourgery ever considered this; they may simply have been struck by the obviously exquisite structure of cancellous bone, which looks so efficient!

Ward (1838) compared the trabecular arrangement in the neck of the femur to a bracket designed to hold a street lamp to its post (Fig. 6.1). We now apply the name *Ward's triangle* to the region of sparse trabecular struts corresponding to the central opening in this "bracket." Ward's comparison is significant because it is an early recognition that bone structure is analogous to structures engineered by people and that bone does not remain where it does not need to be.

We come now to the second key concept, which arose from a legendary conversation between Karl Culmann and Hermann von Meyer. The story goes that one day in 1866 Prof. Culmann, a well-known engineer, visited the laboratory of Prof. Meyer, an equally well-known anatomist, and happened to see a specimen of a proximal human femur that had been sectioned longitudinally to reveal its pattern of trabecular arcades. Culmann was immediately struck by the similarity of this pattern to that of the principal stresses

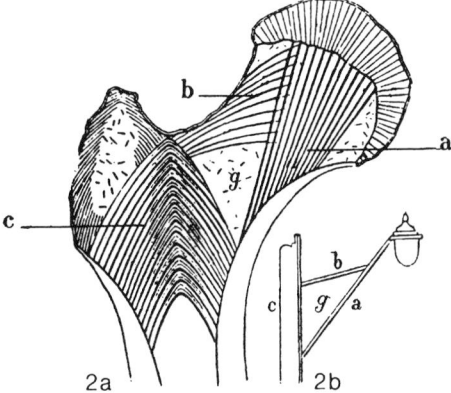

FIGURE 6.1. Sketch showing Ward's analogy between the trabecular arrangement in the neck of the femur and a streetlamp bracket. (From The laws of bone architecture, Koch, JC, *American Journal of Anatomy,* 1917. Reprinted by permission from Wiley-Liss, Inc., a subsidiary of John Wiley & Sons, Inc.)

within a crane structure which he had analyzed. (Fig. 6.2 shows sketches of these two objects.) He is said to have cried out "That's my crane!" (Thompson, 1942, p. 977).[1] The crane was actually named after Fairbairn, the Scots engineer who designed it. Its flowing, bonelike shape is typical of the trial-and-error designs common during the early industrial revolution in Great Britain. It is interesting to note that elasticity theory and the science of strength of materials were rapidly developing during this period of time. Augustin Cauchy had, about 1822, discovered the existence of principal stresses, but decades passed before engineers began to make practical use of the concept. Culmann was one of the first to understand and use the concept of principal stress directions; he was one of several engineers from the continent who led the way in developing a mathematical theory of mechanics, which eventually refined the experimental approaches of British and American engineers. He had apparently been applying his graphical methods for calculating principal stress directions to Fairbairn's crane. In any event, the similarity between the trabecular arcades and the stress trajectories is indeed striking. Clearly, however, the shapes and loads are substantially different for the crane and the femur. Not only that, but the femur is a discontinuous collection of bony struts filled with marrow, while the crane is a homogeneous structure. Principal stresses do not exist for a discontinuous structure. The significance of this was overlooked for a long time (Cowin, 1997). We shall return to this issue several times in this chapter.

This conversation was noted by others interested in bone, including Wolff, and soon afterward he began to publish papers on this subject

---

[1]Some sources have placed this encounter at a scientific meeting, and others have said the femur was actually a metatarsal.

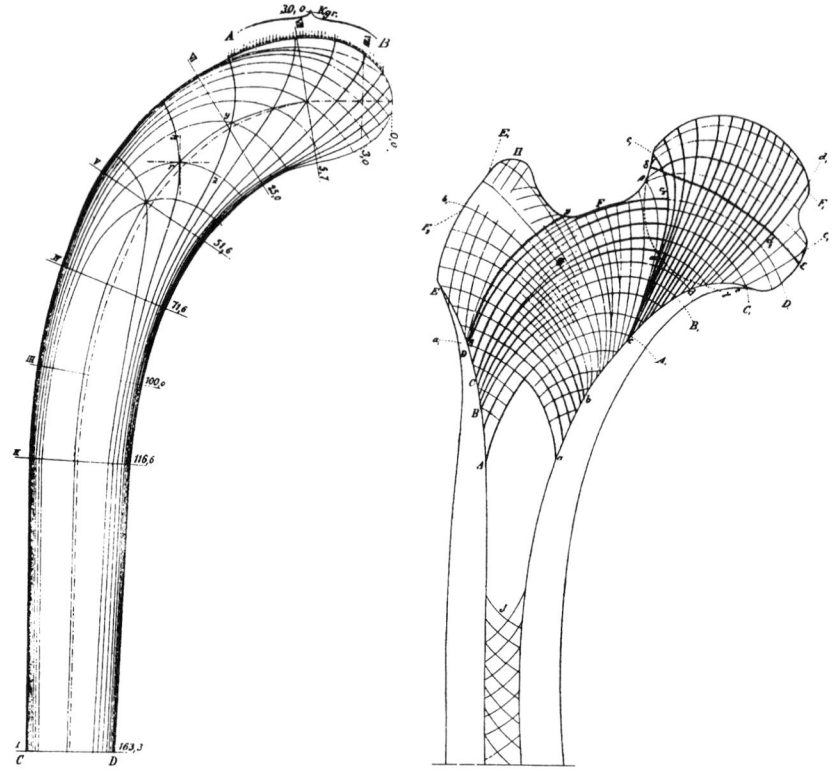

FIGURE 6.2. *Left:* Principal stress trajectories in a Fairbairn crane as calculated by Culmann. *Right:* The trabecular arrangement in a proximal human femur as sketched by von Meyer. (Reproduced with permission from Maquet and Furlong's translation of Wolff's 1892 book.)

(Wolff, 1869, 1870). His interpretations of cancellous architecture quickly rose to dominance. He accepted at face value the implications of the Culmann–Meyer stress-trajectory hypothesis and asserted that there was a mathematically perfect correspondence between trabecular architecture and principal stress trajectories. This similitude is what Wolff's law meant to Wolff, and it is our second key concept.

Wolff went on to try to explain the trabecular structure within many pathologic specimens using this basic idea. In 1892, he published his famous book, *Das Gesetz der Transformation der Knochen* (*The Law of Bone Transformation*), which summarized his earlier writings.[2] It must be

---

[2]This book has been translated into English by Maquet and Furlong. They translated the German word "transformation" as "remodeling," although Wolff was speaking about growth and what we now term modeling.

appreciated, however, that while he spoke of mathematical rules governing trabecular architecture, he was not an engineer or mathematician, and never attempted to formulate a mathematical theory of "Wolff's law." Furthermore, he believed that bone was formed interstitially (i.e., cells pushed out new bone from within the calcified matrix—a physical impossibility), and had no understanding of bone modeling and remodeling as we know them.

Later, Koch (1917), an engineer as well as a physician, resolved some of the issues regarding the comparison of Culmann's stress trajectories for the Fairbairn crane with a human femur. He calculated the stress trajectories in a human femur based on a reasonably accurate load applied to the head and geometry obtained from actual anatomic measurements. In doing so, he was able to demonstrate that trabecular orientations are, after all, similar to mathematical calculations of stress trajectories for a homogeneous structure of the same shape (Fig. 6.3).

We must return now to the nineteenth century, to note the emergence of the third key concept regarding the adaptation of bone to mechanical loading: the idea that this occurs via a self-regulating biologic mechanism. While it may seem obvious to you that this must be the case, that is only because you have been educated in an era when self-regulation is accepted as an essential part of biological organisms. That was not the case in the time of Wolff. It was in his century that Claude Bernard (1813–1878) introduced the idea of homeostasis to physiology: the concept that physiologic systems self-regulate so as to keep a critical internal variable constant. However, it was many years before the full significance of this concept came to be appreciated and accepted. Thus, it was still a fundamentally new idea when, in 1881, W. Roux, influenced by Darwin and others, hypothesized that organisms possess the ability to adapt to changes in their

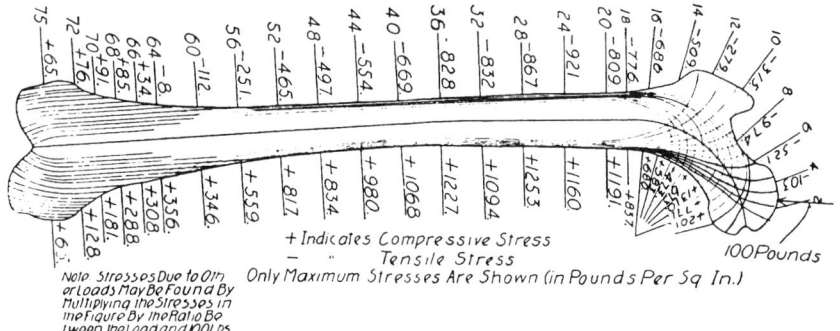

FIGURE 6.3. Koch's computation of the principal stress trajectories and values in a human femur. Note the applied load vector on the femoral head. Koch, 1917.

living conditions. (Note that we are not speaking of evolution here, which involves long-term, genetic changes between one generation and the next, but changes in an individual during its lifetime, even day-to-day or minute-to-minute changes. Also, we are not speaking of adjusting to internal, physiologic changes, but to changes involving interactions with the external environment.) Roux specifically proposed that the ability of bone to align trabeculae with stress trajectories was accomplished by cells forming and resorbing bone according to variations in a *functional stimulus*, in this case, a mechanical stimulus. While his imperfect understanding of biology and mechanics led him to make many mistakes in developing his theory, his concept of self-regulated functional adaptation governed by cell sensitivity to a functional stimulus was extremely important.

In summary then, as Roesler (1987) has pointed out, nineteenth-century researchers provided us with three key concepts regarding bone's ability to adapt to changing mechanical loads: bone structure optimizes strength with respect to the amount of material used; trabeculae line up with principal stress directions; and these things are accomplished by a self-regulating system of bone cells responding to a mechanical stimulus. Today, Wolff's law remains a rather poorly defined "law," but it more or less includes these three core principles.

In the interest of escaping the vagueness and confusion associated with the term "Wolff's law," we propose the following hypothesis about bone's ability to adapt to mechanical loading.

The Mechanical Adaptability Hypothesis

This hypothesis states that *bone structure is regulated so as to minimize fracture risk and bone mass while simultaneously optimizing stiffness.* We call this a hypothesis because we are not sure it is true or how it works. Clearly, minimization of fracture risk is important in most bones, and minimizing bone mass allows the organism to reduce the metabolic energy required for its daily activities, increasing its ability to survive. Following the lead of Currey (1984), we speak of optimizing stiffness because bones need to support loads without bending too much, but they also need some flexibility for shock absorption and to avoid fracture. The optimal stiffness surely varies with the function of the particular bone. For example, Currey (1981) has discussed the divergent functions of three bones. The antlers[3] of the male red deer require impact strength and energy-absorbing power; stiffness is deemphasized. The femoral cortex of the cow requires more stiffness. The tympanic bulla in the ear of the fin whale requires a high acoustic impedance ($= \sqrt{\rho E}$) and therefore very high stiffness. Table 6.1

---

[3]Deer antlers are made of bone. Cow horns, on the other hand, are keratinous tissue, as are fingernails.

TABLE 6.1. Material properties of three different bones

Variable	Red deer antler	Bovine femur	Whale bulla
Work-of-fracture, J/m^2	6190	1710	200
Bending strength, MPa	179	247	33
Elastic modulus, GPa	7.4	13.5	31.3
Density, g/ml	1.86	2.06	2.47
Acoustic impedance, 10^9 kg m^{-2} s^{-1}	3.71	5.27	8.79
Mineral content, % by weight	59.3	66.7	86.4

After Currey, 1981.

shows how the elastic modulus and other attributes vary among the three bones. Although most bones would be more like the cow femur than antler or tympanic bulla in their function, the optimal stiffness obviously varies from bone to bone. It should be understood that this is only a trial hypothesis to serve as a benchmark for exploring this subject. At this point it seems clear that fracture risk and bone weight need to be minimized, but we shall see several alternative possibilities for the optimized variable in the following pages.

A Note on the Use of the Word "Remodeling"

The word "remodeling," in the context of mechanical adaptability, is often used to describe any adaptive change in bone, whether it be remodeling or modeling, as these were defined in Chapter 2. Osteonal "remodeling" always occurs through a process of activation, resorption and formation: this is remodeling as we have previously defined it. "Remodeling" of trabeculae, or of endosteal and periosteal surfaces, may also be A-R-F (activation–resorption–formation) remodeling, or it may involve modeling instead, in which formation and resorption occur independently of one another. Modeling in the formation mode may produce lamellar or woven bone, depending on the circumstances. Generally, woven bone is produced when the addition of more bone is urgent. In this chapter, as elsewhere in this book, the term *remodeling* applies only to the A-R-F sequence. The term *adaptation* will be used in place of the broader definition of "remodeling," that is, to refer to changes that involve modeling or remodeling or both.

Another term that it is useful to mention here is *micromodeling* (Frost, 1986). This term refers to the adjustment of bone matrix microstructure by local factors, including mechanical stress. For example, the regional variations in collagen fiber orientation described in Chapter 4 would be attributed to micromodeling during the formation of the bone matrix. Micromodeling may be accomplished or influenced by cells, or could con-

ceivably occur independently of cells, driven by physicochemical processes. We regard micromodeling as another adaptability mechanism, along
with modeling and remodeling.

Clinical Problems and Mechanical Adaptability

There are two aspects of mechanical adaptability that were not noteworthy
in the historical record but are of particular importance to physicians. The
first is *disuse osteoporosis*: the loss of bone when mechanical loading is
reduced. This type of loss is a systemic skeletal problem when patients are
bedridden or otherwise immobilized; it is also a localized problem when a
bone or limb is immobilized. We shall return to disuse effects later in this
chapter. The other clinical phenomenon in which mechanical adaptability
plays an important role is the self-straightening of a fractured bone when
it has healed in an angulated position (Fig. 6.4). We explore this phenomenon in the next section, before delving into mechanical adaptability in
general, because it serves as a microcosm of this area of research, with clinical observations revealing the phenomenon, prompting a series of
hypotheses regarding the mechanism, and followed by testing of the
hypotheses against further observations.

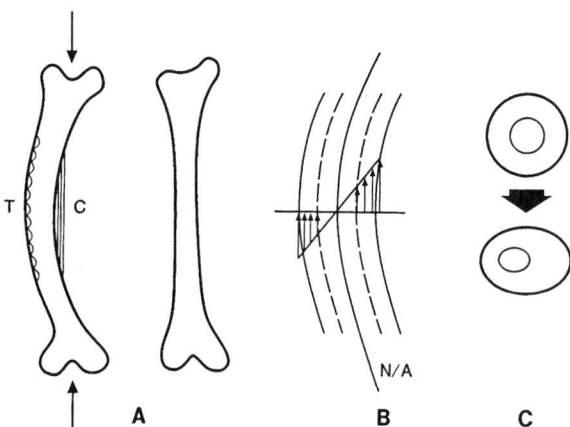

FIGURE 6.4A–C. When a fractured bone heals in an angulated position (**A**, *left*),
bending stresses produced by end loads may provide the stimulus for straightening
(**A**, *right*). However, the assumption that tensile stress (*T*) promotes resorption and
compressive stress (*C*) promotes formation is too simple because the endosteal surfaces have similar stresses (**B**, longitudinal section through the bent bone, *N/A*, neutral axis). This concept results in resorption and formation on the left and right
endosteal surfaces, respectively, producing an asymmetric cross section (**C**).

6.3 Self-Correction of Abnormally Curved Bones

The observation by physicians that fractures which heal in a bent configuration tend to straighten themselves has been used for many years as an example of mechanical adaptability in action. A simple hypothesis to explain this phenomenon was proposed by Jansen (1920) and reiterated by Bassett (1965) and others since. This hypothesis recognizes that if the bent bone is to be straightened, bone must be removed from the convex side and apposition must occur on the concave side; see the left side of Fig. 6.4. When muscle pulls or weight-bearing place the bone in axial compression, there is tensile stress on the convex surface and compressive stress on the concave surface, and its curvature is increased by bending. Then the following associations are implied:

> **tension** . . . . **resorption**
> **compression** . . . . **formation**

Bassett pointed out that a long bone with an angular deformity would be straightened if compressive end loads produced bending stresses that induced these activities on its external surfaces, as shown on the left side of Fig. 6.4. However, Frost (1964a) noted that this explanation did not predict appropriate remodeling responses on the surfaces of the medullary canal. As noted on the right side of Fig. 6.5, there is tension on the endosteal as well as the periosteal side of the bone's left cortex and compression on both surfaces of the opposite cortex. Therefore, the theory predicts that the left cortex will get thinner and the right one will get thicker, rather than both drifting to the right so as to straighten the bone while retaining its symmetric annular cross section.

Frost's Flexural Neutralization Theory

Frost (1964a) proposed an algorithm that would give the desired remodeling result for the case of the bent bone. According to this theory, remodeling is controlled not just by the polarity of the tangential wall stress (i.e., compression or tension) but **by the tendency of the applied end load to alter the relative curvature of the surface.** He suggested that the following associations apply:

> **increased surface convexity** . . . . . . **resorption**
> **decreased surface convexity** . . . . . . **formation**

This scheme allows both the internal and external surfaces of the bone to remodel to reduce the curvature and high stresses associated with excessive bending. We refer to this hypothesis as Frost's *flexural neutralization theory*. Frost proposed that stress levels above a *minimum effective stress* activated the response. (Later, he reformulated his theory in terms of strain rather than stress.)

Frost's theory has been criticized on the grounds that many bones are normally curved, and indeed need to be. Bones become curved during growth under the combined influences of chondral and bony modeling. Limbs that are paralyzed during growth and not subject to normal mechanical forces are not as curved. Rubin (1984) proposed that the observed curvature of various long bones is a mechanism to stimulate continued bone renewal, which might otherwise not occur if strains of sufficient magnitude to cause bone remodeling were "neutralized." Bertram and Biewener (1988) suggested that curved bones make strains more predictable, and that higher stresses in curved bones are a reasonable price to pay to avoid the danger of buckling in straight bones of minimum size. However, Frost proposed that the strain level required to initiate the correction must be above a minimum threshold and so would not prevent all curvature. Instead, curvature would simply be controlled to keep strain levels within some physiologic range.

Another objection to Frost's theory is the difficulty in imagining how bone cells would detect relative changes in concavity or convexity of surfaces much larger than themselves. (This problem is not unlike that of imagining how cells can align trabeculae with principal stress directions!) Is there a related variable that cells might be able to sense?

Stress Gradients and Fluid Flows

Changes in surface curvature induced by bending are a manifestation of the stress gradient perpendicular to the surface. Increased surface convexity is associated with tangential stresses that become more tensile as one moves toward the surface (plus signs in Fig. 6.5). Decreased surface convexity (or increased surface concavity) is associated with stresses which become less tensile nearer the surface (minus signs in Fig. 6.5). An alternative formulation of the flexural neutralization theory is that the polarity of the stress gradient beneath a bone surface affects the cellular activity which occurs there. If one defines a positive surface stress gradient as meaning that the stresses become more tensile nearer the surface, then the following associations may be made:

positive stress gradient . . . . **resorption**
negative stress gradient . . . . **formation**

Now consider the flow of fluids in the bone as it is bent. In general, fluids flow along a pressure (i.e., hydrostatic stress) gradient from more compressed regions toward more tensile regions. Therefore, fluids flow toward surfaces labeled positive and away from surfaces labeled negative in Fig. 6.5. Weinbaum, Cowin, and Zeng (1994) have developed a mathematical theory predicting that physiologic stress gradients should produce fluid flows in the annular spaces between the osteocyte processes and the canalicular walls. As these fluids flow over the mem-

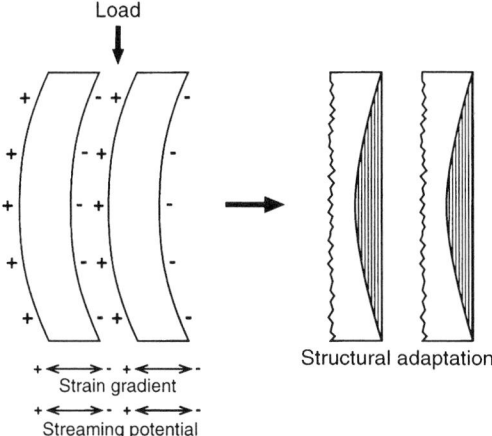

FIGURE 6.5. Bending is associated with stress (and strain) gradients. If the polarity of these gradients is as defined in the text, then they would have + and − signs as shown in the longitudinal section at *left*. The streaming potentials produced by these stress gradients would have similar polarity. If positive and negative signals produce resorption and formation, respectively, the bone would straighten as shown in the *right* diagram.

branes of these processes and interact with the proteoglycan matrix found there, the resulting shear stresses on the cell wall are predicted to be in the range that has been observed to cause the release of intracellular calcium ions in osteoblasts and endothelial cells. It has been hypothesized that this release of intracellular Ca^{2+} leads to the propagation of signals across gap junctions from one cell to another, over considerable distances. However, the mechanisms connecting extracellular shear stresses and communication of a signal from cell to cell remain unclear. In any case, these signals may be communicated through the osteocytic network to bone lining cells on bone surfaces, which may in turn activate modeling or remodeling. It may be that flows toward or away from the surface produce different signals at, and different responses in, the bone lining cells on the surfaces labeled positive and negative in the figure. Work by Owan et al. (1998) supports the theory that stress-generated fluid flow produces the signals sensed by osteocytes and bone lining cells.

Alternatively, the signal perceived by the cells may be electrical (Cowin et al., 1995). Streaming potentials are the source of stress-generated electrical potentials on the surfaces of bones in vivo. Using our definition of stress gradient, a positive electrical potential is associated with a positive stress gradient and vice versa. This has been confirmed by many experiments (e.g., Williams and Breger, 1974). Because other experiments have found

that bone formation is stimulated near a cathode and resorption is stimulated near an anode (Friedenberg et al., 1970), one could hypothesize that
modeling or remodeling are activated by electrical potentials caused by the
stress gradients.

Related Experimental Results

Until recently, no experiments had been performed to test Frost's flexural
neutralization theory. There is now, however, an animal model that has
produced some data suggesting that end loads applied to a curved bone
activate surface responses which reduce the bone's curvature (Torrance et
al., 1994; Hillam and Skerry, 1995). The ulna of the rat is normally curved.
This curvature is developed during growth by modeling drifts (Fig. 6.6).
When additional end loads are experimentally applied to this bone in vivo,
the resorptive portion of these drifts is stopped. The curvature of the bone
stops increasing and instead the bone just gets larger in diameter. These
experiments did not show removal of an established curvature, but they did
show cessation of a developing curvature. Neither did they shed any light
on the mechanisms by which modeling is activated or inhibited on the
bone surfaces, but the rat ulna model appears to be a good one in which
to pursue these mechanisms.

Now that we have examined a small neighborhood of the mechanical
adaptability countryside, we shall expand our horizons. First, we look at
some key experiments to see what more we can learn about how bones
adapt to altered loading conditions. Then, we study several additional theoretical approaches.

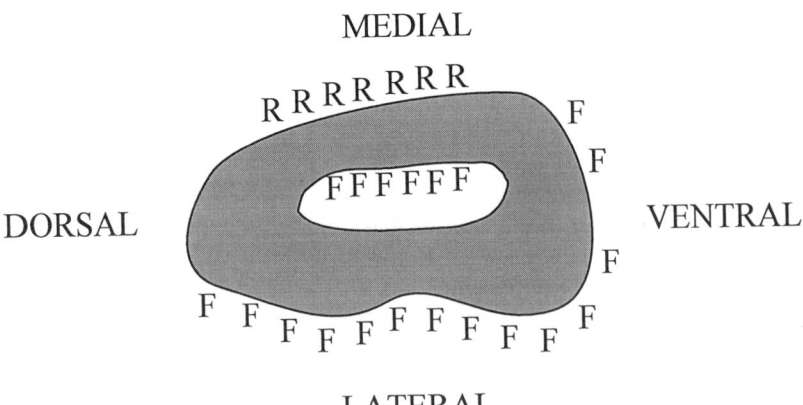

FIGURE 6.6. Normal modeling activity (R, resorption; F, formation) on the midshaft
periosteal and endosteal surfaces of the rat ulna cause it to drift in the mediolateral
direction, making the shaft curved.

6.4 Some Important Experiments

If cells adapt bone to mechanical loads by responding to a mechanical stimulus, it is obviously important to know what that stimulus is and to be able to measure it. We have seen that one suggestion has been fluid flows that are proportional to stress gradients. These would be difficult to measure. Another suggestion has been strain, because if bone cells are attached to bone surfaces they may "feel" the displacements of the surface and respond to them. Strain is also an appealing variable from the point of view of minimizing fracture risk. We saw in Chapter 4 that yield strains are reasonably constant for bone. Therefore, keeping strains near a point well below the yield strain should maintain a margin of safety against monotonic, creep, and fatigue failure. Finally, if stiffness is to be optimized in some sense, the fact that strain is a measure of bone stiffness under a given load (e.g., body weight) may be significant. These are some reasons why strain is of interest as the functional stimulus, and why an important aspect of the mechanical adaptability experimental work has been the measurement of bone strains in vivo.

Measurement of Strain in Living Animals

Until the advent of in vivo strain gauging techniques, the adaptation of bone to mechanical loading could not be effectively analyzed because bone's "mechanical environment" could not be quantitatively assessed at a level commensurate with cellular activity. Lanyon and Smith (1969, 1970) reported the first measurement of in vivo bone strains in animals when they implanted uniaxial strain gauges on the medial tibial cortex in sheep. They found mean longitudinal surface strains of 260 microstrain during walking and a 16% increase when the animals were trotting.

Subsequent to this, Lanyon (1971, 1972) applied strain gauges to the thoracic and lumbar vertebrae in sheep. He reported average peak surface strain ranges in thoracic vertebrae during walking of 142 microstrain ($\mu\varepsilon$), increasing 25% to 187 $\mu\varepsilon$ during trotting. Cochran (1972, 1974) applied strain gauges to the tibia and ulna in dogs, measuring strains in the tibia between 103 and 665 $\mu\varepsilon$, depending on the load and the anatomical position of the strain gage. The mode of the strain, compressive or tensile, was not specified. Shortly after this, Barnes and Pinder (1974) and Turner et al. (1975) added the radius, tibia, metacarpus, and metatarsus of horses to this list. Lanyon et al. (1975) demonstrated that strain magnitudes in the human tibia during locomotion were similar to those in bones of other animals.

Principal Strains and Trabecular Orientations
About this time, Lanyon began to use rosette strain gauges,[4] permitting the measurement of principal strain directions. Applying these gauges to the

---

[4]A group of three gauges oriented at 45° to one another.

calcaneus of a sheep, Lanyon (1973, 1974) demonstrated that the principal orientations of trabeculae coincided with the principal compressive and principal tensile strain directions. Later, Hayes and Snyder (1981) conducted an intensive study of the human patella which concluded that its trabecular orientations are highly correlated with principal stress directions, as determined by finite element analysis. Although this study did not examine in vivo strains on the patella, both studies suggest that Culmann, Meyer, and Wolff were essentially correct about trabeculae aligning themselves with continuum level principal stress (or strain) directions (i.e., what these directions would be in a homogenous structure).

Peak Strains Are Similar Across Species

Subsequently, Rubin and Lanyon (1982) surveyed their own work and that of others and published the data shown in Table 6.2. These data show that when an adult animal's musculoskeletal system is functioning at a high physiologic level (e.g., a dog is running or a bird is flying), peak periosteal strains usually lie between 2000 and 3000 $\mu\varepsilon$. The fact that measured strains were nearly equivalent on different bones and in different animals supported the hypothesis that the skeletons of all animals are adapted to control strain.

There are at least two notable exceptions to the data in Table 6.2. First, strains on the human tibia at a site where stress fractures occur have not been observed to exceed 2000 $\mu\varepsilon$, even during strenuous activity by spe-

TABLE 6.2. Peak functional strains in various animal bones

Bone	Activity	Peak strain	Reference
Horse radius	Trotting	−2800	Rubin and Lanyon, 1982
Horse tibia	Galloping	−3200	Rubin and Lanyon, 1982
Horse metacarpal	Acceleration	−3000	Biewener et al., 1983
Dog radius	Trotting	−2400	Rubin and Lanyon, 1982
Dog tibia	Galloping	−2100	Rubin and Lanyon, 1982
Goose humerus	Flying	−2800	Rubin and Lanyon, 1982
Cockerel ulna	Flapping	−2100	Rubin and Lanyon, 1982
Sheep femur	Trotting	−2200	Rubin and Lanyon, 1982
Sheep humerus	Trotting	−2200	Rubin and Lanyon, 1982
Sheep radius	Galloping	−2300	O'Conner et al., 1982
Sheep tibia	Trotting	−2100	Lanyon et al., 1982
Pig radius	Trotting	−2400	Goodship et al., 1979
Fish hypural	Swimming	−3000	Rubin and Lanyon 1982
Macaca mandible	Biting	−2200	Hylander, 1979
Turkey tibia	Running	−2350	Rubin and Lanyon, 1984

After Rubin and Lanyon, 1982.

cial forces soldiers (Burr et al., 1996). More recent experiments, however, show that strains in excess of 3000 με can be generated during jumping activities, which may explain why "shin splints" are a relatively common injury among basektball players. Second, thoroughbred racehorses generate 5000–6000 με on the surface of their metacarpal bone when running at racing speeds. However, these horses are special animals, bred and trained to run exceptionally fast while they are still juveniles. Strains of this magnitude probably only occur in young animals (or people) in which bone turnover is rapid and many osteons are still mineralizing.

Compressive Strains Are Larger Than Tensile Strains
Because compressive strains in a long bone are usually superimposed on bending strains, the neutral axis is shifted and the peak compressive strains are usually greater than the peak tensile strains (Fig. 6.7).[5] Lanyon et al. (1982) demonstrated that the ratio of (the absolute values of) peak compressive strain to peak tensile strain on the sheep radius varies between 1.45 and 2.34, and tensile strain was never greater than compressive strain during normal locomotion. In fact, tensile strain was always less than, and compressive strain was always greater than, 1000 με. Even after osteotomy of the adjacent ulna, the tensile strain on the radius was less than 1000 με. Because of such tensile-compressive strain differences, Frost (1986) has suggested that strain thresholds for activation of bone adaptation are different on tension and compression surfaces.

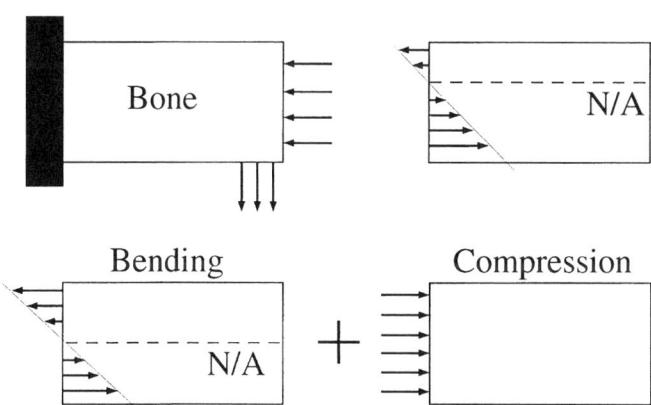

FIGURE 6.7. Most long bones are loaded in combined bending and compression, producing an asymmetric stress distribution in which compressive stresses are greater than tensile stresses.

---

[5]This is not universally true, however. An example would be in the palatomaxillary segment of some lizards, where tensile strains exceed compressive strains (Smith and Hylander, 1985).

Osteotomy Experiments: Surgically Overloaded Bones

Lanyon and his colleagues adapted the earlier osteotomy[6] experiments of Sedillot (1865, 1869) and Wermel (1934, 1935a,b,c,d) to study the way bone adapts to increased load. Goodship et al. (1979) removed a segment of the ulna in **growing** pigs, increasing the load on the radius. The left side of Fig. 6.8 shows a cross section through the radius (A) and ulna (below A, filled with cancellous bone) at the start of the experiment. After the osteotomy, peak strains on the overloaded radius increased from 800 to 1100 µε. (Strains in the radius on the unoperated side simultaneously decreased from 800 to 500 µε, indicating a change in gait had occurred.) Within 2–3 weeks, the increased strains were associated with rapid deposition of **woven** bone on the entire periosteal surface of the intact radius opposite the missing ulna (B). Later, woven bone was not apparent on these surfaces, but such bone was found on the surface nearest the missing ulna (C). After 3 months these responses caused the radius cross-sectional area to be equal to the combined areas of the radius and ulna in the intact limb, and the principal compressive strains in the radius were near normal values. Also, the peak strains on the radius of the intact limb had returned to normal by this time.

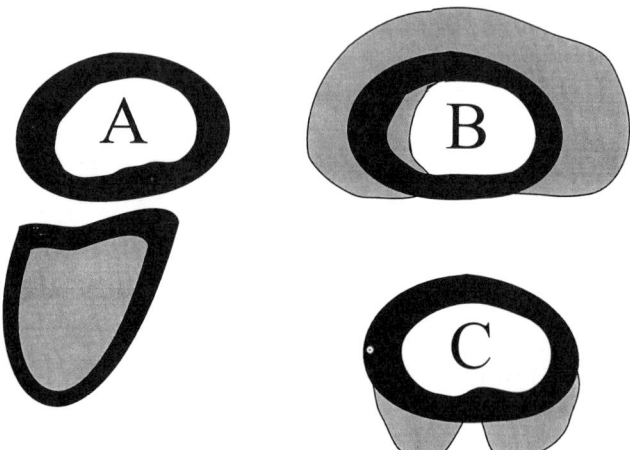

FIGURE 6.8. Results of osteotomy on young pigs. **A** Normal midshaft cross-sectional geometry of the radius and ulna (below) prior to surgical removal of the central ulnar diaphysis. Cortical bone is black; cancellous bone is gray. **B** Appearance of the radius a few weeks after surgery; here, gray material represents woven bone. **C** Three months after ulnar osteotomy the woven bone shown in B has disappeared and new woven bone has appeared facing the missing ulna. (Redrawn with permission from Goodship et al., 1979.)

---

[6]An osteotomy is simply a surgical procedure in which a portion of a bone is cut and removed.

To eliminate the effects of growth and modeling in this experiment, Lanyon et al. (1982) next studied the effects of ulnar osteotomy in **skeletally mature** sheep. Some animals were examined after 6 months and others after 1 year. Lanyon et al. observed that although overstrain of the tensile cortex of the radius was greater than that on the compressive surface immediately after ulnar osteotomy (20% vs. 8%), most new bone formation occurred on the portion of the periosteal surface that was compressed and adjacent to the missing ulna. (Of course, the early flare-up of woven bone on the other surfaces seen in the pig would have been missed in this experiment, if it occurred.) This new woven bone was intensively remodeled with secondary osteons but that in the tensile cortex was not. This adaptation on the radius was consistent with an attempt to replace the missing ulna by increasing the cortical area of the radius and moving it toward the previous location of the ulna. However, it was not consistent with the idea that strain magnitude controlled the adaptive response, because the greatest reaction was found on the surface where the increase in strain was the smallest.

The result was that the strain on the compressed surface subsequent to adaptation was lower than the strain before osteotomy. The observation that bone adapts to produce strains which are lower than normal seemed to be inconsistent with the view that bone reacts to control strain magnitude. Because these results did not seem to fit the notion that absolute strain magnitude somehow controlled the adaptive process, Lanyon and his co-workers suggested that alterations in "strain distribution" (Lanyon et al., 1982), or site-specific strain thresholds (Goodship et al., 1979), may have some role in the control of bone adaptation.

Canine Disuse Experiments

Uhthoff and Jaworski (1978; Jaworski and Uhthoff, 1986) studied the effects of disuse in growing dogs by casting one of their forelimbs for 40 weeks. The metacarpal diaphyseal bone loss in growing dogs was characterized primarily by reduced formation and increased resorption on the periosteal surface, indicating that the normal periosteal bone formation accompanying growth was inhibited or converted to resorption. Relative to the contralateral control bone, bone mass decreased rapidly at first, then more slowly. After 40 weeks of immobilization, it was asymptotically approaching about half the bone mass of the contralateral control bone. When the cast was removed after 32 weeks of immobilization, about 65%–70% of the deficit in bone mass was recovered in 28 weeks. In the diaphysis, this occurred both through endosteal apposition and through additions of woven bone to the periosteal surface. When the experiment was repeated in older dogs, bone loss occurred primarily at the endosteal surface and through increased intracortical porosity (Jaworski et al., 1980). There was no effect on periosteal diameter. Older dogs were less able to recover lost bone following remobilization, recovering only 40% of the lost diaphyseal bone. In older dogs, remobilization stopped the endosteal bone loss and caused apposition of small amounts of lamellar bone to the periosteal surface.

The Porcine Exercise Experiment

Woo and colleagues (1981b) studied the effects of running exercise on the long bones of growing pigs. They found that after 1 year there was no effect on the intracortical bone elastic modulus or failure stress. Exercise produced no woven bone formation on the periosteal surface and no change in periosteal diameter, but endosteal expansion was inhibited; this resulted in increased cross-sectional moment of inertia and whole bone strength.

Avian Isolated Ulna Experiments

To more closely control the mechanical variables in question and thereby perhaps unravel some of the puzzles encountered in their earlier experiments, Lanyon's group next turned to an avian model, the adult turkey, in which the shaft of the ulna was isolated and loaded via pins surgically inserted through metal caps cemented over its ends. Because the animals had no other means of loading their ulnar diaphyses, the only strains were those imposed experimentally by applying load to the pins. Several experiments have been done to study the effects of strain magnitude and distribution, the daily number of load cycles, the effects of static vs cyclic loading, and the effect of strain rate.

Lanyon and Rubin (1984) observed that when no loading was applied over an 8-week period, the ulna suffered endosteal resorption and increased osteonal remodeling. Compressive **static** loading to 2000 µε for the same period of time led to a similar result, but the same strains applied **dynamically** at 1 Hz produced a 24% increase in cortical area, primarily from woven bone formation on the periosteal surface. Subsequently, Rubin and Lanyon (1985) did a dose-response experiment for 1-Hz loading in a bending mode such that peak compressive strains were about twice the magnitude of the tensile strains. When 100 cycles/day were applied at peak strains from 0 to 4000 µε, there was a linear relationship between cortical area after 8 weeks and peak strain magnitude. The linear regression equation was % change in area = −15.3 + 0.015 × (peak microstrain), with $R^2 =$ .69. Setting the left side of this equation equal to zero indicates that an equilibrium loading for maintenance of bone mass (zero change in area) would be 100 cycles/day at about 1000 µε. This was an important benchmark. These investigators did another frequently cited experiment (Rubin and Lanyon, 1984) in which 2050 µε was applied at 0.5 Hz in daily loading rates of 4, 36, 360, and 1800 cycles. It turned out that only 4 cycles/day were sufficient to maintain bone mass at this strain magnitude. After 4 weeks of loading, each of the three higher loading rates produced the same increase in bone mass, about 40% ± 10%, indicating that the woven bone response was an all-or-nothing phenomenon. In this study, the neutral axis for the applied loading was at 90° to that for wing flapping, which may also be why there was no difference between 36 and 1800 cycles/day; that is, the change in strain distribution overwhelmed the effects of cycles/day. In all cases the new bone was periosteal or endosteal woven bone similar to that seen in the Lanyon group's sheep and pig experiments.

The Rubin–Lanyon avian model has proven to be an excellent one for studying the response of cortical bone to graded mechanical loading. It has been underutilized, however, for histomorphometrically assessing the modeling and remodeling results responsible for the bony changes.

Rat Tibia Bending Experiments

Using an in vivo rat model in which the tibia is repetitively loaded in a transcutaneous four-point bending fixture, Turner and Forwood have obtained considerable insight regarding the effects of elevated strains on modeling (Forwood and Turner, 1994; Turner et al., 1994). On the endosteal surface, when strains remained below about 1000 με, no bone formation ensued. Higher strains activated modeling in the formation mode, forming **lamellar** bone. The apposition rate for this endosteal bone formation was linearly proportional to the amount of strain excess over 1000 με, up to at least 2000 με (Fig. 6.9). Strains were higher on the periosteal surface, and adaptive modeling produced **woven** bone, but in this case there was no relationship between production rate and strain magnitude. In this model, as in the avian model, woven bone formation appeared to be an all-or-nothing phenomenon that, once activated, bears no relationship to strain magnitude.

Additional experiments examined the effects of loading on successive days with the magnitude of the overstrain kept constant. One dose of 36

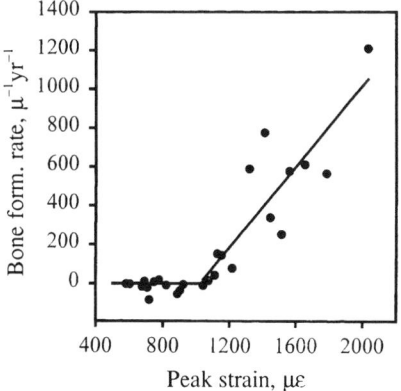

FIGURE 6.9. Apposition rate vs. applied bending load for endosteal surface of rat tibia. (Reproduced with permission from Turner et al., 1994.)

244 6. Mechanical Adaptability of the Skeleton

cycles of bending on a single day, producing 2500 µε on the periosteal surface, was enough to activate lamellar bone formation on the endosteal surface and woven bone formation on the periosteal surface. For both lamellar and woven bone formation, each additional day of loading linearly increased the amount of surface that was modeling. The apposition rate at each point on the forming surface was increased by additional days of loading for woven bone formation but not for lamellar bone formation. Therefore, the data suggest that modeling is a quantum-like phenomenon. On the endosteal surface, each day of loading exceeding the 1000-µε threshold adds a group of osteoblasts and extends the lamellar bone-forming surface, but does not increase the rate at which existing osteoblasts add osteoid. On the periosteal surface, each day of loading adds a group of osteoblasts, extending the woven bone-forming surface. It also seems to increase the rate at which each existing group of osteoblasts makes woven bone, in contrast to the effect of increasing the strain level.

This is very important information. However, one question that is not answered is the extent to which the endosteal-periosteal differences are governed by strain magnitude (strains being higher on the periosteal surface) and the extent to which they are dictated by the fact that one surface is covered by bone lining cells lying adjacent to bone marrow, while the other is covered by fibroblast-like cells in the deep layers of the periosteal membrane.

Summary

These experiments suggest several conclusions:

- Peak periosteal strains during vigorous activities rarely exceed 2000–3000 µε in the long bones of all sorts of animals.
- Trabeculae seem to be aligned with principal strain directions as measured in the adjacent cortical shell, or with principal stress directions computed for a similar homogeneous structure.
- Disuse of a growing long bone diaphysis inhibits periosteal bone formation; disuse of a mature diaphysis results in increased endosteal resorption and intracortical remodeling.
- Mild overloading of a growing long bone diaphysis inhibits endosteal resorption, producing a thicker cortex. Substantially overloading a long bone diaphysis usually activates woven bone modeling on the periosteal surface. However, the locations of these modeling responses may be difficult to rationalize. The periosteal woven bone response appears to be a threshold response which, once activated, is insensitive to increases in load magnitude or the number of cycles applied each day. It may, however, be increased by more days of loading. Endosteal modeling resulting from overload may produce woven or lamellar bone; in the latter case, the apposition rate is linearly proportional to the surface strain excess over 1000 µε. However, increasing the number of days of loading does not change the apposition rate.

These experiments emphasize diaphyseal cortical bone responses, as opposed to trabecular bone architecture, which was the historical focus. In all the experiments that achieved a measurable adaptive response on the periosteal surface, woven bone was formed. Most of the models involve surgical procedures, which may confound the results by adding a regional acceleratory phenomenon[7] to the mechanical effect on the bone. The exigencies of experimental science—getting statistically significant results in a few weeks or months—lead investigators to conduct experiments that may not be very representative of the sublime, gradual, day-in, day-out way in which mechanical adaptation probably works in healthy individuals. In human beings, most cortical bone adaptations are likely to involve more modest load increases and lamellar bone formation. See Bertram and Swartz (1991) for more discussion of these issues.

With these results and limitations in mind, let us now examine some additional theoretical models for mechanical adaptability.

6.5 Some Additional Theories

In this section, we shall describe several additional theories about how the skeleton adapts to mechanical loading. As was the case with the preceding section on experimental work, this section is not meant to be an exhaustive review, but to provide some examples of the principal approaches that have been used. These theories may be organized into four categories, which attempt to explain:

1. Modeling drifts to correct for diaphyseal curvature or osteotomy of a collateral bone
2. Control of directionality in tunneling osteonal BMUs
3. Bone density distribution
4. Trabecular alignment

The first of these includes the flexural neutralization theory, which we have already discussed. We present the remaining examples in sequence by category.

Pauwels' Stress Magnitude Theory: Control of Modeling Drifts

Pauwels (1980)[8] presented an elegant demonstration of how simple assumptions relating surface remodeling to stress can predict the gross distortions in cross sectional geometry that occur in very curved, rachitic

---

[7]Defined (RAP) in Section 2.8.
[8]The cited reference is a compendium of Pauwels' research papers. The actual work discussed here was published in the 1960s.

femurs. Rickets is a disease which, in children, interferes with the normal functioning of the growth plate. As a result, the long bones may grow with abnormal amounts of curvature that modeling drifts fail to correct. The left side of Fig. 6.10 shows three views of a femur from a man who had untreated rickets as a child. The bone has a greatly exaggerated anterior bow. This curvature increased the bending stresses at midshaft, and it was clear to Pauwels that this led in turn to the long, narrow cross-sectional geometry which is shown in the lower right portion of the figure, along with the normal cross-sectional geometry. Both the normal and rachitic cross sections are shown in their correct positions relative to the head of the femur when viewed from above. Pauwels sought a mathematical formula that would predict this adaptation. He explained in detail his method for determining the stresses in the bone, but was less clear regarding his adaptation algo-

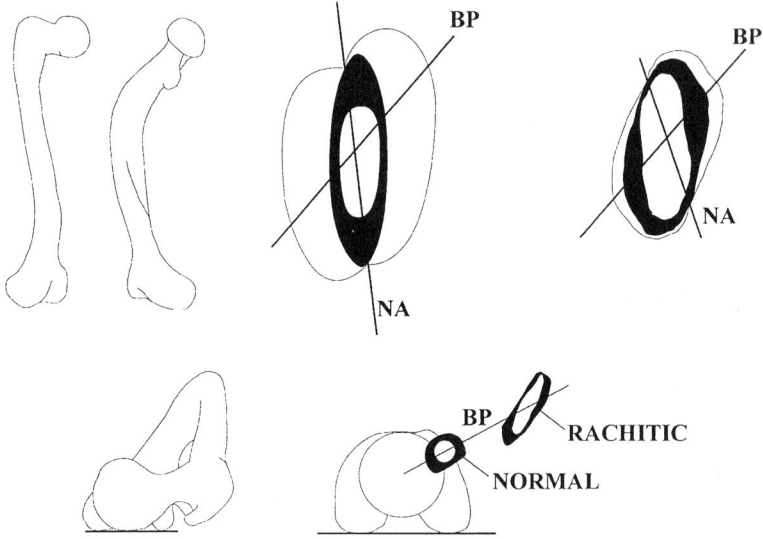

FIGURE 6.10. Effects of mechanically adaptive changes in the femur of a man who experienced untreated rickets. *Left:* Three views of the highly curved femur. *Lower right:* View down the shaft of a normal femur, showing the relative positions of the midshaft cross section for the rachitic bone compared to a normal bone. *BP*, estimated bending plane. Note the extended moment arm of the rachitic femur midsection. *Upper middle:* Pauwels calculated the surface stresses on an elliptical cross section *(black)* loaded by the bending moment shown below. The distance between the curved lines and the surface of the bone section indicates the magnitude of the periosteal stresses. *NA*, neutral axis. Using Eq. 6.1, Pauwels iteratively "modeled" the femur section, producing the shape shown at *upper right,* which is similar to the actual bone section seen below. (Reproduced with permission from Pauwels, 1980.)

rithm. He began with the elliptical cross section shown in the upper middle image (rather than the normal femoral shape), and computed the surface stress, s, from a simulated hip load. He then altered the thickness of the cortex, C, at each point according to the equation

$$C = a + bs^n \qquad (6.1)$$

where a, b, and n were unspecified "arbitrary constants." After this change, the stress and cortical thickness were recalculated, and the process was iterated until, presumably, a steady state was reached. The result of this adaptation is shown in the upper right cross section, which bears a striking resemblance to the actual rachitic cross section. However, it is clear from the figure that, in addition to altering the thickness of the cortex, the calculations moved it radially in and out, creating an effect which implied varying medullary resorption. Neither the rationale nor the algorithm for this aspect of the model was fully explained.

Adaptive Elasticity Theory: Control of Density or Modeling Drifts

These theories have been used to account for both diaphyseal modeling responses and adaptation of bone density to loading. Cowin and his co-workers developed a novel and mathematically rigorous theory of bone adaptation (Cowin and Hegedus, 1976; Hegedus and Cowin, 1976; Cowin and Nachlinger, 1978; Cowin and Van Buskirk, 1978). They viewed internal remodeling as a set of chemical reactions acting to exchange material between the solid bone matrix and the fluids in the void spaces. They employed an extension of continuum mechanics in which the chemical reactions and the bone geometry are forced to obey thermodynamic, mass, and momentum principles. Using this model, it was shown that, under the action of a uniform stress, a nonhomogeneous cylindrical bone would become homogeneous in its density (Firoozbakhsh and Cowin, 1980).

Subsequently, Firoozbakhsh and Cowin (1981) showed that for uniaxial stresses and certain constraints of the remodeling coefficients, their model was equivalent to that of Pauwels for surface remodeling. Immediately thereafter, Cowin and Firoozbakhsh (1981) presented a somewhat simpler model for surface remodeling that explored the implications of a generalized strain error hypothesis for mechanical adaptation. The starting point was the equation

$$M = C_{ij} (e_{ij} - e_{ij}^{\circ}) \qquad (6.2)$$

where M is the net rate of addition to the bone surface (M is negative for resorption), e_{ij} is the strain tensor, and e_{ij}° is a reference or "set point" strain tensor. This theory is much more general than most formulations of strain-governed adaptation, which typically consider only one component of strain, or some other scalar quantity. The C_{ij} are "surface remodeling rate coefficients" that can be positive or negative, depending on whether net formation or resorption occurs. Although the theory does not specify the biological mechanisms that are embodied in the C_{ij}, it allows explicit computation of adaptive results for complicated circumstances (such as realistic loading which superimposes compression, bending, and torsion).

Obviously, if this theory is to have any practical application, the values of the remodeling rate coefficients must be determined. To that end, Cowin's group (1985) compared the predictions of their model to the results of five animal experiments (the canine disuse experiments, the sheep and pig osteotomy experiments, and the exercised pig experiment described earlier). The C_{ij} tensor contains coefficients for each component of strain. The authors were able to simplify matters considerably for the specific applications of the theory, so that there were just two coefficients relating surface apposition rate, M, to longitudinal stress at the surface. By adjusting these coefficients separately for each experiment, excellent reproductions of the changes in bone cross section were reproduced. However, widely varying values for the coefficients in each experiment were required, possibly because the coefficients vary with species and age as well as bone surface, or because the theory is based on an incorrect choice for the mechanical stimulus.

While developing theoretical models somewhat along the lines of Cowin's, Hart and co-workers (Hart et al., 1984a,b; Hart and Davy, 1989) have formulated their models in terms of variables that quantify (or at least relate to) processes of cell differentiation and cell function. They assume that some set of variables representing the bone strain history combines with genetic, hormonal, and metabolic factors to determine the regions of bone surfaces where modeling or remodeling will occur, the numbers of active osteoblasts and osteoclasts per unit area in these locations, and the average activities of individual cells. Their approach melds and extends engineering representations of mechanical influences on cells initially developed by Cowin and co-workers with formulations of cell responses initially developed by Frost (1964b) and Martin (1972, 1985, 1995). Their approach was able to predict some of the results of canine osteotomy overload experiments (Burr et al., 1989a,b; Oden et al., 1995).

What Controls Osteonal Tunneling Directions?

Osteons are not precisely longitudinally oriented in a long bone diaphysis, but spiral around and through the cortex at a shallow angle (Cohen and Harris, 1958). Lanyon and Bourn (1979) showed a good correspondence between this orientation and the principal strain directions. The mechanism by which such directionality is achieved has received little attention.

While the activation of osteonal BMUs may be achieved by relatively brief, isolated signals, directional control seems to require a signal that is at least intermittent. The only theory found in the literature to explain this behavior is that of Martin et al. (1974). This theory asserts that the stresses in the bone surrounding the osteonal remodeling cavity play an important role in organizing and directing the cellular activities in the BMU. Like the external surfaces of a whole bone, the walls of the cavity produced by an osteonal BMU experience a variety of stresses. Most importantly, stress concentrations will occur in the vicinity of the cutting cone and reversal region. These high stresses may produce conditions that can control the cutting cone's direction of travel.

The Stresses Around an Osteonal BMU: Wolff's Law Inside Out
Figure 6.11 shows schematic diagrams of an osteonal BMU and a simple spherical cavity in a loaded bone; the bone is assumed to be loaded in simple compression, and the BMU is assumed to be aligned with the principal compressive stress direction. When a simple spherical cavity lies in a compressive stress field, p, the equatorial wall stress is increased to $3p$, and the wall stress at the pole of the cavity is $-p$ (Peterson, 1974). The $-$ sign indicates that this polar stress is tensile. Thus, the surfaces of the void that are parallel to the load direction experience tangential stresses three times greater than normal, and the surfaces perpendicular to the load carry a stress of opposite polarity. In the case of a cavity having the elongated geometry of the BMU, the stress concentration factor can be expected to be reduced, but the reversed polarity of the stress at the apex of the cutting cone will remain. Consequently, a BMU aligned with a compressive stress field has its osteoclasts working on a surface that carries a tensile stress and its osteoblasts working on surfaces which carry compressive stress.

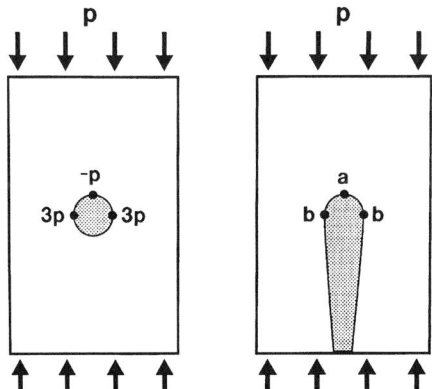

FIGURE 6.11. *Left:* Stress concentration and stress of opposite polarity at apex of a spherical cavity in a uniaxial compressive stress field, *p. Right:* Diagram of a BMU cavity aligned with the compressive stress direction in a long bone. The tangential stress on the resorption surface *(a)* would be tensile; that on the refilling surface *(b)* would be compressive.

Now consider the stress gradients surrounding an osteonal BMU. There is a positive gradient at the resorption surface and a negative gradient at the initial refilling surface. (It is left as an exercise to show this.) Thus, the cellular activities within the osteonal BMU bear the same relationships to stress gradients as do the activities of cells remodeling the external surfaces of a curved bone shaft carrying end loads in Frost's flexural neutralization theory. Thus, it is again hypothesized that a positive strain gradient activates osteoclasis at the resorption front and a negative strain gradient activates bone formation at the beginning of the refilling region.

Steering in the Principal Compressive Strain Direction
Having seen that cell location and function may be governed by the mechanical environment of the osteonal cavity, one may now ponder how these considerations could lead to directional control of the osteon. Suppose, as depicted in Fig. 6.12, that the osteonal BMU becomes misaligned with the principal compressive stress direction. In that case, the location of the maximum positive strain gradient shifts from the apex of the

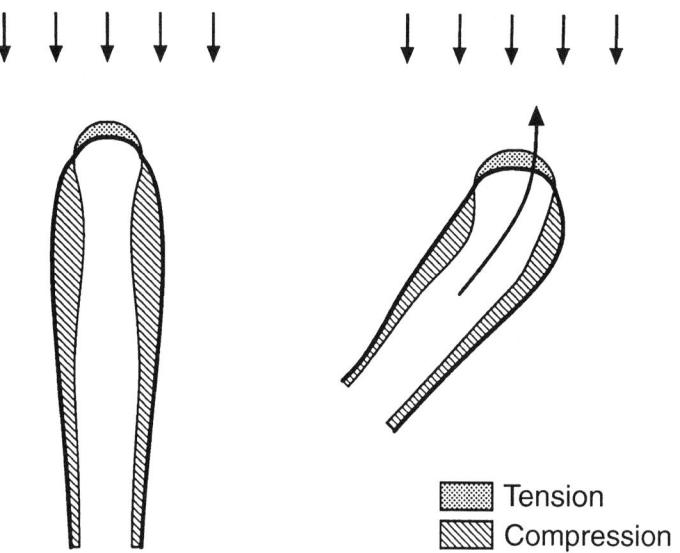

Tension
Compression

FIGURE 6.12. *Left:* Surface stresses in a BMU aligned with the compressive stress direction. Thickness of *hatched region* inside the cavity approximates the compressive stress magnitude; thickness of the *stippled region* above the cavity approximates the tensile stress. *Right:* Corrective steering of a BMU that has become misaligned with the compressive stress direction. The peak tensile stress has moved to the left side of the resorption surface; if resorption is enhanced there, the BMU will turn in that direction.

cutting cone to a point associated with the principal compressive stress direction. One would then expect the most vigorous resorption to occur at this point, so that the osteon would turn in the desired direction. Thus, the principles that we have outlined provide a mechanism for automatically steering osteons in the principal compressive strain direction.

If the cortices of long bones are loaded primarily by bending and compressive end loads, and to a lesser extent by torsional loads, steering of osteons in the compressive stress direction fits nicely with the observation that osteons spiral around the shaft at a shallow angle. However, the theory encounters two difficulties in that it does not explain the tendency of BMUs to dig through the thickness of the cortex, nor does it explain the behavior of osteonal resorption cavities on the tensile side of bones subjected to bending. Before investing too much effort in considering these problems, a more definitive picture is needed of the patterns of osteonal alignment in the cortices of bones habitually loaded in different ways.

Adaptive Finite Element Models: Control of Density

Carter et al. (1987b) and Huiskes et al. (1987) introduced the finite element approach to the development of predictive relationships between mechanical loading and trabecular bone density. For an example of this general method in its later form, see Beaupre et al. (1990a,b). We present here a simplified version of the Carter–Beaupre approach, but one that captures its essentials. This theory assumes that the mechanical stimulus is some time-averaged aspect of stress or strain. A *daily stress stimulus*, Ψ, is defined as

$$\Psi = (\Sigma \, n_i \, \sigma_i^m)^{1/m} \tag{6.3}$$

where n_i is the number of cycles/day of a particular kind of loading, i, which produces a continuum level "effective stress," σ_i, at the location in question. This effective stress has assumed different forms, each a scalar quantity having units of stress, for example, $\sqrt{2EU}$, where E is elastic modulus and U is strain energy density. Because stress, strain, and strain energy are all proportional to one another, the results are all similar. The summation is over i, that is, over all the significant kinds of loads that are applied each day. Next, a finite element model is used to determine the values of σ_i at each location in the bone, each "location" corresponding to an element in the model. An error function is then defined as

$$e = \text{constant} \times (\Psi - \Psi_{AS}) \tag{6.4}$$

where Ψ_{AS} is an "attractor state," or equilibrium value, of the stress stimulus. Originally, the apparent density of each element was assumed to change according to e. In the Beaupre version of the model, the net rate of addition or removal of bone on an external or internal surface (the apposition rate, M, μm/day) is made a linear function of the error:

$$M = c_1 + c_2 e \tag{6.5}$$

where c_1 and c_2 are coefficients containing the biological mechanisms by which bone cells respond to the mechanical stimulus. For internal remodeling, the bone formation rate (volume of bone added or subtracted per unit volume per day) can be written as

$$V_F = dB_v/dt = M k S_v \tag{6.6}$$

where B_v is the bone volume fraction, S_v is the internal surface area per unit volume, and k is the fraction of this surface that is actively remodeling. Because apparent density, d, is the product of B_v and the density of the calcified bone matrix, eq. 6.6 allows one to track the changes in d that occur in each element. These density changes in each element are converted to modulus changes using the Carter–Hayes correlation:

$$E = \text{constant} \times d^3 \tag{6.7}$$

(recall Eq. 4.12). The finite element model is solved again with the new modulus values to obtain a new stress stimulus for each element. The model is iterated to find the distribution of bone density (or external geometry), which minimizes the error function.

The Value of m

To use the model, a value must be obtained for the exponent m in the stress stimulus equation. Beaupre et al. (1990b) did this in the following way. Using data from two of Rubin and Lanyon's isolated avian ulna experiments, they plotted two combinations of applied strain magnitude and cycles per day that resulted in bone maintenance, that is, neither losing or gaining bone. These are the points on the left (4 cycles at 2000 $\mu\varepsilon$ and 100 cycles at 1000 $\mu\varepsilon$) in Fig. 6.13. They added to these two points a third, which represented the strain measured by Lanyon on the tibia of a human (400 $\mu\varepsilon$), and assumed that this person's tibia was loaded 10,000 cycles/day and was in an equilibrium state. These three points gave a slope of 4.88; based on other studies (Whalen et al., 1988), this was reduced to 4. Figure 6.14 shows a typical pattern of bone density in the proximal femur predicted by a model like this when the loading consists of either one or three point loads applied to the femoral head and corresponding abductor loads applied to the greater trochanter. The two key similarities between the simulation result and real bones are the path of dense bone leading from the medial cortex up to the head of the femur, and the area of low density in the middle of the femoral neck known as Ward's triangle (it has an oval shape in the three load simulation).

Huiskes and co-workers (1987) have used adaptive finite element models to study the responses of bone to the mechanical consequences of

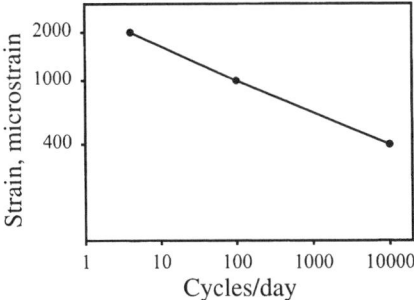

FIGURE 6.13. Applied peak stress vs. cycles/day for equilibrium of bone mass. (Replotted from Beaupre et al., 1990b.)

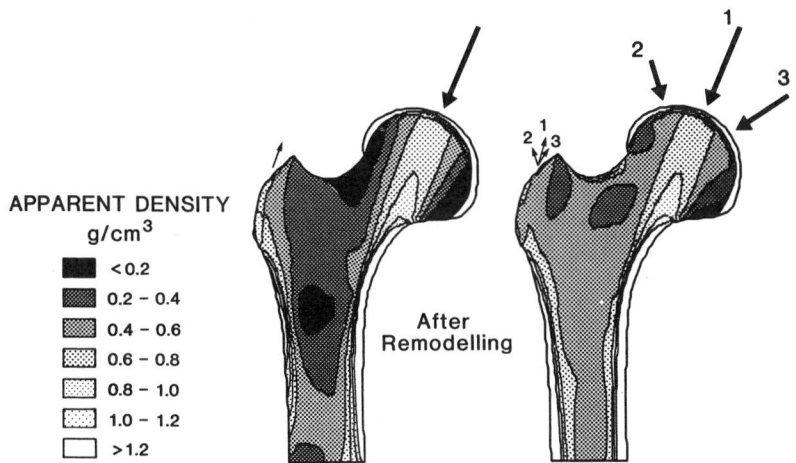

FIGURE 6.14. Results of a typical model of the kind used by Carter and co-workers. Less dense areas are indicated by *darker shading* as seen in scale at left. The model at *left* was loaded by a single set of hip and abductor forces *(arrows)*; that at *right* was loaded by three different loads (i.e., different *i* in Eq. 6.3), representing abduction, adduction, and normal hip positions. (Reproduced with permission from Carter et al., 1987b.)

implanting a joint prosthesis. They used strain energy density as the mechanical stimulus to which the bone cells respond, and showed that such models could predict patterns of postimplantation bone density change adjacent to total hip prostheses. It has been shown that the Carter and Huiskes approaches are equivalent if the coefficients are properly chosen (Carter et al., 1989; Jacobs, 1994).

Self-Trabeculating Models: Control of Density and Trabecular Alignment

If one makes an iterative finite element model like those we have been describing, in which the density and then the modulus of each element is adjusted in proportion to its stress or strain or strain energy density (SED), and lets it run indefinitely, a very interesting thing often happens. The model may not approach an equilibrium state in which there is a smooth variation in element densities from one region to another. Instead, almost all the elements may either go to the minimum or maximum allowable density. This behavior has been termed "0–1" behavior because each elemental "square" in the model becomes either empty or full. The full elements are arranged to form a network of trabecula-like struts that carry loads from one part of the model to another.

Figure 6.15 shows one of the first such models in the orthopaedics literature, produced by Weinans et al. (1992). In this model, the mechanical stimulus was strain energy density (SED). It shows a model that has fixed elements at the bottom and carries a set of vertical point loads across the top which increase from right to left. The "1" or bone-filled elements are

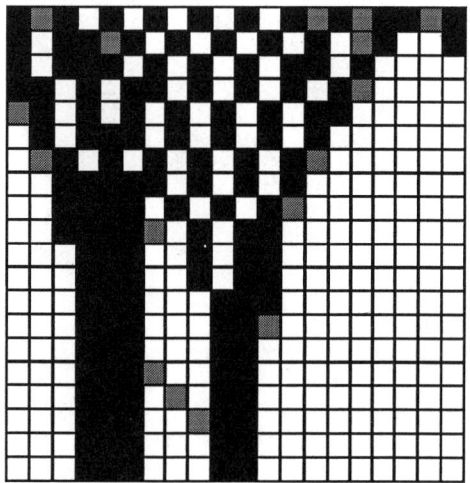

FIGURE 6.15. An example of an early self-trabeculating finite element model. A compressive load was applied to the upper surface, diminishing from left to right. The density of the material in each element is represented by a gray scale. Most elements have either maximum (black) or minimum (white) density. Two walls of cortical bone have developed below a metaphyseal region. Note the "checkerboarding" effect described in the text. (Reproduced from J Biomechanics, Vol. 25, Weinans et al., The behavior of adaptive bone-remodeling simulation models, 1425–1441, 1992, with kind permission from Elsevier Science Ltd., The Boulevard, Langford Lane, Kidlington OX5 IGB, UK.)

black and the "0" or soft tissue-filled elements are white. The bottom portion of the model has evolved into two asymmetric cortices, and the top portion has become a network of "trabeculae." The figure conveys the self-trabeculating concept well, although it is a relatively crude, two-dimensional model.

The 0–1 behavior, in which the model develops a trabecular architecture on its own, was very exciting to see, for if this behavior truly represented a physical phenomenon, rather than some sort of numerical processing behavior peculiar to a computer model, it could provide insight about cancellous bone structure and the mechanical adaptability hypothesis. However, there is another feature of Fig. 6.15 that is very disturbing. Notice that many of the bone-filled regions contact one another only at their corners, that is, at a single node of the finite element mesh. This effect is known as "checkerboarding," and it signals that the finite element model is not representative of reality because there would be infinite stress at such contact points. Engineers designing optimized composite materials saw such behavior as well, and learned that this it can be avoided if the modulus of each element is adjusted not on the basis of its SED (or stress or strain) alone, but based on a mean value for several neighboring elements.

Mullender et al. (1994) made a similar discovery when their subsequent models incorporated a biologically motivated feature that introduced a similar smoothing effect. They assumed that the SED is the mechanical stimulus sensed by osteocytes and communicated through their processes. Each cell then responds according to a sum of stimuli: that produced at its location, and those received from osteocytes in other locations. They also assumed that the strength of the stimulus, S_E, decays exponentially with the distance communicated. The stimulus received at a given location is of the form

$$S_E = \Sigma \left[(U_i - k) \exp(-d_i/D) \right] \tag{6.8}$$

where U_i is SED at the location of an osteocyte labeled i, located a distance d_i away, k is an equilibrium value of SED, and D is a decay constant. The summation is over i, that is, all the surrounding osteocytes. When such a model is run, starting from a homogeneous mesh of elements having an intermediate density, the element densities again migrate to extreme values, so that 0–1 behavior persists and the model self-trabeculates, but checkerboarding does not occur (Fig. 6.16).

An even more interesting aspect of such models is that the trabeculae which develop align themselves with the principal stress directions or, more correctly, what the principal directions would be if computed for a homogeneous structure of the same shape and bearing the same loads (Fig. 6.16). Higher loads produce more and thicker trabeculae. Furthermore, if the model orients its trabeculae to loads directed one way, and then the load direction is changed, the model will realign its trabeculae to the new load direction. This realignment occurs even though the mechanical stimulus is a scalar that contains none of the information about the individual

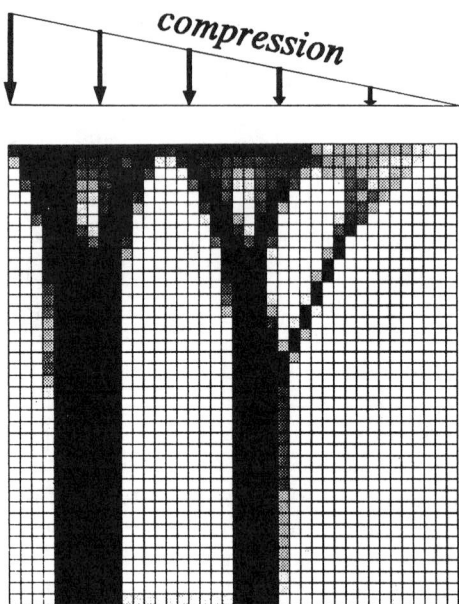

FIGURE 6.16. Self-trabeculating model similar to that of Fig. 6.15 but with a finer mesh. Checkerboarding has been eliminated by using Eq. 6.8. (Reproduced from *J Biomechanics,* Vol. 27, Mullender et al., 1389–1394, 1994, with kind permission from Elsevier Science Ltd., The Boulevard, Langford Lane, Kidlington 0X5 IGB, UK.)

components of the stress tensor which an engineer would ordinarily use to compute the principal stress directions! We return to this point later.

The models we have described so far assumed that osteocytes not only sensed the mechanical stimulus, but altered the density at their location as well. Although osteocytes may make minor changes in the adjacent bone matrix, they cannot resorb or form bone to any significant degree. These processes must be carried out by osteoclasts and osteoblasts, and they must occur on bone surfaces, not in the bone midsubstance. Smith et al. (1997) modified the model of Mullender et al. (1994) to incorporate this requirement. The elements of their model represent only the trabeculae and have a constant density and elastic modulus. The model also assumes that the cells located on the surfaces of trabeculae are bone lining cells that can activate resorption or formation, depending on the sign of the error signal they receive from neighboring osteocytes and bone lining cells. Otherwise, the model is similar to previous models, with all cells sensing SED and communicating this information to nearby cells according to Eq. 6.8. If the signal received by a lining cell is positive (the nearby SED is greater than the attractor state value), formation occurs by adding a new element (representing a packet of new bone) to the surface. If the signal is negative, the element beneath the lining cells is removed from the mesh ("resorbed"), and the model's stresses are recomputed. This model, which is more realistic from the point of view of the actual biologic processes, continues to adjust itself to changes in load direction or magnitude just as the earlier models did.

Mathematical Stability Considerations

Harrigan and Hamilton (1994) have analyzed the mathematical nature of models such as Carter's. The central fact that they have illuminated is that the stability of such models depends on the relative magnitudes of the exponents in the two principal equations:

$$E = k \rho^n \tag{6.9}$$

and

$$\Psi = (\Sigma \, n_i \sigma_i^m)^{1/m} \tag{6.10}$$

It turns out that $n < m$ is a necessary and sufficient condition for the simulation solution to have global stability and uniqueness. Because most models use $n = 2$ or 3 and $m = 4$ or 5, their solutions meet these conditions. Of course, that does not mean that they are necessarily correct mechanically or biologically!

Bone Tissue Is a 0–1 Material

The 0–1 behavior seen in these finite element models is contrary to the assumption that apparent density may vary continuously in continuum level models of a large region of bone like the proximal femur. On the other hand, 0–1 behavior is very appropriate for a finite element model that represents the structure of individual trabeculae in a much smaller region of cancellous bone. At this scale, the elements in the model should represent either bone or marrow. Although the density of bone matrix within trabeculae may vary because of variations in mineralization, it can be argued that these variations are small compared to continuum-level porosity variations. As we have seen, the mineralization of bone tissue depends primarily on the rate of remodeling and normally varies only a few percent. Thus, bone is inherently a 0–1 material. To the extent that it is not, simulation of density variations would call for tracking the mineralization of new packets of bone produced by remodeling.

Related Engineering Models

The more recent versions of the self-trabeculating models are three dimensional and have fine enough meshes that the individual trabeculae are several elements across, eliminating some of the concerns stemming from the crudeness of the early models. The engineering literature on optimization of structures contains similar models. The analyses of Strang and Kohn (1986) suggested that optimal structures should contain highly oriented fibers rather than a continuously varying interior, not unlike trabecular struts of bone. Other papers describe finite element-based models very similar to those of Weinans and Mullender that optimize the stiffness of a structure relative to its weight. For example, the top of Fig. 6.17 shows an optimized design for a cantilever "aircraft support beam" (Reiter and Rammerstorfer, 1993). It is fixed along its left end and supports a downward load at its lower right corner. There is a striking similarity between this structure and that of the inside of a vulture's wing bone, shown in the bottom of Fig. 6.17.

A Conundrum Solved?

We noted in Section 6.2 that there is a fundamental problem with the essential concept of Wolff's law: that is, that trabeculae align themselves with the average principal stress directions in the loaded bone. At the continuum level of analysis, there is no problem with this because principal stress directions can be computed by overlooking the discontinuities in the cancellous bone structure. Then, the alignment of trabeculae with principal stress trajectories means that the principal axes of the fabric tensor (recall

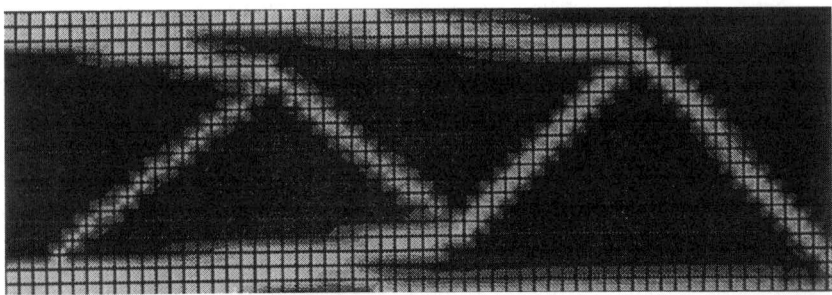

FIGURE 6.17. *Top:* Optimized design of an "aircraft support beam." (Reprinted from P. Pedersen, Ed., *Optimal Design with Advanced Materials,* 1993, p. 31, with kind permission from Elsevier Science–NL, Sara Burgerhartstraat 25, 1055 KV Amsterdam, The Netherlands.) *Bottom:* Photograph of trabecular struts within a vulture's wing. (Reproduced with permission from Thompson, 1942.)

Section 4.5) are parallel to the principal stress directions (Cowin, 1986). The problem is that bone cells are incapable of understanding continuum mechanics or of overlooking the discontinuities which form the trabecular surfaces on that they may reside. Thus, the "principal stress directions for the proximal femur" are, at their level, entirely fictitious (Cowin, 1997).

How, then, could one expect bone cells to "know" such fictitious directions, and form trabeculae so that they line up with them? The self-trabeculating models suggest an answer to this perplexing question. Perhaps bone cells are able to do this not because they "know" what the fictitious stress directions are, but because it turns out that those directions coincide with the trabecular orientations achieved by keeping the strain energy density at a constant value. In other words, by shaping and aligning trabeculae so as to control the local strain energy density, it may happen that the trabecular orientations in a real bone are parallel to the principal stress trajectories in a facsimile bone of the same shape and loading, but made of a continuous substance. The self-trabeculating models support this proposition, but it would be nice to have a mathematical proof that it is true.

Synthesis

While the foregoing theories vary in approach, it is possible to boil them down further than the four categories that we initially defined. The flexural neutralization and osteon alignment theories both propose that resorption and formation are determined by the stress gradient beneath the surface in question. They are essentially the same theory applied to two different structures. Both concepts are also compatible with the theory that osteocytes sense bone strain by reacting to fluid flow over their processes, because such flows would be driven by stress gradients.

Similarly, the bone density optimization and trabecular alignment theories are essentially the same; the latter is just formulated at a smaller scale. Also, the adaptive elasticity theories may be moved into this camp, because they also relate formation and resorption to stress or strain magnitude. Thus, we have just two kinds of theories, with the mechanical stimulus being stress gradient in the first and some scalar measure of stress magnitude (e.g., strain or strain energy) in the second. Both theories are consistent with osteocytes being the mechanical sensors, and both seem to be dependent on communication across the osteocytic network, for sensing a stress gradient in one case and signal averaging in the other (to avoid "checkerboarding" in a computer model and perhaps some equivalent instability in reality). It is clear that stress magnitude does not work as the mechanical stimulus in theories of the first kind, but it is not clear that stress gradients would not work as the stimulus in theories of the second kind. Given the inhomogeneity and anisotropy of the bone tissue, stress gradients may be ubiquitous in bone, and the osteocytic network may be able to communicate information about both stress magnitude (or strain energy density) and stress gradients. Clearly, more work needs to be done to synthesize a single, general theory from the menu currently in the literature.

Frost's Mechanostat Theory

We shall describe one more theory about mechanical adaptability, one that does not fit into the four categories listed. Frost (1986, 1987b) introduced this algorithm, called the *mechanostat theory*. Frost's flexural neutralization theory sought to predict whether the modeling or remodeling would be resorptive or formative, once activated. This theory seeks to predict **when** such activation will occur in response to increased or decreased strain magnitudes.

The Equilibrium Strain Range
The mechanostat theory begins with the concept that there is a *minimum effective strain* which must be exceeded to excite an adaptive response. However, instead of simply postulating that strains below a certain threshold will evoke no response, Frost suggests that there is an *equilibrium range* of strain values which will evoke no response. Strains above this range evoke a positive adaptive response (increased bone), and strains below this range cause a negative adaptive response (loss of bone). Others have suggested similar physiologic ranges, sometimes called *"dead zones"* or *"lazy zones."* In this framework, the minimum effective strain becomes the upper limit of the equilibrium strain range, or the strain beyond which an augmenting adaptive response is initiated. There would also be a lower strain limit defining the strain below which bone loss would occur.

Frost's mechanostat theory asserts that modeling and remodeling have opposite responses to loading that pushes strains above or below the equilibrium range:

Increased loading....... modeling is increased and remodeling is inhibited.
Decreased loading...... modeling is inhibited and remodeling is increased.

The logic behind this algorithm is as follows. While the inhibition of remodeling by high strain may seem "backwards" because of the long-held concept that increased loads or strain stimulate "remodeling" to increase bone strength, it must be remembered that we are using the strict definition of remodeling, which does not include modeling. Now, think first of cortical bone. Osteonal remodeling increases porosity as a result of the remodeling space and the new Haversian canals created. It is therefore desirable that osteonal remodeling should be depressed by overload and stimulated by disuse. In trabecular bone, remodeling can theoretically decrease porosity, but trabecular and endosteal BMUs in adults are usually characterized by incomplete replacement of the bone removed. Therefore, the mechanostat theory makes sense there as well. Other than through longitudinal growth (endochondral ossification), the primary means of adding bone to the skeleton is modeling in the formation mode. Therefore, it is also reasonable that mechanical overload should activate modeling. One limitation of the algorithm is its vagueness regarding the possibility that overload would activate modeling in the resorption mode. Endosteal resorptive modeling could be compensated by periosteal formation, or just not be part of the adaptive picture. These aspects of the algorithm are not clear, and become an issue in its experimental verification.

Experimental Verification
We may compare the mechanostat against the results of several of the experiments described previously. When Woo et al. (1981b) exercised growing pigs, there was no effect on periosteal diameter, but endosteal expansion was inhibited. This result is consistent with the mechanostat theory if the endosteal surface in young pigs is remodeling (inhibited by overuse) rather than modeling. The failure of the periosteal surface to show increased bony apposition would make sense if the rate of apposition were high already because the animals were growing. This is a reasonable supposition because pigs form plexiform bone at the periosteal surface during growth, and the function of this kind of primary bone formation seems to be to add bone quickly.

Disuse osteopenia in the casted forelimbs of growing dogs (Uhthoff and Jaworski, 1978; Jaworski and Uhthoff, 1986) was produced both by the failure of the periosteal surface to expand and by expansion of the medullary cavity. This result is consistent with inhibition of modeling on the periosteal surface and stimulation of remodeling on the endosteal surface. However, bone loss was also seen on the periosteal surface; according to the mechanostat, this should have been produced by remodeling, but this is unclear. Remobilization of the young dogs reversed both these trends: woven bone was apposed to the periosteal surface, while the slow apposition of lamellar bone to the endosteal surface of normal young dogs was also restored. In older dogs, in which modeling is less responsive, disuse enhanced endosteal resorption and increased intracortical porosity, consistent with stimulated remodeling (Jaworski et al., 1980; Jaworski and Uhthoff, 1986). There was virtually no effect on periosteal diameter, consistent with the observation that this envelope almost always is restricted to modeling in the formation mode. Remobilization (Jaworski and Uhthoff, 1986) caused a small amount of bone to be added to both the endosteal and periosteal surfaces, but only partially restored bone mass in the diaphysis. The endosteal apposition was consistent with refilling of the remodeling space caused by an inhibition of remodeling. Much less bone was apposed periosteally than that in younger dogs, and the new bone was more highly organized, demonstrating the relative slowing of modeling processes in adults.

The isolated avian ulna model (involving adult animals) of Rubin and Lanyon (1985) also provides a good test of mechanostat predictions, and demonstrates that bone surfaces respond differently depending on whether the strain stimulus is above or below the equilibrium range. Disuse was associated with increased intracortical remodeling and endosteal resorption, as in immobilized adult dogs. Mechanical overload was associated with insignificant osteonal remodeling, a reduction in endosteal resorption, and abundant periosteal modeling producing woven bone. Similar observations concerning the effects of overload were obtained in the osteotomy studies of Lanyon and colleagues, as we have described. (It is noteworthy that increased osteonal remodeling has been observed in these experiments, but only after the woven bone formed by modeling had been present for some time and, presumably, needed remodeling to repair fatigue damage.)

Another relevant experimental model is that of Jee and co-workers (Jee and Li, 1990; Li et al., 1990; Jee et al., 1991), who unloaded one hindlimb of rats by bandaging, thus overloading the opposite limb. The increased resorption and reduced bone formation in the trabecular bone of the tibial metaphysis of the unloaded limb was consistent with increased remodeling and inhibited refilling, resulting in a 50% reduction in bone mass and a new equilibrium state after 26 weeks. Overloading resulted in increased bone mass caused by reduced bone trabecular remodeling and increased periosteal modeling in the formation mode. Schaffler (1990) has obtained similar results in dogs with one limb immobilized in a cast.

Thus, while some questions remain, the available experimental data support the essential features of the mechanostat theory. Reduced mechanical loading increases activation of new BMUs, increasing the number of new osteons and producing trabecular and endosteal resorption. Periosteal modeling in the formation mode is inhibited. Conversely, periosteal modeling is commonly activated by overload, which also inhibits activation of new BMUs, depressing formation of new osteons and limiting loss of bone from trabecular and endosteal surfaces. However, there must be an important limitation to the overload aspect of the mechanostat: if the overload produces damage, we saw in Chapter 5 that remodeling is activated to repair it.

Relationship of Mechanically Adaptive Responses to Other Control Factors

The response of bone to mechanical factors can be expected to depend on the balance that occurs between mechanically prompted physiologic signals and those from systemic or local nonmechanical factors. The nature of competing systemic (or global) and local signals to bone cells has been discussed by both Lanyon (1981, 1984; Lanyon et al., 1986) and Frost (1986). The idea is that various factors within the body compete for control of bone cells. For example, bone adjacent to marrow, particularly fatty marrow, has a tendency toward net resorption regardless of the hormonal or mechanical environment (Johnson, 1964; Frost, 1986). Thus, local "modulators" compete with or modify other signals, which undoubtedly accounts for some of the envelope specificity in bone remodeling (e.g., periosteal vs. endosteal).

Conversely, global factors may modify local responses, as when female distance runners lose rather than gain bone from excessive training-induced *amenorrhea*, a simulacrum of menopause. Exogenously supplied estrogen reverses the bone loss in these runners (Cumming, 1996). Similarly, a study of male basketball players found that in a rigorous training regimen they lost bone mass unless given supplemental dietary calcium; calcium loss through perspiration resulted in a physiologic calcium demand that was met by bone resorption (Klesges et al., 1996). Skeletal health has been likened to a three-legged stool, the legs being diet, hormonal status (estrogen, vitamin D), and exercise. All three legs must be robust for the stool to stand. It is apparently also possible for local skeletal disturbances to have global effects. When marrow is destroyed within a bone, its replacement involves filling the medullary canal with woven bone, which is subsequently resorbed and replaced by new marrow. Bab et al. (1985) found increased bone formation in the mandibular condyle of rats after removal of bone marrow from the tibia. Thus, bone messengers intended to govern local responses may "escape" to the systemic circulation and produce global remodeling responses.

Frost (1985) has suggested that certain osteoporoses result from an increase in the lower limit of the physiologic strain range, so that normal mechanical loading is perceived as a disuse state and bone is lost. It is possible that estrogen affects this "set point," so that it changes at menopause. If so, postmenopausal bone loss may occur because the normal mechanical stimulus becomes low relative to the cells' altered reference value. Conversely, the anabolic effects of low doses of parathyroid hormone (PTH) suggest it may move the set point in the other direction (Gunness and Hock, 1993). Thus, alterations in modeling and remodeling may occur because of changes in strain, changes in the physiologic strain range set points, or combinations of these factors.

Mechanical Adaptability and Damage Repair

We have hypothesized that mechanisms exist to optimize the stiffness of bone while minimizing fracture risk and bone mass. We have examined several possibilities for such mechanisms, such as the ability of osteocytes to sense strain and communicate such a mechanical stimulus to bone lining cells and periosteal cells. These cells have the capability of activating modeling or remodeling to change the architecture of a bone in ways that affect its stiffness, fracture risk, and weight. However, as we discussed in Chapter 5, these cells also may activate remodeling to repair fatigue or creep damage. What is the relationship between damage repair and mechanical adaptability?

The answer has to do with the trade-off between weight and fracture risk from creep and fatigue. Let us digress again to Culmann and nineteenth-century engineering for a useful analogy. In bridges, fatigue can be an important element of failure. After some disasters, early American engineers began to build their bridges of elements large and robust enough that stresses were very low and fatigue failure would not occur. Not having the benefit of the mechanical theory being developed by Culmann and others in Europe, they overdesigned, wasting material and effort. As a young engineer, Culmann toured America and recognized this problem (Timoshenko, 1953). Later, he used theory to design bridges in which stresses were low enough to avoid fatigue failure, but not unnecessarily low. Now, imagine what would happen if a way could be found to repair fatigue damage in a bridge soon after it occurs. Then the stress level could be allowed to rise, and the bridge could be made even lighter, saving further on materials and expense. That is the advantage that bones have over bridges. Damage repair by remodeling allows bones to be lighter, even though they carry stresses that would otherwise lead to fatigue failure within the organism's lifetime.

Damage repair is not part of mechanical adaptability in the sense that it shapes elements of the skeleton to control stiffness and stresses, but it is an important part of the picture insofar as it allows higher stresses to be tolerated. The same may be said of the elements of bone tissue structure that limit and control microdamage. In fact, all these matters are components of the system that has evolved to optimize bone stiffness while minimizing fracture risk and weight. This system works similarly in the skeletons of many different animals, from hummingbirds to elephants.

6.6 Mechanical Adaptability in Cartilage

There is evidence that chondrocytes also respond to mechanical signals and that this capacity is instrumental in the development of cartilaginous structures. These mechanical effects could affect the development of bones from their cartilaginous models in the embryo and the functioning of the growth plate. We describe next a pair of theories directed at these phenomena, based on assumptions about the overall effects of stress on cartilage.

The Carter–Wong Chondral Calcification Theory

As reviewed by Pauwels (1980), several early writers proposed that the differentiation of cells to produce different kinds of connective tissue (bone, cartilage, tendon, etc.) was differentially promoted or inhibited by various kinds of mechanical stress (tension, compression, shear). This concept also included changes in chondrocytes leading to the calcification of cartilage. Carter and co-workers (Carter, 1987, Carter et al., 1987a) developed this concept into a theory and explored its predictions with finite element models using methods similar to those described for bone. The basic premise of the theory is that the responses of chondrocytes to shear stresses lead to calcification of adjacent cartilage, while compressive dilatational (i.e., hydrostatic) stresses hold chondrocytes in a condition that preserves the cartilage in an uncalcified state.[9] This hypothesis is consistent with the aqueous nature of cartilage; that is, uncalcified cartilage can support hydrostatic loads much better than shear loads. To assign scalar values to the dilatational and shear portions of the local stress state, two variables are calculated from the three principal stresses (σ_1, σ_2, σ_3):

$$\text{Dilatational stress} = D = (\sigma_1 + \sigma_2 + \sigma_3)/3 \qquad (6.11a)$$

$$\text{Octahedral shear stress} = S = [(\sigma_1 - \sigma_2)^2 + (\sigma_2 - \sigma_3)^2 + (\sigma_3 - \sigma_1)^2]^{1/2}/3 \quad (6.11b)$$

It is proposed that the stimulus to calcify is an ossification index, I^*, given by

$$I^* \propto \Sigma\, n_i\, (k\, S_i + D_i) \qquad (6.12)$$

where k is a coefficient that determines the balance between the two kinds of stress in driving calcification. As in the bone adaptation models, the subscript i denotes different kinds of daily activities or loading categories, and the summation is over all such categories. If a finite element model is made of the cartilaginous anlage of a long bone like the femur, loads are applied to its ends to simulate joint and muscle forces, the elements in the model are made to calcify and become stiffer when Eq. 6.12 reaches a threshold, and the model is iterated, interesting things happen (Carter et al., 1987a). First, a region in the center of the diaphysis calcifies; this corresponds to the primary ossification center in the developing embryonic bone. Because of shear stresses in the cartilage next to the calcified region, the calcified region spreads toward the ends of the diaphysis. As it does so, shear stresses increase in the ends of the bone, and these regions start to calcify; they correspond to the secondary calcification centers in the embryonic epiphyses. As these three ossification centers continue to grow, the regions of uncalcified cartilage shrink to four zones, corresponding to the articular cartilage on the joint surfaces and the growth plates at each end of the bone. These regions persist as cartilage because a stable mechanical situation dominated by hydrostatic stress develops in these sites. Thus, this theory holds that the development of a bone from a cartilaginous model in the embryo is not simply "genetically programmed," but is in large part driven by the effects of mechanical stresses on the cells. This happens in much the same way that cells in the embryo are influenced by other *epigenetic factors* such as chemical gradients established by neighboring cells. Presumably, the growth plate calcifies and closes when the hormonal changes associated with puberty override the mechanical influences that have stabilized it.

---

[9]All stress states can be broken down into two parts: a hydrostatic or dilatational component, which produces a volume change but does not distort, and a shearing component, which distorts shape.

This principle has also been used to explain the morphology of synovial joints (Carter and Wong, 1988), the calcification patterns in the human sternum (Wong and Carter, 1988) and the characteristic pattern of calcification in a fracture callus (Carter et al., 1988).

Frost's Chondral Modeling Theory

While the Carter–Wong theory offers an explanation for the de novo development of the growth plate, Frost's (1990) chondral modeling theory offers mechanical reasons for changes in the orientation or shape of the growth plate as it functions. A much older theory known as the Heuter–Volkmann law holds that the physis grows faster when placed in tension and slower when normal amounts of compression are exceeded. This law is based primarily on the effects of static loads seen in clinicians' efforts to control the rate of growth of the whole growth plate. Frost's theory, on the other hand, is directed at understanding differential rates of growth within a growth plate, based on dynamic loading. It holds that higher stresses in either tension or compression accelerate growth, up to a point (Fig. 6.18). If peak compressive stress becomes too high, the growth rate will decrease. (If peak tensile stress is too high, the growth plate will be torn apart.) Now consider the implications of this with respect to the directionality of growth. If the dynamic joint force is not directed normal to the plane of the growth plate,

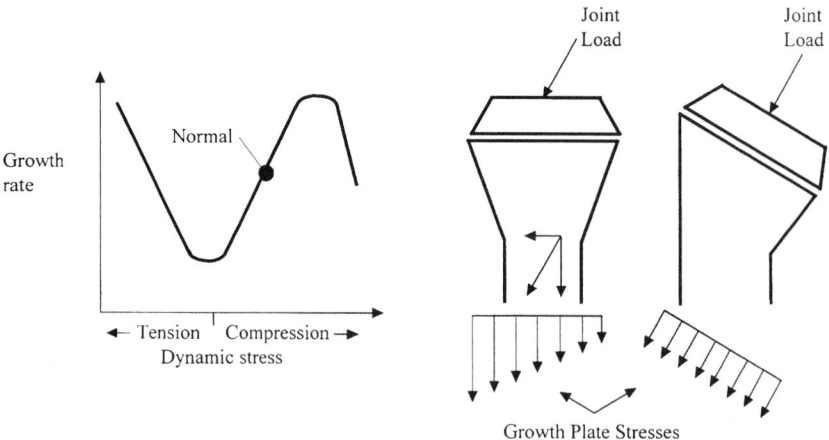

FIGURE 6.18. In the chondral modeling theory, growth rate increases with either tensile or compressive dynamic loading *(left)*. Excessive compressive stresses inhibit growth, however. A misaligned joint force *(middle)* creates a stress gradient across the growth plate and a corresponding growth gradient, resulting in rotation of the growth plate and elimination of the malalignment *(right)*.

but at an angle to it, bending stresses are superimposed on its normal compression. As a result, the stresses on one side of the physis are reduced relative to the other, creating a growth rate gradient across it. This gradient rotates the plane of the growth plate so as to realign it normal to the joint force. If, however, the dynamic compressive stresses increase too much on one side of the physis, growth there is slowed rather than accelerated, and the orthogonality of the growth plate and the joint load will get worse instead of better. Frost maintains that this theory explains why mild problems with genu varum and valgus get better on their own and more severe ones do not. He also asserts that the curvature of long bones is established by asymmetric physeal growth as well as modeling in the diaphysis.

A more complete theory, based on the actual molecular responses of chondrocytes to various stress conditions, is needed to provide a solid foundation for the Carter–Wong and chondral modeling theories, which appear to be two sides to the same coin. Together, the effects attributed to these theories may ultimately explain much of the diversity in joint morphology to be described at the end of Chapter 7. In concluding this chapter, let us now consider the relationship between mechanical adaptability and evolutionary changes in the skeletons of animals.

Box 6.1 Technical Note
Teleology and Adaptation, Optimization and Evolution

An engineer who has designed a structure can say, "The purpose of this strut is to reduce bending moments on this beam," but we cannot analogously say that the purpose of a strut of trabecular bone is to reduce the bending moment on an adjacent trabecula. The difficulty lies with the word *purpose*. The attribution of purpose to physiologic mechanisms is called teleology, and in this text we have tried to avoid it. We might say, instead, "a consequence of this trabecular strut is that bending moments on this adjacent trabecula are reduced." This is not just an esoteric point of semantics, it is important in understanding the nature of biological adaptation, whether it be somatic or evolutionary.

In the course of natural selection a group of organisms produces a great many genetic variants, some of which are better adapted in the sense that they will leave more progeny in the next generation. By this mechanism, species evolve, and adaptations such as bone remodeling come into being and are preserved. The process is a passive and stochastic one: the concept of **purpose** does not enter into it.

There are two sides to biological adaptation—the organism and its environment. When the environment is noisy and unpredictable, the adaptive process functions less efficiently. As we saw in Chapter 1, the mechanical environment of vertebrate bones is provided to a large extent by their own muscles; much of this loading is predictable and controllable. But the outside environment also directly loads bones, and these forces may be less predictable and controllable. If a creature walks on level ground all day, and at the end of the day steps unexpectedly into a hole, which of these is mechanically most significant in driving adaptation, the 20,000 cycles of gait or the one cycle of tripping that generates twice as much skeletal stress? Lanyon (1993) has favored the idea that skeletal adaptation is primarily an error-driven process, and it is the unusual loads that are most significant, but this important question has not been subjected to much research.

Organisms that live in water inhabit a mechanical environment which is seemingly more predictable than the terrestrial one, and so should be able to mechanically 'fine-tune' (a teleological term!) their skeletons better than land dwellers. But the buoyancy of water works against the adaptive advantage of a lighter skeleton in that a heavy skeleton balances low-density fat, resulting in an organism with an average density close to that of water, thereby facilitating aquatic locomotion. Conversely, birds (that is, flying birds, not terrestrial birds like chickens and ostriches) find exceptional adaptive value in a minimal-weight skeleton, which improves the efficiency of flight. Other factors contribute to the skeleton's mechanical milieu as well, including pregnancy and other seasonal fluctuations in body weight, antler growth, hibernation, migration, and fighting among males. These factors must complicate the optimization of skeletal structure.

This chapter is devoted to discussing the proposition that the skeleton contains adaptive mechanisms that optimize its structure in some sense. It would be a mistake, however, to hypothesize that **evolution** is directed toward maximizing efficiency of structures. Evolution is directed toward maximizing genetic representation in subsequent generations, which in turn is associated with efficiency, but only rather loosely. Indeed, the physiques of many ancient relict species contain features that are hardly paragons of efficiency. Consider the antlers of a moose. They are heavy and cumbersome, and get caught in branches, and require great amounts of calcium and phosphorus to construct. With respect to locomotion or food gathering, their contribution to efficiency is negative; yet they are favored by evolution, apparently because the moose with biggest antlers is the most likely to be genetically represented in the next generation.

6.7 Mechanical Adaptability and Evolutionary Adaptability

There has been considerable debate among engineers, biologists, and clinicians concerning the balance between genetics and mechanical adaptability in deciding the structures of bones. Some have argued that the anatomy and histology of a bone such as the human femur are determined primarily by genetic information expressed by various cells as they form the skeleton and maintain it in the adult. Others have come to believe that mechanical factors strongly influence the expression of the genetic plan carried by the cells. In closing this chapter, it is appropriate to consider the relationships between mechanical adaptability and evolutionary change as determinants of skeletal structure. This discussion is based on the ideas of Bateson (1963, reprinted in Bateson, 1987).

Somatic Vs. Evolutionary Adaptation

The adaptation of bone to alterations in mechanical loading is an example of somatic change, that is, a change in body form or function that does not entail a genetic change and is therefore not passed on to succeeding generations. Somatic change is distinct from evolutionary change, which does involve a modification of the genome of the organism. A simple example of somatic change is obtaining a sun tan during a beach vacation; obviously, this does not alter the genetic code for skin color, and the tan is not passed on to the vacationer's children. Nevertheless, somatic change plays an important role in evolution.

Bateson pointed out that the ability to adjust to environmental changes by somatic modification of body systems is essential to survival because evolutionary change operates relatively slowly, through chance mutations. In addition, somatic adjustments allow the organism to survive when a genetic mutation, which might ultimately have survival value, disrupts one or more body systems. Conversely, genetic adjustments to the somatic system itself allow the homeostatic range to be restored when a mutation has driven it to an extreme. Thus, evolution involves constant interaction between slow, haphazard genetic changes and rapid, predictable somatic changes. In short, somatic flexibility allows time for adaptive modifications of the genome to occur, and also helps the organism tolerate their adverse side effects.

A Nonskeletal Example

Consider, for example, what happens when the vacationer mentioned above goes instead to visit Machu Picchu, the ancient Incan ruins in the Peruvian Andes. After flying to Lima, the tourist's next stop is the city of Cuzco with an elevation of 3700 m (12,000 feet). Picking up her backpack and heading for her hotel, she soon finds her heart pounding and her breath coming in gasps because of the thin air. Clearly, these are responses to the trouble her cells are having in getting enough oxygen. Her increased volume and rate of breathing and pumping blood are somatic changes allowing her to cope with the environmental changes. By the time she leaves Cuzco 2 days later to continue her journey to Machu Picchu itself, carrying her backpack no longer requires her heart and respiration rates to be adjusted upward so much because her hematopoietic system has produced more red blood cells. This second level of somatic adaptability requires more time to put in place, but much less metabolic energy than pumping blood and air faster.

However, more red blood cells per unit volume of blood (called *hematocrit*) is not the best physiologic adjustment for living permanently at high altitude, either. "Lowlanders" who go to high altitudes may suffer severe cardiovascular problems caused by the effects of increased hematocrit on the physical properties of their blood. Some individuals can adjust to living at high altitudes, but many cannot. However, groups of people have lived at high altitudes (principally in the Andes and Himalyan mountains) for many generations and appear to be free of altitude sickness. It is possible that these populations have experienced genetic changes, through differential reproductive success, which have adapted them to high altitudes in ways that avoid the problems associated with increased hematocrit (Cruz-Coke, 1978).

Somatic Change in the Skeleton

Comparing bone's mechanical adaptation to Bateson's altitude adaptation example, the correlate to increased heart and respiration rates is an increased remodeling rate to repair load-induced microdamage. The correlate to increased hematocrit is periosteal and endosteal modeling to decrease the stresses in the bone. These skeletal responses can have problematic sequelae, just as the respiratory ones can. The increased rate of remodeling requires metabolic energy, and it introduces remodeling space porosity, which can increase strains and the damage formation rate. The formation of additional bone mass also requires metabolic resources, and it increases the bone mass that must be carried about each day, at a very significant energy cost.

Somatic Vs. Evolutionary Effects

How does one know when skeletal variations are of somatic or genetic (evolutionary) origin? The influence of skeletal somatic change is clear in the experiment of Woo and co-workers, in which the exercise level of pigs was increased, or that of Uhthoff and Jaworski, in which the forelimbs of dogs were immobilized. It is not at all clear, however, when one walks among the skeletons of existing and extinct animals in a museum. For example, the long bones and crania of Neandertals were much more robust than those of modern *H. sapiens*. Trinkaus (1983) has explained this as a somatic change: Neandertals lived during a very cold period in Europe and may have had to forage long distances for food. Trinkaus suggests that the bones adapted to this elevated level of exercise by becoming more robust. Data showing that bone turnover rates were lower in Neandertals than in modern populations (Abbott et al., 1996) support this hypothesis if the mechanostat theory is correct and if addition of bone to periosteal and endosteal surfaces prevented microdamage from accumulating as a result of the higher activity levels. It is also possible, however, that the robustness of the Neandertal skeleton represents a genetic difference from modern *H. sapiens*. For example, if the limits of the equilibrium strain range were genetically shifted to lower levels, ordinary loads would stimulate more bone formation through modeling in growing Neandertals, and suppress remodeling in adults, fitting the observations equally well. Alternatively, some aspect of just the periosteal tissue could have become genetically altered in Neandertals, so ordinary signals to periosteal cells resulted in enhanced bone formation and larger skeletons. The observation of a heavier cranium, hardly useful for foraging long distances for food, seems to fit with this possibility and further supports the hypothesis that the skeletal robustness of these creatures was not somatic. The likelihood of this seems increased now that the analysis of mitochondrial DNA (Krings et al., 1997) makes it unlikely that Neandertals are a part of our own genetic heritage (to the relief of some!). In any case, this example illustrates the difficulty in distinguishing between somatic and evolutionary skeletal changes on the basis of skeletal appearance alone.

In fact, Bateson's point was that all evolutionary change involves endless somatic and genetic adjustments, which together achieve adaptations that would be impossible if only one or the other existed. Thus, it is shortsighted to debate which caused what, exactly. It is more important to learn how somatic and genetic change interact in the skeleton, as they do in other organ systems. This kind of knowledge should ultimately be very useful in treating many bone injuries and diseases.

6.8 Summary and Further Reading

Wolff's law and the mechanical adaptability of bones to mechanical loading are not only of great interest to basic scientists interested in the skeleton; they are of great clinical utility as well. Orthopedists see the effects of these phenomena in their daily practices. We have reviewed the history of these concepts in some detail to gain a better understanding of how the current ideas evolved. We saw that there are three key concepts, including optimization of strength with respect to weight, alignment of trabeculae with principal stress directions, and regulation by systems of cells able to respond to a controlling stimulus. We then reviewed some of the more important experiments in this area, and saw that they have emphasized cortical rather than trabecular bone. Although disuse and overload effects are clear, it may be not be appropriate to relate the latter to normal development because woven rather than lamellar bone formation occurred.

We also reviewed some of the theories regarding mechanical adaptability of cortical and trabecular bone, and found that the putative ability of osteocytes to sense strain or strain energy is emerging as an important element of such theories. The discovery that control of strain energy within a loaded structure leads to 0–1 architectures with load-aligned trabeculae has been a key development in recent work. Finally, we asserted that damage repair is an important adjunct to mechanical adaptability because it allows higher strains to be tolerated, and that damage repair and mechanical adaptability are somatic changes which extend the organism's ability to cope with environmental changes without committing to an irreversible genetic change.

Much of the historical material and interpretation at the beginning of this chapter came from two reviews by Roesler (1981, 1987); these are essential reading for anyone interested in this subject. As Roesler notes, Wolff's monograph, *Das Gesetz der Transformation der Knochen*, has often been cited but seldom read. A short time with Maquet and Furlong's translation of this book (Wolff, 1892) gives one a much better concept of the insights and limitations of "Wolff's law." Reading the original papers of the experiments and theories that we have described will, of course, give a much greater appreciation for the finer points of this field. Frost's perspective remains unique in its attempt to balance biologic, mechanical, and clinical observations. Finally, *Bone Structure and Remodeling*, edited by Odgaard and Weinans (1995), is an excellent collection of papers by key investigators in this field, including a thought-provoking and entertaining exchange of biologic and engineering views of the subject by Currey and Huiskes.

6.9 Exercises

6.1. In the Lanyon–Rubin avian model for studying the effects of mechanical loading on bone, a diaphysis is isolated as shown in Fig. 6.19. Pins extend from the metal caps over the bone ends, as shown. Assume a load is applied between the ends of the pins as shown. Calculate the longitudinal strain on the upper and lower surfaces of the bone assuming it is cylindrical with the following characteristics:

Outer diameter = 20 mm
Inner diameter = 10 mm
Load = 30 N
Moment arm (d) = 40 mm
Elastic modulus = 20 GPa

6.2. Assume that a diaphyseal cross section carries a bending moment M. Let the cross section be circular with inner and outer diameters of 30 and 20 mm, respectively, and assume that the neutral axis (NA) passes through the centroid. Draw the cross section and write an equation

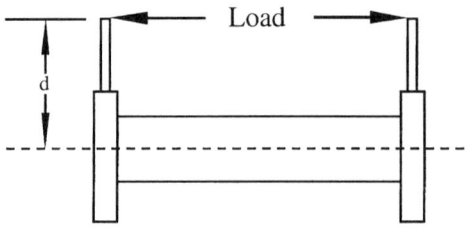

FIGURE 6.19. Isolated diaphysis loaded by projecting pins to be analyzed in Exercise 6.1.

for the stress, σ, as a function of angular position Θ measured from the neutral axis. Assume that modeling adds or removes bone on the periosteal and endosteal surfaces at a rate proportional to the local longitudinal stress on the surface:

$$M_{F/R} = -K\ \sigma(\Theta) \tag{6.13}$$

where the apposition rate $M_{F/R}$ is negative for tensile stress and positive for compressive stress and the rate constant K is the same for each kind of modeling and each surface. Sketch a line adjacent to the original periosteal surface that would represent the addition or removal of new bone at each point around the circumference after some arbitrary period of time. Do the same for the endosteal surface. (Scale your sketch so that the maximum thickness of bone added is about half the original cortical thickness.) What has happened to the bone's cross-sectional geometry? Would this change be advantageous in some way?

6.3. Reanalyze the bone in Exercise 6.2 assuming that $M_{F/R}$ is always formative and proportional to the **magnitude** of the longitudinal surface stress. How does the shape of this new bone distribution compare with that seen in Goodship's pig osteotomy experiment (Fig. 6.8B)? Discuss how the shape of the periosteal new bone distribution would change if M were proportional to the fourth power of the stress.

6.4. Reanalyze the bone in Exercise 6.2 assuming that $M_{F/R}$ is proportional to the radial **gradient** of stress. Assume that a positive stress gradient produces resorption ($M_{F/R} < 0$) and a negative stress gradient produces formation ($M_{F/R} > 0$). (A positive stress gradient is defined as one in which the stress becomes more tensile or less compressive, as one approaches the surface from within the bone matrix.) Let the rate constant be the same for formation and resorption, and for the periosteal and endosteal surfaces. Discuss the variations of the results in Exercises 6.2 through 6.4 and how they compare with the experimental results described in this chapter.

6.5. Draw a diagram to show that the stress gradients at the polar (resorption) and equatorial (reversal region) surfaces of an osteonal BMU aligned with a compressive stress direction are positive and negative, respectively, using the sign convention stated in the text and Exercise 6.4.

6.6. To see how 0–1 behavior develops in an adaptive model, try the following.[10] Use the Voigt model described in section 4.6 as a miniature "finite element model" containing just two elements. Let each element have the same cross-sectional area, $A = 5$ mm^2, and let this model be loaded by a force $F = 40\ N$. The strain, ε, is the same in both elements.

[10]This problem is adapted from an example given in Weinans et al. (1992).

the apparent density and modulus of each element are ρ_i and E_i, respectively, where $i = 1, 2$ identifies the elements. Assume that the time rate of change of apparent density is

$$d\rho_i/dt = B[(U_i/\rho_i) - k] \qquad (6.13)$$

where U_i is strain energy density $(U_i = E_i\varepsilon^2/2)$, B is a coefficient, and k is an equilibrium value of U_i. Assume also that the modulus-density relationship is $E_i = 2.5\ \rho_i^3$ (ρ_i in g/cm^3, E_i in GPa). Derive an expression for the strain as a function of F, A, E_1, and E_2; this can be used to calculate U_i. With these tools, set up a spreadsheet and iterate the model through a series of time steps to see what happens to the density of each element. Try starting with $\rho_1 = 1.9$ and $\rho_2 = 1.8$ g/cm^3. You will find you have to restrict the densities to a reasonable range; for example, 0–2.0 g/cm^3. Explore the behavior of this system by varying the starting densities, Eq. 6.13, and the modulus-density equation. How can you change the latter to avoid 0–1 behavior?

6.7. Consider the diaphysis of an adult's long bone subjected to a constant amount of daily torsion. Suppose that the apposition rate on the periosteal and endosteal surfaces is proportional to the difference between the average shear stress, σ_s (recall Eq. 4.4), and a set point or equilibrium value, σ_{SE}. If the difference is positive, formation occurs, and if it is negative, resorption occurs. Suppose that σ_{SE} is the same for both the periosteal and endosteal surfaces. What would the consequences of this be? [Such a model has been discussed in detail by van der Meulen and co-workers (1993; van der Meulen and Carter, 1995).]

6.8. In Exercise 6.7, a stable situation could be produced by introducing a "dead zone" in place of σ_{SE}, that is, a range of stress values within which the apposition rate remains zero. Sketch a graph of apposition rate (positive and negative values) vs. shear stress showing the dead zone; label the periosteal and endosteal response zones. What geometric feature of the bone would determine the width of the dead zone? Can you think of an assumption besides a dead zone that would permit stability?

6.9. An archer's skeleton, leather quiver strap still around his spine, was recovered from an English ship sunk in 1545. This man's left forearm bones and the tips of the first three fingers on his right hand were obviously thickened (Encyclopaedia Britannica, 1997, Vol. 29, p. 541). Think about the forces that produced these bony responses in the radius and ulna and in the phalanges. Considering the arm and finger responses separately, do you think the original periosteal responses involved lamellar or woven bone? Why? Should woven bone responses always be categorized as "pathologic" rather than "adaptive"?

7

Synovial Joint Mechanics

> [The bone ends] are covered with a smooth elastic crust, to prevent mutual abrasion; connected with string ligaments, to prevent dislocation; and enclosed in a bag that contains a proper fluid deposited there for lubricating the two contiguous surfaces.
>
> William Hunter (1743)

7.1 Introduction

Up to this point, this book has primarily discussed bone, but we now put aside this tissue and direct our attention to the soft tissues of the skeletal system. In this chapter we explore the mechanics of synovial joints, and in the last chapter, tendons and ligaments. However, we must not forget what we have learned about bone, because its presence heavily influences the characteristics of the new tissues we are going to study.

Synovial or diarthroidal joints are those that move freely and contain synovial fluid, like the knee and elbow joints, as opposed to joints that do not, such as the "tether" joints between the vertebral bodies in the spine. Figure 7.1 is a schematic diagram of a typical synovial joint. The synovial fluid is manufactured by cells in the *synovial membrane*, which lines the *fibrous joint capsule*. The joint capsule is aided by ligaments (not shown) and the surrounding musculotendinous tissue in holding the two bones of the joint

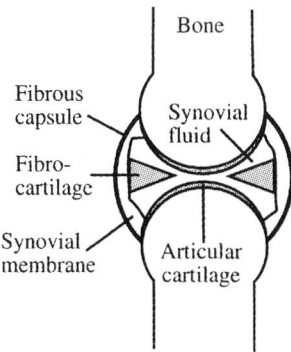

FIGURE 7.1. Schematic diagram of a typical synovial joint.

together (i.e., "stabilizing" the joint). The bone surfaces within the joint capsule are covered by *articular cartilage*, also called *hyaline cartilage*. The purpose of this tissue is **not** to serve as a shock absorber, as some textbooks state, but to provide a suitable surface for lubrication and wear prevention. Some joints also contain *fibrocartilage disks* that help maintain the other joint components in appropriate positions and reduce stress by extending the load-bearing surfaces. Such disks are generally found in joints where there is substantial rotation or torsion but incongruous joint surfaces (e.g., the acromio-clavicular joint). In the knee, these disks are called *menisci*. An important compositional difference between articular cartilage and meniscus is that the latter contains type I instead of type II collagen (Mow et al., 1992).

Functions of a Joint

Synovial joints provide essentially frictionless motion between limb segments while transmitting relatively high loads between them. Joints in animals are the equivalent of "bearings" in machines. Of course, these bearings do not support continuously rotating shafts, but components that start, stop, and oscillate. A similar engineering example might be the bearing that supports the oscillating tub in a washing machine.

We saw in Chapter 1 that the forces acting in the joints between bones are quite large relative to the external loads being supported; the stresses in the bones, tendons, and ligaments are therefore relatively high as well. The same may be said of the stresses in cartilage. The contact pressures exerted on the cartilage in various normal joints typically range between 2 and 11 MPa (Afoke et al., 1990).

Joint Diseases

Joint disease and osteoporosis are the two most common skeletal pathologies in modern society. One or the other seems to give almost everyone problems eventually. Their human and economic costs are enormous, and are increasing as other diseases are controlled and people live longer. Although these conditions do not kill directly, they do so indirectly in most painful and debilitating ways. (Immobilization itself, from the pain of arthritis, can be life-threatening to an elderly person.) The causes of both osteoporosis and arthritis are equally poorly understood, but the difficulties with understanding them are different. Mechanically, osteoporosis is rather simple: there is not enough bone tissue, and therefore the bone structure is weak and fractures occur. The complexity is in the bone remodeling biology that leads to diminished bone structure. In the case of an arthritic joint, on the other hand, both the mechanics of joint function and the cell-level biology are poorly understood. However, it seems clear that most kinds of arthritis involve feedback loops between mechanical damage processes and inflammatory processes (Fig. 7.2). For example, mechanical overuse may create fatigue damage in cartilage. This damage may incite an inflammatory process that activates bone and cartilage cells to not only repair the damage, but also to attempt to reshape the tissues so as to improve the mechanical situation. If too much begins to happen at once, however, the circumstances become chaotic. Enzymes intended to get rid of a damaged tissue also destroy healthy tissue. The size and shape of a joint can be altered rather easily in young people by bone and cartilage modeling, but after growth stops, these processes become sluggish and, apparently, inaccurate. New cartilage and bone production intended to reduce stress instead result in misshapen "spurs" (called *osteophytes*) around the periphery of the joint, which often get in the way and cause pain.

Alternatively, an infection or a defect in the patient's immune system may cause an inflammatory response in a joint when mechanical loading is normal. This response can lead directly to tissue destruction, which so jeopardizes mechanical properties that normal loads produce damage-level strains. Thus, there are many possibilities for the development of "vicious circles" in which mechanical and cellular processes in the bone and cartilage tissues, which normally operate in a controlled way at modest levels, become rampant and unstable.

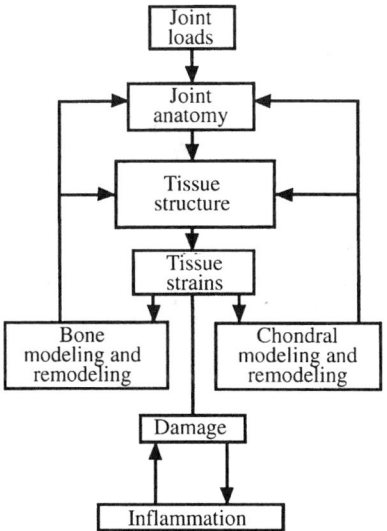

FIGURE 7.2. Block diagram of potential feedback loops in synovial joint physiology. Joint loads and anatomy determine tissue stresses, which produce tissue property-dependent strains that influence bone and chondral modeling and remodeling. If tissues are damaged, inflammatory responses as well as modeling and remodeling of bone and cartilage, may be activated, influencing joint anatomy and tissue properties. In some instances, the system may be stable and self-correcting; in others, it may become unstable and self-destructive.

Understanding joint mechanics is essential to understanding various forms of arthritis. The way synovial joints work is not entirely clear, but much is known and it is a fascinating story.

7.2 Mechanical Properties of Cartilage

Initial Points

There are three important initial points to be made about the mechanical role of cartilage.

Cartilage Is 70% Water

First, the mechanical properties of cartilage are very much dependent on its porous structure and proteoglycan content. The cartilage matrix of collagen and proteoglycan molecules constitutes about 30% of the tissue. The rest is essentially all water. As was emphasized in Chapter 2, proteoglycans are extremely hydrophilic, so that the tissue is essentially inflated with water that is held in place by electrochemical interactions with the proteoglycans. If anything interferes with the proteoglycan content (e.g., the release of destructive enzymes), or with the ability of chondrocytes to make normal cartilage (e.g., a genetic defect), the mobility of the water increases and the mechanical functioning of the cartilage will deteriorate.

The mechanical properties of both bone and cartilage are closely related to their porous structures, but cartilage is a very different kind of porous material compared to bone. First of all, the void dimensions of bone are 50–300 μm across; those of cartilage are about 50 Å. The solid matrix of bone is relatively macroscopic and rigid; that of cartilage is microscopic and flexible. Think of a porous pumice stone (bone) compared to a wad of steel wool (cartilage), each containing a different kind of liquid. The liquid in the porous stone has microscopic molecules, but that in the steel wool has molecules only slightly smaller than the steel fibers themselves. In addition, these big liquid molecules tend to adhere strongly to the steel fibers. It is easy to see why the resistance of these two systems to deformations of varying rates is quite different.

Reverting back now from the steel wool metaphor to the actual cartilage, we stress again the significance of the interactions between chemical and mechanical factors. The numbers of negative charge groups in the proteoglycans will rise and fall in proportion to the pH of the synovial fluid. Also, increased NaCl concentration in the synovial fluid provides positively charged ions that are attracted to and shield the proteoglycan charges. For these and other reasons, the chemistry of the cartilage has substantial effects on its mechanical properties.

Cartilage Is Highly Viscoelastic

Because cartilage is a very porous, water-filled material, its mechanical properties are viscoelastic. In particular, its elastic modulus varies greatly with the rate of loading. At very slow rates of loading, the modulus is only a few MPa, but at physiologic rates of dynamic loading it may approach 500 MPa (Higginson and Snaith, 1979; Radin et al., 1970; Unsworth, 1981; Yamada, 1973). This range of behaviors should be borne in mind when reading the rest of this chapter, because many of the descriptions of cartilage stiffness measurements refer to behavior at slow rates of loading. For example, indentation tests are usually quasistatic, and the "equilibrium" or "aggregate" modulus refers to the stiffness associated with an infinitely slow rate of loading.

Cartilage Is Not a Shock Absorber

Finally, while it is much less stiff than cortical bone, cartilage does not serve to "cushion" or reduce impulsive forces in joints. It cannot do this because it is so thin that its capacity to absorb energy is insignificant compared to (a) eccentric contractions of muscles and (b) energy absorption in the bones on either side of the joint. (Exercise 7.1 asks you to demonstrate this to yourself.) Instead, the role of cartilage is to provide a self-renewing, well-lubricated, load-bearing surface.

Structure of Articular Cartilage

Figure 7.3 is a schematic diagram of the structure of articular cartilage in relation to its foundation of subchondral bone. The collagen fibers in the cartilage have been described as forming arches (the "arcades of Benninghoff") (1925) such that their orientation is parallel to the surface in the *superficial tangential zone* and perpendicular to it in the deeper zones. The chondrocytes tend to line up in columns parallel to the collagen fibers in the middle and deep zones. As in the growth plate, the deepest layer of

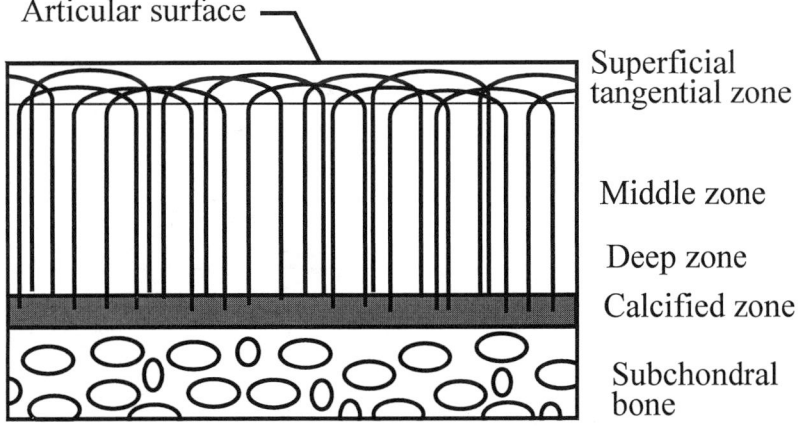

FIGURE 7.3. Schematic diagram of collagen fiber orientations in articular cartilage.

cartilage calcifies.[1] The junction between calcified and uncalcified cartilage is called the *tidemark* and shows up as a line in stained histologic sections. The calcified cartilage layer may minimize the stress concentration that would exist at the bone–cartilage interface by virtue of their different stiffness. Beyond these general observations, the structure of cartilage remains poorly understood, but even this degree of understanding reveals much about the origins of the anisotropy and inhomogeneity in cartilage's mechanical properties.

Permeability

Because cartilage consists of an organic matrix and water, it is useful to think of it as a *biphasic* system, with water being one phase and the collagen–proteoglycan matrix the other. The volume fraction of the water phase can be called the porosity of the cartilage. Although the water molecules are restrained to a significant degree by the proteoglycan molecules, they may nevertheless migrate through the tissue under the influence of a pressure gradient. This may be modeled using Darcy's law:

$$Q = k \, dP/dx \qquad (7.1)$$

where Q is the flow rate per unit cross-sectional area, dP/dx is the pressure gradient, and k is the *hydraulic permeability coefficient* (sometimes called the *apparent permeability*). Typical values of k are shown in Table 7.1. Permeability diminishes in the deepest layers of the articular cartilage, and also varies with the composition of the liquid phase. Clearly, the viscoelastic resistance of the tissue to deformation depends on its permeability. On the other hand, the permeability is a function of deformation. We return to this concept later.

TABLE 7.1. Properties of lateral femoral condyle cartilage in several different species

Variable	Human	Bovine	Canine	Monkey	Rabbit
Aggregate modulus, MPa	0.70	0.89	0.60	0.78	0.54
Poisson's ratio	0.10	0.40	0.30	0.24	0.34
Permeability, $10^{-15} m^4/Ns$	1.18	0.43	0.77	4.19	1.81

From Mow et al., 1991.

---

[1] In fact, one may think of articular cartilage as a miniature growth plate that exists throughout one's lifetime. Its chondrocytes manufacture cartilage very slowly, however, so that instead of adding length to the parent bone they simply replace the cartilage worn away at the articular surface.

Indentation Testing

It is difficult to form mechanical test specimens of cartilage because it is only found in thin (2–3 mm) layers. Therefore, many of the early mechanical tests involved indentation of the surface as a means of gaining compressive stiffness data. Such experiments were based on the equation derived by Boussinesq (1885) for a hemispherical indenter applied to an elastic half-space (Fig. 7.4). If an indenter of radius r penetrates to a depth d under load P, the elastic modulus is given by

$$E = P(1-v^2)/2dr \qquad (7.2)$$

where v is Poisson's ratio of the indented material. Assuming that the cartilage was indeed of infinite depth and that it was incompressible (so that $v = 0.5$), Sokoloff (1966) obtained $E = 2.3$ MPa for the elastic modulus of articular cartilage. Later, when a mathematical solution was obtained for indentation of an elastic layer on a rigid foundation (i.e., cartilage on relatively rigid bone) by a plane-ended cylinder, Hayes et al. (1972) found that E was only 1.6 MPa. As more work was done, it became clear that this value, too, was probably inaccurate because cartilage is not truly incompressible and fails to meet the assumptions of the tests in other ways as well (e.g., it is anisotropic, inhomogeneous, and viscoelastic). Table 7.1 shows more recent data on the compressive modulus and Poisson's ratio of articular cartilage for various species. These data were obtained from compression tests on machined specimens using techniques similar to those described later in this section.

Tensile Tests

Woo and co-workers (1976) pioneered the tensile testing of cartilage. They sliced articular cartilage into miniature "dumbbells" having a thickness of 0.25 mm and clamped them into a tensile test apparatus. Strain was measured by videotaping the specimen as it was pulled apart and measuring, in each video frame, the distance between two lines drawn across the undeformed specimen.

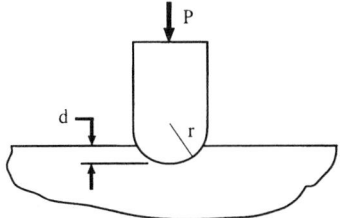

FIGURE 7.4. Diagram of indentation experiment. An indenter having a hemispherical end of radius r is pressed into the surface of the tissue for a distance d by a load P.

When cartilage is loaded, the way the force is shared between the fluid and solid phases is highly dependent on the nature of the loading. In tension, it is primarily the solid phase of the cartilage that carries the load; in particular, it is the collagen fibers. Figure 7.5 shows a typical tensile stress-strain relationship for a collagenous soft tissue. It has a "toe" region with an initially low modulus, followed by a region of increasing stiffness. The sketches on the right side of the figure illustrate how the toe region is attributable to straightening of the "fibers" in the material; in the case of cartilage, it is probably the collagen fibrils more than the proteoglycan molecules that are behaving in this way.

Cartilage produces the same sort of curve seen in Fig. 7.5 except that its upper portion remains nonlinear because the collagen fibers are not brought into full alignment. Its stress-strain relationship can therefore can be described by an equation of the form

$$\sigma = A\,(e^{B\varepsilon} - 1) \tag{7.3}$$

where σ is stress, ε is strain, and A and B are coefficients. To find the modulus at any point on the curve, one merely differentiates to obtain

$$E = d\sigma/d\varepsilon = B(\sigma + A) \tag{7.4}$$

Figure 7.6 shows the shapes of such stress-strain curves for bovine articular cartilage as a function of depth below the surface and orientation relative to the split line direction (recall Fig. 2.16). Table 7.2 shows the elastic modulus at a stress of 2.5 MPa for each position and orientation. Note that the tensile modulus is greater parallel to the split lines and nearer the articular surface, where the load direction corresponds to the collagen fiber orientation. Ultimate tensile stresses for articular cartilage range from 9 to 18 MPa, and ultimate tensile strains from 60% to 120%.

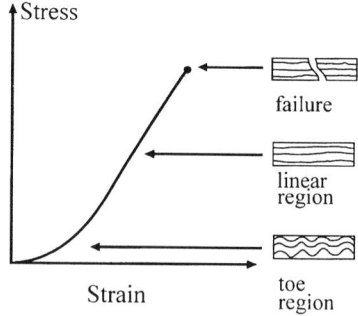

FIGURE 7.5. Typical tensile stress-strain curve for collagenous soft tissue. The diagrams at *right* indicate how the collagen fibers straighten in the toe region, become stretched in the linear region, and finally break at failure. (Reproduced with permission from Buckwalter and Mow, 1992.)

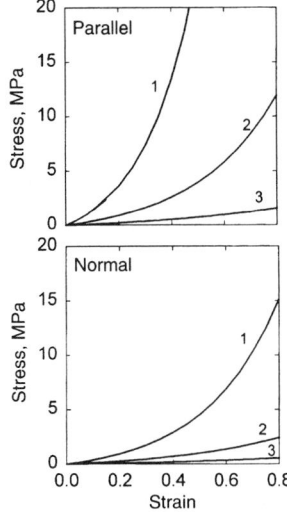

FIGURE 7.6. Stress-strain curves for bovine cartilage specimens as functions of orientation and depth. (Plotted from data from *J Biomechanics*, Vol. 9, SL-Y Woo et al., Measurements of nonhomogeneous directional mechanical properties of articular cartilage in tension, 785–791, 1976, with permission from Elsevier Science.)

TABLE 7.2. Data for bovine cartilage

Direction	Slice # (relative depth)	Tensile modulus, MPa
Parallel to split lines	1	23.0
	2	11.2
	3	5.0
Normal to split lines	1	12.2
	2	6.6
	3	3.6

From Woo et al., 1976, with permission from Elsevier Science.

Biphasic Theory of Articular Cartilage

When a cartilage specimen is loaded in tension or compression, drops of fluid can be seen to exude and collect on its surfaces. This is a clear sign that the cartilage liquid phase flows under the influence of pressure gradients produced by mechanical deformation of the solid phase. Clearly, movement of the fluid phase produces viscoelastic behavior of the tissue. On the other hand, the permeability of the system is also altered when the solid phase is deformed. Mansour and Mow (1976) studied this behavior in 200-μm-thick specimens of cartilage clamped between two porous metal platens. Through these porous platens a pressure gradient was imposed across the specimen: the applied pressure, P_A. The resulting flow was measured and the permeability coefficient k calculated. By compressing the specimen to different strain levels (ε), the relationship between permeability and tissue strain could be studied. Typical results (Fig. 7.7) show that an interaction occurs between the deformations of the fluid and solid phases when cartilage is loaded. The curves can be fit to the equation

$$k = A\, e^{\alpha \varepsilon} \tag{7.5}$$

where A and α both decrease as the applied pressure increases. To understand this interplay, Mow and co-workers developed what has become known as the *biphasic theory* of cartilage mechanics. This theory assumes that cartilage consists of a soft, porous, elastic solid, the pores of which communicate and are filled with water. Both phases are assumed to be incompressible. The details of this theory are beyond the scope of this book, but several examples are presented showing how it describes the compressive viscoelastic behavior of cartilage.

Confined Compression Creep Test

Mow et al. (1980) conducted creep tests of cylindrical specimens of cartilage using the scheme shown in Fig. 7.8. A cylindrical specimen of cross-sectional area A is cored from a joint surface and placed in a cylindrical metal well with solid walls and bottom. A porous metal loading platen is placed on top of the specimen. A constant load, F, is applied at time $t = 0$ and the specimen deformation is recorded until an equilibrium deformation is reached. In this situation, creep is controlled by the exudation of fluid through the porous loading platen against the specimen's top surface. The movement of fluid to this interface is in turn controlled by the permeability, k, of the specimen. The equilibrium deformation is controlled by the

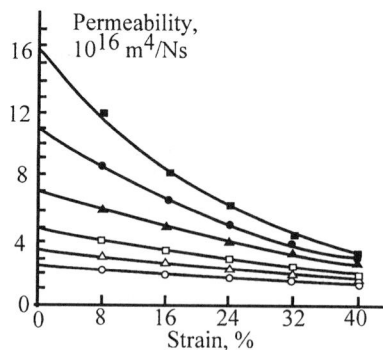

FIGURE 7.7. Permeability vs. compressive strain for cartilage. The family of curves shows how permeability decreases as increasing pressure is applied. From *top* to *bottom*, the pressures were 0.17, 0.34, 0.69, 1.03, and 1.72 MPa. (Reproduced with permission from Skalak et al., *Handbook of Bioengineering*, 1987a, McGraw-Hill: New York. With permission of The McGraw-Hill Companies.)

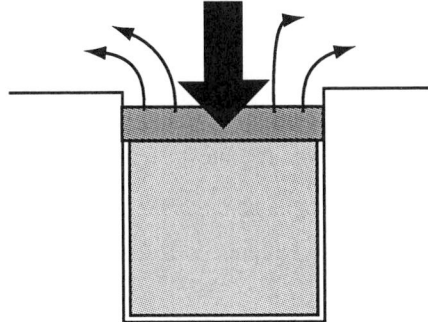

FIGURE 7.8. Schematic diagram of the experimental setup used for creep compression tests of cartilage. Fluid is allowed to flow through the porous loading platen.

equilibrium or *aggregate modulus* of the specimen, H_A. If the initial height of the specimen is h, and its displacement under stress $\sigma = F/A$ is $u(t)$, the biphasic theory solution for the ratio of the deformation to the initial specimen height as a function of time is (Woo et al., 1987a)

$$u/h = (\sigma/H_A)\{1 - 2\ \Sigma\ [\exp(-\Lambda H_A kt/h^2)/\Lambda]\} \tag{7.6}$$

where t is time, $\Lambda = (n + 1/2)^2 \pi^2$, and the summation is from $n = 0$ to ∞. Notice that as $t \to \infty$, this equation simplifies to $\sigma = H_A\ (u/h)$. Figure 7.9 and Table 7.3 show the results of fitting this equation to creep data for articular cartilage and two other kinds of cartilaginous tissue: nasal cartilage and knee meniscus. The distinct differences in the properties of these tissues are striking. Notice the inverse relationship between the aggregate modulus and the equilibrium deformation. Also, the cartilage with the greatest water content has the greatest permeability.

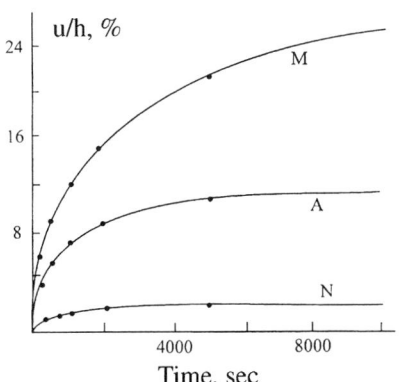

FIGURE 7.9. Strain (u/h) vs. time for articular cartilage *(A)*, meniscus *(M)*, and nasal cartilage *(N)*. (Reproduced with permission from Skalak et al., *Handbook of Bioengineering*, 1987a, McGraw-Hill: New York. With permission of The McGraw-Hill Companies.)

TABLE 7.3. Compressive properties of three types of bovine cartilage

Tissue	Aggregate modulus (H_A), MPa	Permeability (k), $10^{15} m^4/Ns$	H_2O by weight, %
Nasal cartilage	5.64	0.49	75.6
Articular cartilage	0.85	4.67	80.9
Meniscus	0.41	0.81	73.9

From Woo et al., 1987a.

Confined Compression Relaxation Test

In this experiment, the apparatus is the same as that depicted in Fig. 7.8 except that the instantaneous load application is replaced by a ramp displacement like that shown in Fig. 7.10. After a specified maximum displacement is reached, it is maintained while the reactive stress produced in the specimen is monitored. As Fig. 7.10 shows, there is a nonlinear increase in stress until maximum displacement is reached, then the stress "relaxes" to an equilibrium value. Fluid flows from the specimen early in the test (e.g., at A and B), but later the strain redistribution is accompanied by entirely internal fluid movements (e.g., at C and D).

The biphasic theory predicts the stress during both the compression and the relaxation phases of the experiment. The equation for the latter is (Woo et al., 1987a)

$$\sigma = -(H_A V_0/h)\{t_0 - 2 \Sigma [K^{-1} \exp(-Kt) - 1]\} \tag{7.7}$$

where V_0 is the displacement rate, t_0 is the time at which the displacement is stopped (B in Fig. 7.10), and $K = n^2 \pi^2 k H_A/h^2$. The other symbols retain their former definitions, and the summation is from $n = 1$ to ∞. Note again that as $t \to \infty$,

$$\sigma = H_A V_0 t_0/h = H_A \varepsilon \tag{7.8}$$

because $V_0 t_0/h$ is the strain, ε. These equations fit experimental relaxation data quite well, and the values of H_A and k that achieve these fits agree well

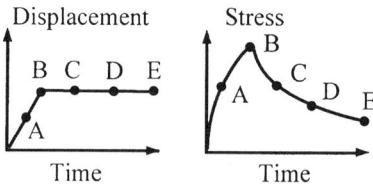

FIGURE 7.10. *Left:* Displacement applied to cartilage specimen in compression relaxation test. *Right:* Stress-time curve for the test. Letters *A–E* show equivalent positions on the two curves. (Reproduced with permission from Skalak et al., *Handbook of Bioengineering,* 1987a, McGraw-Hill: New York. With permission of The McGraw-Hill Companies.)

with those obtained from creep experiments. There are advantages to both methods of obtaining the aggregate modulus. The relaxation test takes less time to reach equilibrium, but it requires an apparatus to achieve a displacement curve like that shown in Fig. 7.10.

In the course of testing various kinds of specimens, it has been learned that the aggregate modulus of cartilage is inversely proportional to water content but increases in proportion to hyaluronic acid content. For example, diminishing the water content from 85% to 75% increases the modulus from about 0.3 to 1.0 MPa, and a 30% increase in hyaluronic acid content can double the modulus. There is also a reduction in modulus with age, but this is not a predictive relationship because the individual variability at every age is very great.

Inhomogeneity of Modulus
Schinagl and co-workers (1996) loaded excised specimens of bovine articular cartilage in compression and then measured the displacements within the tissue as a function of depth below the articular surface. They allowed 40 min to elapse after applying the load to allow the displacement field within the cartilage to reach an equilibrium state. Displacements were then measured by using a microscope to record the positions of chondrocytes before and after application of the load. (The DNA in the cells was marked with a fluorescent dye to make them easier to locate.) The results showed that displacements were greater near the articular surface and diminished exponentially through the thickness of the cartilage; thus, the modulus in the upper layers was a fraction of that in the depths of the cartilage.

Effect of Strain Rate on Tensile Tests
Figure 7.11 shows a graph of the effects of varying strain rate on the stress-strain relationship of bovine articular cartilage tested in tension (Woo et al., 1987a). As with bone, the stiffness of cartilage increases as it is deformed more

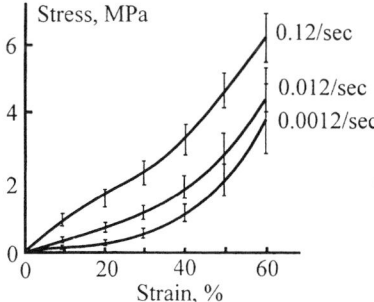

FIGURE 7.11. Stress-strain curves for cartilage at three different strain rates. (Reproduced with permission from Skalak et al., *Handbook of Bioengineering*, 1987a, McGraw-Hill: New York. With permission of The McGraw-Hill Companies.)

quickly. Notice that the curve for the fastest strain rate has an inflection point, but the others do not. This phenomenon is thought to be caused by resistance to fluid flow. At slow strain rates, there is time for fluid to flow without offering much resistance to the applied deformation. At high strain rates, fluid pressure resists the loading early in the test. This pressure also expresses fluid from the tissue, reducing its participation in the production of stress and allowing the slope of the stress-strain curve to decline (i.e., to inflect). The biphasic theory produced the following equation to describe this kind of a test:

$$\sigma = E_S\, \varepsilon + (b^2\, d\varepsilon/dt)/(H_A k)\, [A_0 - \Sigma[A_n \exp(-4\alpha_n^2\, H_A\, k\, t/b^2)/\alpha_n^2] \quad (7.9)$$

where ε is strain and $d\varepsilon/dt$ is strain rate, E_S is the elastic modulus of the solid matrix, and A_n and α_n are constants. The summation and other symbols are as in Eqs. 7.6 and 7.7. This result describes only the low strain portion of Fig. 7.11 (i.e., $\varepsilon < 10\%$) because the biphasic theory assumes infinitesimal strains and linearity. At higher strains the collagen fibers have been stretched to the point that their participation in resisting the applied deformation is of increasing importance. It is postulated that the three curves have similar slopes at their right ends because by then the stretched-out collagen fibers are carrying virtually the entire load, and fluid effects are insignificant.

Effect of Strain Rate on Compression Tests

The tensile elastic modulus of the specimens in Fig. 7.11 is in the neighborhood of 1–10 MPa, depending on strain magnitude and rate. This value is similar to the aggregate modulus for compression of articular cartilage at low strain rates. There is evidence that the compressive modulus is at least an order of magnitude higher when dynamic loading is involved. Figure 7.12 shows how this modulus (measured at 10% strain) increases as a function of strain rate. Even higher moduli have been recorded for the dynamic

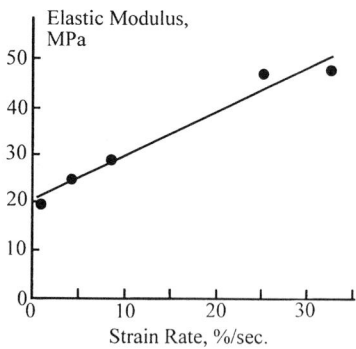

FIGURE 7.12. Compressive elastic modulus vs. strain rate for bovine articular cartilage. (Reproduced with permission from Radin et al., 1970.)

modulus of cartilage when the dynamic loading is superimposed on creep strain. (Imagine the cartilage in your knees creeping into a compressed state while you stand waiting for a traffic light to change, then being dynamically loaded as you step off the curb.) Under these conditions, the cartilage behaves as a linear elastic material with a modulus of 100 to 500 MPa, depending on the amount of creep strain (Higginson and Snaith, 1979).

In summary, the mechanical properties of cartilage are largely determined by fluid flow through the collagen–proteoglycan matrix. Thus, permeability is an important factor, and the tissue is highly viscoelastic. The importance of this kind of behavior becomes clear in the next section. Unlike most engineering bearings, synovial joints do not involve lubricants acting on rigid, inert surfaces. Instead, the lubricant flows into, and interacts with, bearing surfaces that change their shape to accommodate both loads and lubrication.

7.3 Lubrication of Joints

Synovial joints have extremely low friction coefficients, about equal to an ice skate on ice. Just how this happens is not fully understood, but much can be deduced on the basis of research on joints themselves and on engineering lubrication theory, which has a long and interesting history.

Friction

Friction is produced between sliding surfaces because they cannot be perfectly smooth, but have projections that interlock (Fig. 7.13). Frictional forces arise from interdigitation of the asperities on the two surfaces. The compressive force between the sliding surfaces must be overcome to enable the interdigitating asperities slide up and over one another. In addition, adhesive forces may exist between the two surfaces.

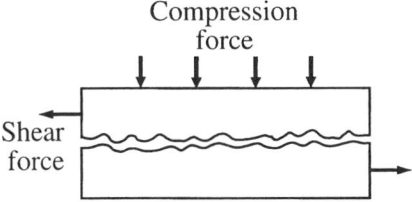

FIGURE 7.13. Representation of two surfaces sliding on one another. The required shear force must exceed the frictional force resulting from interactions between the asperities of the two surfaces.

Thus, to slide more easily past one another, the two surfaces must be separated slightly against the action of the compressive force. The primary purpose of lubrication is to reduce interlocking and adhesion between the asperities on the two surfaces. The surfaces of articular cartilage seem to have ridges not unlike those in the figure, at a microscopic level. The roughness of these surfaces has been observed to increase with aging and joint disease. However, drying of the tissue during preparation for microscopy may exaggerate such features, and there is currently uncertainty about the magnitude of the surface ridges on cartilage.

The two basic laws of friction are known as Amontons' laws[2]:

- The friction force is directly proportional to the applied load.
- The friction force is independent of the apparent area of contact.

The friction force is defined as the shear force required to pull one surface across another. The friction coefficient, μ, is just the ratio of the friction force, F_F, to the applied load, or normal force, N, which pushes the surfaces together:

$$F_F = \mu N \qquad (7.10)$$

Static friction coefficients usually are greater than sliding friction coefficients. Typically, friction coefficients between unlubricated surfaces are about 0.3 and those in lubricated machinery bearings are 0.05 or less (Table 7.4). Friction coefficients in synovial joints are among the smallest ever measured, about 0.001–0.005.

The reason that the friction force is independent of the bearing surface area is that when the area increases, the number of barriers to sliding increases, but the compressive force across each one decreases. Therefore, making an articular surface larger reduces the continuum level or macroscopic stresses in the tissues of the joint, but in principle it does not affect the frictional force.

Wear

Wear occurs in a bearing because projections become broken by crashing into one another or tear off because of adhesion, or because foreign bodies create additional "mountains" that temporarily lodge on one surface and gouge the opposite surface. These phenomena occur in synovial joints as well, causing the cartilage surfaces to wear away over time. Foreign bodies play little part in wear of normal joints, but can play a major role in prosthetic joints, where wear particles become foreign bodies and accelerate the wear problem. They can also migrate to the periphery of the joint, contact the synovial membrane, and cause an inflammatory response. This response may in turn produce enzymes that cause bone resorption or pain. Reduced friction is associated with reduced wear.

---

[2]The French scientist Guillaume Amontons presented his classic paper on this subject to the Academie Royale in December of 1699.

TABLE 7.4. Typical friction coefficients

Materials and conditions	μ
Rubber on concrete, wet or dry, static	1.0
Brake material on cast iron, clean and dry, static	0.4
Brake material on cast iron, lubricated with mineral oil, static	0.1
Graphite on steel, static	0.1
Hickory on dry snow, waxed, 4 m/s, −3°C, dynamic	0.18
Hickory on dry snow, unwaxed, 4 m/s, −3°C, dynamic	0.08
Ice on ice, 4 m/s, 0°C, dynamic	0.02
Articular cartilage in human joints, dynamic:	
Human knee; Charnley (1960)	0.005–0.02
Porcine shoulder; McCutchen (1962)	0.02–0.35
Canine ankle; Linn (1967, 1968)	0.005–0.01
Unsworth et al. (1975)	0.01–0.04
Malcom (1976)	0.002–0.03

Values from *CRC Handbook of Chemistry and Physics* (1984) and Mow and Soslowsky (1991).

Types of Lubrication

There are several different mechanisms by which lubrication can be achieved. Each of the following mechanisms is thought to play a role in the lubrication of synovial joints.

Boundary Lubrication

> All things and everything whatsoever however thin it be which is interposed in the middle between objects that rub together lighten the difficulty of this friction.

So said Leonardo da Vinci, and while this translation may not win a prize for punctuation, it is perhaps as good a description as any of the idea behind *boundary lubrication*. Placing a "boundary" layer (i.e., an extremely thin layer, perhaps one molecule thick) of almost any liquid-like material between the bearing surfaces holds the surface projections slightly apart, reduces adhesion, and keeps the projections from interlocking. However, it does not produce the lowest possible friction coefficients because there is still some contact between the surfaces.

Squeeze Film Lubrication

A distinct improvement over boundary lubrication can be realized if a total separation of the bearing surfaces can be achieved by creation of a pressurized layer of fluid between them. One way to achieve this is as follows. When you run and your leg swings forward, its knee joint is minimally loaded, allowing the cartilage surfaces to separate and minimizing the friction. When this leg enters its stance phase, body weight pushes the femoral condyles down through the synovial fluid toward the tibial side of the joint. When contact is made, frictional forces greatly increase. The viscosity of the fluid resists the descending femur and increases the amount of time required for the two sides of the joint to meet. In the meantime, friction remains low. The bearing surface separation produced by such load-and-unload cycles is called *squeeze film* lubrication.

Hydrodynamic Lubrication
A way to achieve prolonged fluid pressure to hold up the body weight and separate the joint surfaces is to use parallel rather than perpendicular relative motion between them, that is, sliding motion of one surface across the other (Fig. 7.14). The fluid next to each surface sticks to it. As one surface moves across the other, friction between adjacent fluid molecules drags the intervening fluid into the space between them, creating hydrodynamic pressure that forces them very slightly, but completely, apart. The friction force is essentially only that between the fluid molecules. For this kind of lubrication to work, the bearing or joint must be moving fairly rapidly. Thus, it can function in synovial joints only during flexion or extension motions.

Hydrodynamic lubrication was discovered by Beauchamp Towers when he was asked in 1882 by a royal commission in England to study the problem of friction in machinery (Dowson, 1979). Soon after, in 1886, Osbourne Reynolds published a theory explaining how hydrodynamic lubrication worked. While the application of this research was to be in the field of conventional engineering, Reynolds ended his classic paper with the speculation that hydrodynamic lubrication was important in biologic as well as engineering bearings. It is worth quoting from his paper because he squarely identifies the issue in joint lubrication that remains uncertain today: how can a reciprocating bearing maintain a

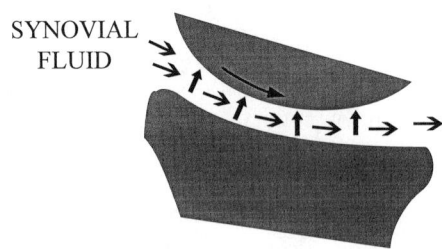

FIGURE 7.14. Hydrodynamic lubrication results in complete separation of the surfaces.

state of hydrodynamic lubrication?

> The question as to whether this action can be continuous or not turns on whether the action tends to preserve the matter between the surfaces at the points of pressure as in the apparently unique case of the revolving journal, or tends to sweep it to one side as is the result of all backwards and forward rubbing with continuous pressure. The fact that a little grease will enable almost any surfaces to slide for a time has tended doubtless to obscure the action of the revolving journal to maintain the oil between the surfaces at the point of pressure. And yet, although only now understood, it is this action that has alone rendered our machines, and even our carriages possible. The only other self-acting system of lubrication is that of reciprocating joints with alternate pressure on the separation ... of the surfaces. This plays an important part in certain machines, as in the steam engine, and is as fundamental to animal mechanics as the lubrication action of the journal is to mechanical contrivances.

It is clear that Reynolds not only foresaw the possibility that hydrodynamic lubrication occurs in synovial joints, but also that there would be difficulty in explaining how the system could work if it involved "rubbing with continuous pressure."

To gain a more tangible feel for the criteria that must be met if a joint is to be hydrodynamically lubricated, we shall derive Reynolds' equation and use it in an example problem. The first step in this process is to recall the definition of viscosity. In an elastic solid, stress is linearly proportional to strain and the slope of the stress-strain curve is the elastic modulus. In a Newtonian fluid, stress is linearly proportional to strain rate, and the slope of the line is called viscosity (Fig. 7.15). If the stress-strain rate relationship is nonlinear, the fluid is non-Newtonian. If the slope of the curve decreases with increasing strain rate, the fluid is said to be *thixotropic*. It will be seen that synovial fluid is thixotropic, and that this is advantageous for joint lubrication.

We consider a simplified joint such as that shown in Fig. 7.14. We assume that there is a flow of fluid between the joint surfaces and that:

- The joint surfaces are impermeable to the fluid.
- The fluid is incompressible and Newtonian.

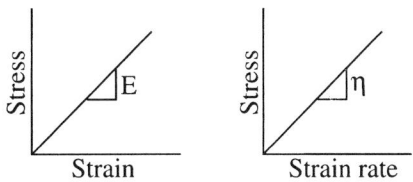

FIGURE 7.15. The correspondence between elastic modulus in solids and viscosity in fluids.

- The fluid's viscosity is homogeneous and isotropic.
- The pressure gradient in the vertical direction is much less than that in the flow direction.

First, we obtain a continuity of flow condition. Consider the column of fluid of width dx sketched in Fig. 7.16A. Let Q be the flow rate (per unit width of the joint space) into the left side of the column, and $Q + dQ/dx$ be the flow rate out the right side. Because we do not want any fluid sources or sinks in the column, we require that the flow in equal the flow out, and this reduces to

$$dQ/dx = 0 \qquad (7.11)$$

Now consider the balance of forces on a cube within the column (Fig. 7.16B). Assuming the flow is associated with a pressure gradient in the x-direction, the force on the left side is $p\ dydz$ and that on the right side is $(p + dp/dx)\ dydz$. There are also shear stresses on the top and bottom surfaces of the cube because one side of the joint is moving relative to the other. If the shear stress is τ, then the shear force on the bottom (acting to the left) is $\tau\ dxdy$ and that on the top (acting to the right) is $(\tau + d\tau/dz)$ dxdy. Equating the forces in each direction gives

$$p\ dydz + (\tau + d\tau/dz)\ dxdy = (p + dp/dx)\ dydz + \tau\ dxdy = 0 \quad (7.12)$$

which reduces to

$$dp/dx = d\tau/dz \qquad (7.13)$$

If the fluid is Newtonian, then the shear stress is related to the velocity gradient in the z-direction by the viscosity, η, so that

$$dp/dx = d/dz\ (\eta dv/dz) \qquad (7.14)$$

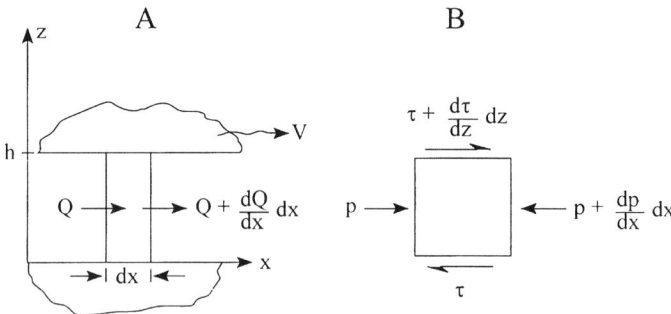

FIGURE 7.16A,B. Diagrams for derivation of Reynolds' equation. **A** A fixed column of fluid with thickness dx defined within the space between the joint surfaces. The upper surface is moving with velocity V relative to the lower surface. **B** Free-body diagram of a small cube of fluid within the column in **A**.

where v is velocity. Because the pressure, p, is not a function of z, we may integrate this equation to obtain

$$\eta \, dv/dz = dp/dx \, z + C_1 \tag{7.15}$$

where C_1 is the constant of integration. Assuming, as we have, that η is not a function of z either, we integrate again to get

$$\eta v = (z^2/2) \, dp/dx + C_1 z + C_2 \tag{7.16}$$

where C_2 is a second integration constant. To find C_1 and C_2, we use the conditions $v = 0$ at $z = 0$ and $v = V$ at $z = h$, where $h(x)$ is the distance between the joint surfaces. This gives

$$C_1 = \eta V/h - (h/2) \, dp/dx \quad \text{and} \quad C_2 = 0$$

Substituting these into Eq. 7.16, one has

$$v = (z^2 - zh)/2\eta \, dp/dx + Vz/h \tag{7.17}$$

The gradient of this velocity in the z-direction is

$$dv/dz = (z - h/2)/\eta \, dp/dx + V/h \tag{7.18}$$

Integrating the flow velocity over the joint separation, h, gives the flow rate Q in the x-direction per unit width of the joint in the y-direction:

$$Q = -(h^3/12\eta) \, dp/dx + Vh/2 \tag{7.19}$$

Substituting this into the continuity equation, Eq. 7.11, we obtain

$$d/dx \, [(h^3/12\eta) \, dp/dx] = V/2 \, dh/dx \tag{7.20}$$

which can be integrated with respect to x to yield

$$h^3 \, dp/dx = 6 \, V\eta h + C_3 \tag{7.21}$$

where C_3 is the integration constant. To find a value for this constant, we assume that there is some point along x where the pressure separating the two surfaces is a maximum. At that value of x, $dp/dx = 0$. Letting $h = h^*$ there, we find that $C_3 = -6V\eta h^*$. Making this substitution, we finally obtain the one-dimensional version of Reynolds' equation:

$$dp/dx = 6V\eta \, (h - h^*)/h^3 \tag{7.22}$$

In this equation, V is the relative velocity of the two bearing surfaces, h is the distance between them, and h^* is the value of h where p has its maximum value. Equation 7.22 captures the fact that hydrodynamic lubrication depends on the product of three factors: the viscosity of the lubricant, η, the velocity of one joint surface relative to the other, V, and variations in the lubricant film thickness, h.

Example Problem. To illustrate the use of Reynolds' equation and to test the feasibility of hydrodynamic lubrication between articular surfaces, a simple problem will be solved. Suppose the two bearing surfaces of a joint may be approximated by planes, the upper one of which is fixed in space at a slight angle to the other (Fig. 7.17). An x-y coordinate system is established as shown. The lower surface slides to the left with velocity $-V$. The value of x at the left end of the upper surface is x_0 and that at the right end is $x_0 + w$. The slope of the upper surface is s and the minimal distance between the two surfaces is h_0. Both s and h_0 are very small. In general, the distance h between the surfaces is given by $h = sx$; therefore, $dx = dh/s$. Making this substitution, Reynolds' equation becomes

$$dp = -6\eta V s^{-1} \left[(h - h^*)/h^3\right] dh \tag{7.23}$$

Integrating, one has

$$p = -6\eta V s^{-1} \left[-(1/h) + (h^*/2h^2) + C_4\right] \tag{7.24}$$

where C_4 is another integration constant. C_4 and h^* may be found using the following boundary conditions. To the left and right of the region between the two surfaces, the pressure is assumed to be zero. Thus,

$$\text{At } x = x_0, \quad h = h_0 \text{ and } P = 0$$

$$\text{At } x = x_0 + w, \quad h = h_1 \text{ and } P = 0$$

Making these substitutions and solving the resulting pair of equations yields

$$h^* = 2\, h_0 h_1/(h_0 + h_1) \quad \text{and} \quad C_4 = 1/(h_0 + h_1)$$

When these equations are substituted into Eq. 7.24 and the leading minus sign is taken inside the square brackets, one has

$$p = 6\eta V \left[(1/h) - h_0 h_1/(h_0 + h_1)\, h_2 - 1/(h_0 + h_1)\right]/s \tag{7.25}$$

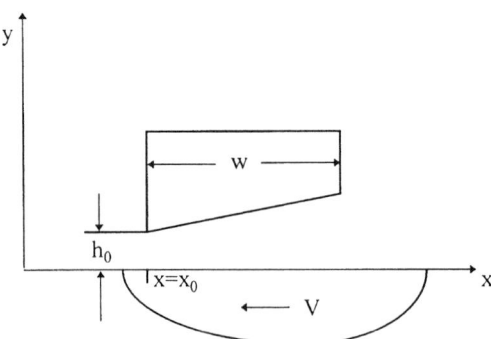

FIGURE 7.17. A flat lower surface moves to the left with velocity V relative to a fixed, canted upper surface having width w and slope s.

By substituting $h = sx$ into this equation and doing some algebraic manipulation, it can be put in the form

$$p = 6\eta V (sx - h_0)(h_1 - sx)/[(h_0 + h_1) s^3 x^2] \qquad (7.26)$$

This equation clearly shows that the pressure between the bearing surfaces is a parabolic function of x with zeros at each end (where $sx = h_0$ and h_1, respectively).

Figure 7.18 shows the pressure p as a function of position between the two bearing surfaces and the cant angle between them for the following parameters:

$$V = 0.25 \text{ m/s} \quad \eta = 0.05 \text{ Pa-s} \quad h_0 = 5 \text{ μm} \quad w = 5 \text{ cm}$$

Note first that the pressures are the same order of magnitude as the contact pressures between joint surfaces, suggesting that hydrodynamic lubrication may indeed be able to separate the surfaces under physiologic load conditions. Note also the extremely small cant angle that is required for hydrodynamic lubrication to be effective. Separation pressures drop off rapidly if the angle is less than a half minute or more than half a degree because of the presence of s^3 in the denominator. This kind of behavior is largely independent of the bearing surface geometry.

The integral of p across the joint (from left to right in Figs. 7.17 and 7.18) must equal the joint force if the separation between the two joint surfaces is to be maintained. Equation 7.22 shows that this integral (the joint force) must equal the product of η, V, and a geometric function of the interbearing fluid space. Another way of saying this is that a load-normalized "viscovelocity factor,"

$$\frac{\text{viscosity} \times \text{velocity}}{\text{joint force}}$$

should be constant for a given joint geometry.

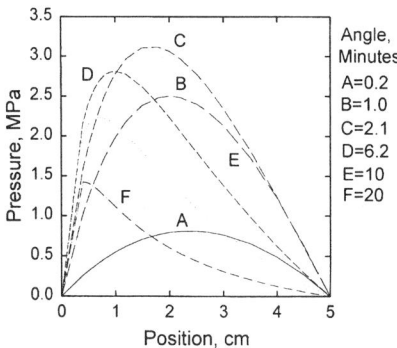

FIGURE 7.18. Effect of cant angle on pressure distribution for hydrodynamic lubrication of the system in Fig. 7.17.

Plotting the friction coefficient (i.e., the ratio of frictional force to joint force) vs. this ratio for bearings under various conditions gives Fig. 7.19. At low velocities (or viscosities, or high loads), friction is relatively high because lubrication is by boundary or "thin-film" lubrication. As the velocity increases, hydrodynamic lubrication begins and friction falls. Then, however, as velocity continues to increase, friction begins to rise again because of "internal friction" in the lubricant. As the strain rate in a Newtonian lubricating fluid increases, the shear stress increases as well, increasing the frictional force. Thus, such lubricants limit the remarkable advantage obtained from the hydrodynamic lubrication mechanism.

To prevent the deterioration of lubrication efficiency at high velocities, one may take advantage of the fact that some liquids containing long, chainlike molecules have a peculiar kind of viscosity that decreases as the rate of shearing increases (Fig. 7.20). If such a *thixotropic* liquid is used as the lubricant, the friction force does not rise so fast at high sliding speeds. It turns out that healthy synovial fluid is a very thixotropic lubricant. Furthermore, measurements have shown that the degree of thixotropy is decreased with age and by various joint diseases (Fig. 7.21). This result suggests that the efficiency of lubrication, and thus the avoidance of wear, declines with age and the appearance of disease. Whether the diminishment of thixotropy is a primary cause of joint degeneration is unknown, but it would seem to be a contributing factor.

Elastohydrodynamic Lubrication
This term refers to situations in which the bearing surfaces are compliant enough (i.e., have a low elastic modulus) so that they deform significantly under the pressures produced in the lubricating fluid as it passes over them. It seems clear that this applies to articular cartilage, but the difficulty in mathematically modeling the situation has prevented a credible demonstration of elastohydrodynamic effects in synovial joints.

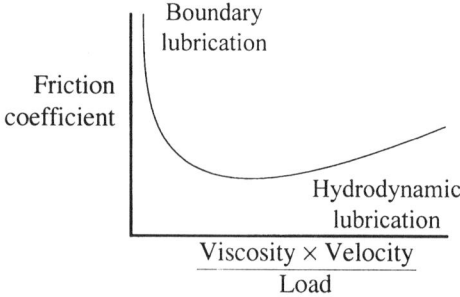

FIGURE 7.19. Schematic diagram of the effect of the viscovelocity factor on the measured friction coefficient.

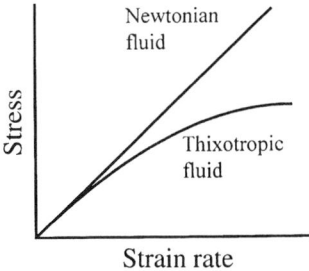

FIGURE 7.20. In a thixotropic fluid, the slope of the stress-strain rate curve decreases at the strain rate increases.

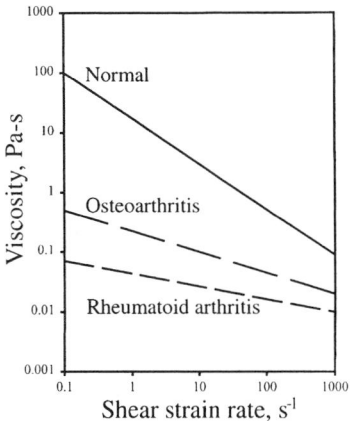

FIGURE 7.21. Synovial fluid viscosity vs. strain rate for representative normal, osteoarthritic, and rheumatoid arthritis joints. Water would be a horizontal line at 1.8×10^{-3} Pa-s. (Data from Cooke et al., 1978.)

Weeping Lubrication

The source of the water that fills the pores of cartilage is the synovial fluid in the adjacent joint space. However, the hyaluronic acid molecules are too large to pass into the pore spaces. Therefore, the fluid that is contained in the pores of cartilage is actually a filtrate of synovial fluid. Lewis and McCutchin (1959) postulated that, just as this fluid is expressed from a cartilage specimen during a compression test, it is also exuded into the load-bearing region when two joint surfaces are pressed together during joint flexion. This flow of fluid under pressure was hypothesized to contribute to separation of the joint surfaces. As in the case of elastohydrodynamic lubrication, it has been difficult to demonstrate convincingly that such an effect really occurs.

Boosted Lubrication

This theory of joint lubrication was proposed independently by Maroudas (1967) and Walker et al. (1968). The idea is quite simple. The large hyaluronic acid molecules of the synovial fluid that get left behind when synovial fluid is transported into the cartilage is hypothesized to form a gel-like layer on the articular surface. This "concentrated lubricant" is postulated to play a major role in the lubrication process. The problem is to understand where and when on the joint surface this happens. Weeping lubrication claims that fluid flows out of the loaded surface, not in. Squeeze-film lubrication has the synovial fluid being squeezed out to each side of the contact region. The proponents of boosted lubrication claim, on the other hand, that because of the microscopic ridges and grooves on the surface, when the two cartilage surfaces are pressed together, pools of synovial fluid are trapped and the water is forced into rather than out of the cartilage. These conflicting ideas have generated intense debate.

Synovial Joint Lubrication

Before trying to draw some conclusions about which of these lubrication mechanisms are most important in synovial joint lubrication, it is useful to first explore some of the early research in this area.

Pendulum Experiments

After Reynolds' comments, there seems to be no discussion of hydrodynamic lubrication in joints again until MacConaill proposed it in 1932. A few years later, simple experiments were done in which joints were set to swinging by attaching one limb segment to a pendulum, and the rate of decay of the swinging motion was studied (Jones, 1934, 1936). The fact that the amplitude of the motion decayed nonlinearly led to the conclusion that viscous damping was at play, providing support for the theory that a layer of fluid separated the joint surfaces, and hydrodynamic lubrication existed in the joint. Many years later, Charnley (1960) pointed out that the nonlinear decay could be for another reason: that the soft tissues about the joint resist large motions more than small motions. Soon, Barnett and Cobbold (1962) did an experiment that showed that this indeed happens (Fig. 7.22). A dog's ankle joint was rigged as the fulcrum of a pendulum, showing that the resistance to motion became less, and the decay in motion more linear, when the adjacent soft tissues were removed. However, to properly interpret this experiment, it needs to be understood that the joint was rigged "upside down" so that the pendulum weight compressed the joint surfaces together continuously. Thus, this experiment did not examine the possibility of short-term hydrodynamic lubrication following an unloaded state.

Another pendulum experiment suggests that hydrodynamic lubrication does occur in joints after all if the joint load is not too great. For boundary lubrication, the friction coefficient, μ, is independent of the joint contact force,

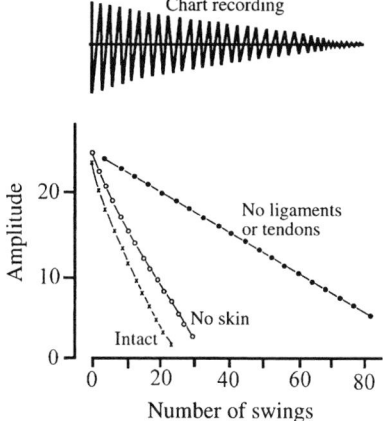

FIGURE 7.22. Experimental results from an early pendulum test of synovial joint lubrication. The *upper trace* shows oscillation decaying over time. The *lower graph* shows the nonlinearity of the decay was a function of the adjacent soft tissues. (Redrawn with permission from Barnett and Cobbold, 1962.)

N. When the joint surfaces are separated by a layer of fluid, as in hydrodynamic lubrication, the friction coefficient will **decrease** as the joint load increases because of the relationship $\mu = F/N$ and the fact that, in this case, the frictional force, F, is largely independent of the normal force, N. Figure 7.23 shows how the friction coefficient for the distal interphalangeal joint of a human finger decreases with increasing force across the joint (expressed as

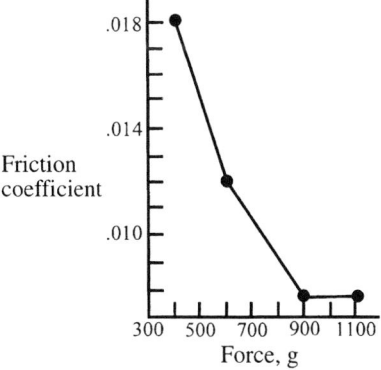

FIGURE 7.23. Friction coefficient vs. force across the joint for a human finger joint. (Redrawn with permission from Barnett and Cobbold, 1962.)

mass), strongly suggesting that the lubrication mechanism is not simply boundary lubrication. However, the plateauing of the curve when the force exceeds 900 g indicates that sufficient joint force can reduce the fluid layer between the articular surfaces to a thin film (i.e., boundary lubrication).

How Much Separation Is Needed?

Dintenfass (1963a,b) argued that elastohydrodynamic lubrication was probably occurring because the required pressures would deform cartilage significantly relative to the fluid film thickness. The minimal fluid layer necessary to produce hydrodynamic lubrication should be greater than the surface roughness of cartilage, which has been measured by several investigators at 1–6 μm (e.g., Thomas et al., 1980). Several analyses of elastohydrodynamic lubrication as a mechanism for joint lubrication have yielded fluid film thickness of less than 1 μm, however (Mow and Soslowsky, 1991). To circumvent this objection, Dowson and Jin (1986) argued that the physiologic pressures produced in joints would flatten the unevenness of the cartilage surface, permitting "microelastohydrodynamic lubrication" to work, but this idea has not been widely accepted. More recently, attempts to refine the measurement of surface roughness of cartilage have suggested that the in vivo roughness is much less that that measured on dried specimens, and that the measurement probe may "cut through" a superficial layer of smooth material and register instead the subsurface structure (e.g., proteoglycan and collagen fiber bundles). Thus, just plain hydrodynamic lubrication may be an adequate theory after all.

What Are the Lubricating Ingredients?

The principal argument against hydrodynamic lubrication in joints, and in favor of boundary lubrication, is the obvious fact that these bearing surfaces are not in continuous motion. Indeed, they are stationary much of the time and, it could be argued, often move slowest when carrying the highest loads. Experiments conducted so that only boundary lubrication could operate show remarkably low friction coefficients for cartilage on cartilage, supporting such a hypothesis. It has been suggested that the boundary lubricant could be (1) lipids in the cartilage (because fat solvents increase the friction coefficient), (2) a hyaluronan gel precipitated from synovial fluid, or (3) a lubricating glycoprotein called *lubricin* (Little et al., 1969; Swann et al., 1981). Later, a phospholipid called dipalmitoyl phosphatydilcholine (DPPC) was put forth as the active boundary lubricant in synovial joints (Hills, 1989)

Back to the Biphasic Theory

Some of the many questions about the role played in lubrication by the movement of fluid into and out of the cartilage have been clarified by extension of the biphasic theory to the problem of lubrication, as opposed to the mechanical properties of cartilage (Mow and Soslowsky, 1991).

Analysis indicates that when a parabolically shaped porous indenter is slid across a layer of cartilage at physiologic speeds, fluid is expressed from the cartilage into the leading-edge region and fluid is imbibed back into the cartilage near the trailing edge (Fig. 7.24). Fluid "weeping" from the cartilage begins well in advance of the arrival of the load, and as the load increases, so does the flow. Then, elastic recoil of the cartilage allows fluid to reenter the tissue when the load is past, presumably leaving a layer of hyaluronic acid gel on the surface. Thus, the theory predicts that both the weeping and boosted lubrication theories have a role to play, as do boundary and hydrodynamic lubrication. It must be understood, however, that the values of many of the variables in the theoretical model are poorly defined, and much remains to be done in taking full advantage of the potential of the model to elucidate the etiology of various joint diseases.

Diversity of Joint Architecture

In concluding this chapter, we should take care to point out that there is great variety in the anatomy of synovial joints (Adams, 1993). The implications of this with respect to lubrication are unclear, but it certainly raises the prospect that different combinations of the proposed lubrication mechanisms may predominate in different kinds of joints. To make this point more tangible, we briefly describe four human synovial joints having very different anatomical conformations. Often, there seems to be a trade-off between the need for range of motion and the need for stability under load.

Temporomandibular Joint

Motions of the mandible during chewing and speaking are extremely complex, with up-down, side-to-side, and in-out motions that must be regulated with great precision. Moreover, the two temporomandibular joints are linked by the mandible and the skull, so that motion in one joint produces motion in the other joint. It is clear that a highly constrained joint would be inadequate for these needs. The articular surface of the mandible does not

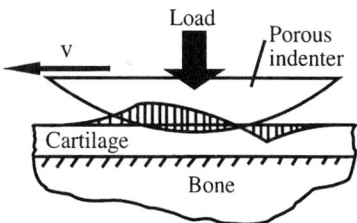

FIGURE 7.24. Flow out of the cartilage in front of an advancing porous indenter (at *left*) and back into the cartilage behind it (at *right*). (Redrawn with permission from Mow and Soslowsky, 1991.)

directly bear on the articular surface of the temporal bone; instead elastic fibrous articular disks (menisci), 2- to 3-mm thick, lie between them. Rolling and sliding of the mandible on the disk, and the disk on the temporal bone, allow considerable freedom of motion. Stability is provided primarily by muscles and to a lesser extent by ligamentous and bony restraints.

Trapeziometacarpal Joint of the Thumb
The base of the thumb articulates with two interlocking saddle-shaped joint surfaces; that is, each surface is concave in one direction and convex in the orthogonal direction. The two surfaces are not precisely congruous. This geometry allows considerable freedom of motion when the thumb is unloaded. When the thumb is loaded (e.g., in a pinch grip) the two saddles push together and the joint geometry contributes to stability. Relative motion of the two joint surfaces is entirely by sliding because a rolling motion is not possible with this geometry. Because friction is clearly increased in sliding contact, this restriction has lubrication implications.

Hip Joint
The ball-and-socket joint of the hip consists of the spherical femoral head tethered to the socket shaped acetabulum of the pelvis by a short ligament, seemingly a situation in which joint contact is continuous. However, the convex part is actually larger in diameter than its concave articulation, so that at low loads joint contact is confined to the periphery. As loads increase, the acetabulum undergoes elastic deformation and the two surfaces become congruent. When the joint is again unloaded, the residual force in the acetabulum pops the femoral head partially back out of the joint. During gait, the femoral head is not only rotating within the acetabulum, but is moving in and out of the acetabulum with each step. This movement may support squeeze-film lubrication in this large, heavily loaded joint. The elbow joint has a similar incongruity.

Tibiofemoral Joint at the Knee
The distal femur terminates in two rounded condyles that articulate with the much flatter surface of the proximal tibia. This arrangement allows for a combination of rolling and sliding as the joint moves, but contributes little to joint stability, which depends on dynamic muscle action and on ligamentous restraints. These stability mechanisms are often inadequate, and soft tissue injuries about the knee are common. The menisci interposed between the joint surfaces presumably to help distribute forces across the joint, and may help to provide the extraordinarily narrow fluid spaces necessary for hydrodynamic lubrication.

These few examples give a glimpse of the diversity of joint architecture. Obviously, caution is required in applying Reynolds' equation, or any other theory, to synovial joints. It seems likely that several different lubrication mechanisms function to varying degrees in different joints.

7.4 Summary and Further Reading

We have described the mechanics of cartilage and joint lubrication in synovial joints as they are currently understood. Arthritic diseases involve interactions between mechanical damage and inflammatory responses within a joint and the adjacent bone. The mechanics of cartilage are dominated by its high water content and the matrix of proteoglycan and collagen molecules, which retain the water during compression and resist tensile stresses, respectively. Because of its high water content, cartilage is highly viscoelastic, and its permeability is an important variable. The biphasic theory has been extensively used for interpreting the results of compression tests and predicting the mechanical behavior of cartilage. The lubrication of synovial joints is still poorly understood, but is thought to involve both boundary and hydrodynamic lubrication mechanisms. The flow of synovial fluid in and out of the cartilage on the joint surfaces seems to be important as well.

For additional information on this subject, there are several books: *Tribology of Natural and Artificial Joints*, by Dumbleton (1981), *Biomechanics of Diarthroidal Joints*, edited by Mow et al. (1990), and *Mechanics of Human Joints* edited by Wright and Radin (1993). There are also several reviews by Mow and his collaborators (Mow et al., 1986; Mow and Mak, 1987; Mow and Soslowsky, 1991, Mow et al., 1991), and by Woo et al. (1987a). An early paper on the biphasic theory is that by Mow et al. in 1980. More recent papers on the "triphasic" extension of the theory (the third element being ions) are by Lai et al. (1991) and Gu et al. (1993). A clearly written engineering textbook on lubrication theory is *Basic Lubrication Theory* by Cameron (1981), and *History of Tribology* by Dowson (1979) provides fascinating context for this chapter. Finally, a second edition of the book *Basic Orthopaedic Biomechanics*, edited by Mow and Hayes (1997), devotes four excellent chapters to synovial joints.

7.5 Exercises

7.1. Determine the relative amounts of strain energy absorption in the cartilage and bone of the proximal half of the tibia when it is loaded uniformly over the articular surface by an arbitrary compressive force. Simplify the situation by ignoring the fibula and assuming that the tibia consists of three parts: a cylindrical diaphyseal segment of cortical bone, a metaphyseal segment of cancellous bone in the form of a truncated cone, and a disk-shaped cartilage layer (Fig. 7.25). The dimensions shown in the sketch are a = 10 mm, b = 30 mm, c = 50 mm, e = 4 mm, f = 70 mm, and g = 130 mm. Assume the diaphysis is a hollow circular cylinder having inner and outer diameters as shown, and the cartilage and cancellous regions are uniformly filled with tissue. Assume also that stresses are uniformly distributed throughout each

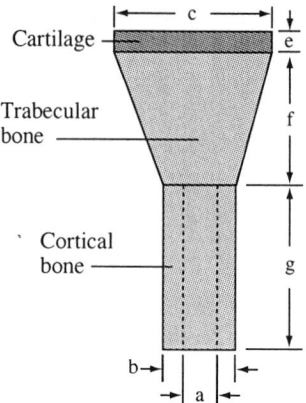

FIGURE 7.25. Model for the proximal half of the human tibia.

individual segment of the model and each tissue is linearly elastic with the following elastic moduli:

Cortical bone	20,000 MPa
Cancellous bone	200 MPa
Cartilage	20 MPa (Radin et al., 1970) or
	200 MPa (Higginson and Snaith, 1979)

Calculate the total strain energy in each segment of the model as a percent of the total energy in the model. What is this percentage for the cartilage using the two different dynamic modulus values?

7.2. Calculate the average stress and strain values in the cartilage and cortical bone in Exercise 7.1 if the uniform compressive load is the body weight of a 70-kg person (i.e., approximately 700 N).

7.3. In Fig. 7.21, viscosity is plotted as a function of strain rate. Show that when a fluid is sheared between two surfaces, one of which is moving with velocity V relative to the other, the strain rate is equal to the velocity gradient across the fluid. Then calculate the relative velocity between the joint surfaces associated with a mean strain rate of 1000 s^{-1} if they are separated by a distance of 1 μm (10^{-6} m). Compare this to an estimate of the relative velocity in the knee joint.

7.4. Suppose a hydrodynamically lubricated joint consists of a moving horizontal surface on one side and a stationary, exponentially curved surface on the other, as depicted schematically in Fig. 7.26. The joint is $W = 0.04$ m wide and the equation for the upper surface is

$$y = b_m e^{kx} \tag{7.27}$$

where k is a constant and the minimum distance between the joint surfaces is $b_m = 5$ μm. The viscosity of the synovial fluid is 0.05 Pa-s. The

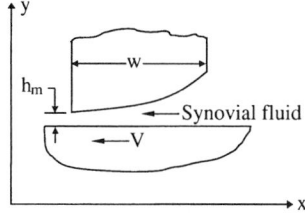

FIGURE 7.26. Diagram for exponentially curved joint surface problem (not to scale).

relative velocity between the joint surfaces is 0.25 m/s. Using Reynolds' equation, obtain an equation for the fluid pressure, p, as a function of position between the two surfaces.

7.5. In Exercise 7.4, find the value of k that maximizes the hydrodynamic pressure. Plot the corresponding surface.

7.6. In Exercise 7.4, calculate the force exerted on the joint surfaces by p, assuming that the depth of the joint surfaces (into the page in Fig. 7.26) is 0.04 m. How does this force compare to a body weight of 700 N?

7.7. In the experiments that oscillated joints as pendulums, it was necessary to find a relationship between the decay of the oscillation and the friction coefficient in the joint. Solve this problem; that is, show that the friction coefficient is approximately

$$\mu = (a - b)/4R$$

where $a - b$ is the change in amplitude during one swing and R is the distance from the center of rotation of the joint to the articular surface. Other features of the problem are shown in Fig. 7.27. (Hint: use energy methods. If you really get stuck, look in Barnett and Cobbold, 1962!)

7.8. Using a flat-ended cylindrical indenter of radius R = 2.5 mm, Obeid and co-workers (1994) measured the mechanical properties of human

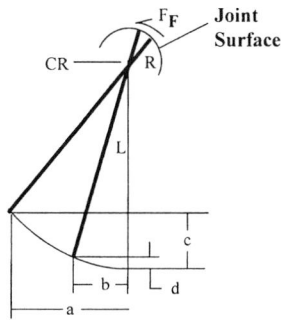

FIGURE 7.27. Diagram for pendulum experiment. F_F is the friction force against the articular joint surface; CR is the center of rotation. (Redrawn with permission from Barnett and Cobbold, 1962.)

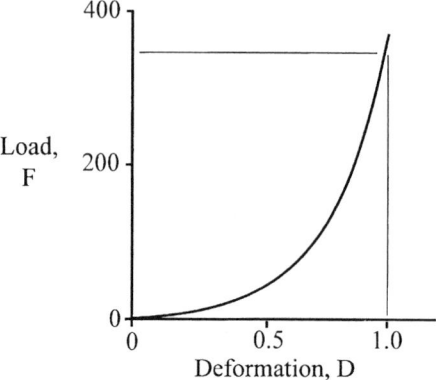

FIGURE 7.28. Load vs. deformation curve for indentation test of human articular cartilage. (Redrawn with permission from Obeid et al., 1994.)

articular cartilage from the tibial plateau in the knee joint. A typical load-deformation curve was as shown in Fig. 7.28 when the load was applied in about 0.5 s. The theoretical relationship between load, F, and the maximum deformation, D, produced by such an indenter (Sneddon, 1965) is

$$F = 2RED/(1 - v^2) \qquad (7.28)$$

where E is the elastic modulus and v is Poisson's ratio. Calculate the average (secant) modulus for the specimen shown in the figure as it was loaded to 350 N and 1 mm deformation. Assume that Poisson's ratio is 0.4. How does your result compare with typical values of cartilage's aggregate modulus? Explain the difference. How would the behavior shown in Fig. 7.28 change if the same deformation were achieved in 0.1 s, as when a person jumps?

8

Mechanical Properties of Ligament and Tendon

The movement of animals is like that of automatic puppets...For they have functioning parts that are of the same kind: the sinews and bones. The latter are like the pegs and the iron in our example, the sinews like the cables.

Aristotle (384–322 B.C.), De Motu Animalium

8.1 Introduction

Ligaments and tendons, the flexible structures that bind together the musculoskeletal system, are extraordinarily strong in resisting tensile loads. For example, the digital flexor tendon from the foreleg of a horse is strong enough to support the weight of two large automobiles. The Anglo-Saxon word for tendon is "sinew," which also means "strong" or "tough." As Aristotle realized, without ligaments and tendons to stabilize and animate our skeletons, they would be mechanically useless. We now turn our attention to these important tissues.

Both tendon and ligament are composed of dense fibrous connective tissue, but they differ in morphology and function. Ligaments bind one bone to another to restrict their relative motions. Tendons provide the connecting link from a muscle to a bone. Although textbooks sometimes refer to ligaments as if they were distinct and easily identified anatomical structures, this is often untrue. Many ligaments represent thickenings or specializations within a joint capsule, and their margins may be blurred and indistinct. Moreover, there is considerable individual variation in ligaments, and a ligament that is well defined in one person may be obscure or absent in another. Similarly, textbooks often treat tendon as if it were a uniform generic material, when in fact there is considerable variation in the properties of different tendons that represent adaptations to specialized function. These variations are unlikely to be entirely genetic in origin; many may also derive from the different ways in which each individual habitually uses a particular part of their musculoskeletal system. As with bone, cells may remodel ligaments and tendons to adapt them to the imposed loads.

8.2 Functional Considerations

Ligament

Ligaments are tough bands of fibrous connective tissue that bind bones together and support organs. Ligaments originate and insert on bone. From a skeletal point of view, their principal function is to maintain correct bone and joint geometry. Ligaments, together with their associated joint capsules (composed of similar material), are often referred to as *passive joint stabilizers*, and together with the articular contours they determine a joint's range of motion. Ligamentous damage occurs when the joint is forced beyond this functional range. Lax or torn ligaments that permit joint motion beyond the normal range can precipitate cartilage degradation.

A second and less appreciated function of ligamentous tissue is *proprioception*. Several studies have demonstrated the presence of stretch-sensitive *mechanoreceptors* in the ligaments of animals and humans (Schutte et al., 1987; Vangsness et al., 1995; Petrie et al., 1997). Studies in cats have demonstrated that the mechanoreceptors in the joint capsule and ligaments of the knee trigger muscle contractions that protect the knee from extremes of motion (Palmer, 1958; Freeman and Wyhe, 1966). Recent evidence suggests that similar reflex arcs help to protect the human knee (Borsa et al., 1997; MacDonald et al., 1996). Reconstructive procedures may restore the structural role of ligaments, but traumatic loss of proprioception is most likely permanent.

Tendon

Tendons are muscle-to-bone linkages that transmit the forces developed by muscle contractions across joints to stabilize them or produce motion. They originate in muscle, cross at least one joint (sometimes several), and insert to bone. The role of tendons in proprioception is well established. Mechanoreceptors in tendons have been studied since Laporte and Lloyd first described their significance in 1952. These receptors, called *Golgi tendon organs*, sense tension and provide feedback control that alters muscle activity when resistance to movement is encountered.

Energy storage is another important function of tendon. During walking, phasic transfer of energy between potential and kinetic forms during the up-and-down motions of each stride allows efficient locomotion. During running, this mechanism breaks down, and energy efficiency is achieved instead by temporarily storing energy as elastic strain energy in stretched tendons and returning this energy to the system later in the gait cycle. It has been estimated that as much as 50% of the energy required for running is stored in the Achilles tendon, the quadriceps tendon, and the ligaments of the foot at initial contact and then returned at push-off (Alexander and Bennet-Clark, 1977). Short-term energy storage in isolated tendons occurs with efficiencies approaching 95% (Alexander, 1988; Gillis et al., 1995).

8.3 Structure and Composition

Ligament

In its natural state, ligament is 55% to 65% water; its other components are shown in Table 8.1 as percentages of dry weight. Collagen (primarily type I with some type III) comprises approximately 70% to 80% of the dry weight and elastin another 10% to 15%. Unlike cartilage, proteoglycans comprise a very small percentage of ligaments and tendons. (This is consistent with the observation that they do not support compressive loads.) These relative proportions vary according to ligament location and function.

Collagen, the major organic constituent of bone, is by far the most abundant extracellular protein contained in ligament. Fibrillar collagen gives ligament its high tensile strength and is synthesized by specialized cells called *fibroblasts*. The collagen molecule in ligament is a glycine-rich triple helix, composed of two alpha-1 chains and one alpha-2 chain. As with any other protein, collagen synthesis begins with gene transcription in the nucleus followed by translation of the mRNA in the cytoplasm. Translated products are routed to the endoplasmic reticulum where the procollagen molecule is assembled and then extensively modified before being secreted into the extracellular matrix. Even after secretion several enzymes continue to modify the procollagen molecule in preparation for assembly. Finally, the processed molecules, now simple triple-helical subunits, spontaneously self-assemble into collagen *microfibrils* (4 nm in diameter), which in turn assemble into sequentially larger strands of protein called *subfibrils* (20 nm in diameter) and *fibrils* (50–500 nm in diameter). Collagen fibrils can be visualized with electron microscopy and have a characteristic striated pattern resulting from the ordered assembly of the individual collagen molecules within the microfibrils. At the top of the structural hierarchy are collagen fibers (100–300 μm), which can be seen using light microscopy (Fig. 8.1).

Fibroblasts also possess the machinery required to break down and remove old collagen, an important function in response to injury. To do this, fibroblasts synthesize and secrete *neutral collagenases*, enzymes designed to cleave the collagen triple helix. Evidence suggests that collagen turnover is a continual process throughout life. However, in healthy tissues, collagen turnover is much slower in ligaments and tendons than in bone (Neuberger and Slack, 1953).

TABLE 8.1. Major components of ligament, tendon, and skin as percent dry weight

Component	Ligament	Tendon	Skin
Collagen (mostly as type I)	70–80	75–85	56–70
Elastin	10–15	<3	5–10
Proteoglycans	1–3	1–2	2–4

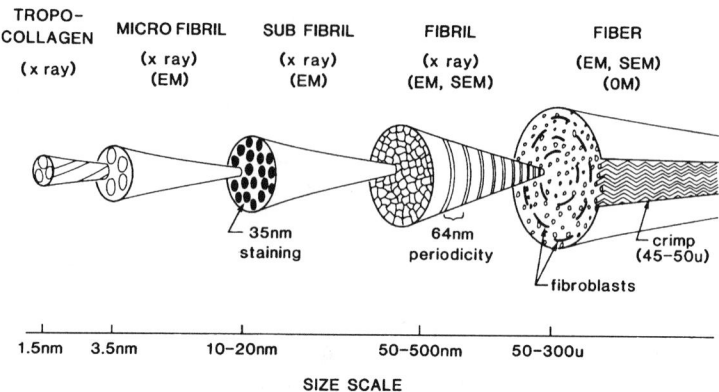

FIGURE 8.1. Structure of a collagen fiber in a ligament. Parenthetical notes at *top* indicate the kind of imaging that revealed the structure: X-ray diffraction, transmission or scanning electron microscopy (EM, SEM), or optical microscopy (OM). (Reproduced with permission from Kastelic J, Deformation in tendon collagen, *Symp Soc Exp Biol,* Vol. 34, p. 397, 1980, Cambridge University Press.)

In the absence of load, the collagen fibers demonstrate a wavy *crimp pattern*. As the structure is stretched, the crimp pattern disappears and the fibers become straight (see Fig. 7.5). The collagen fibers may be grossly aggregated into bundles that are oriented along the principal axis of stress experienced by the ligament. As a joint progresses through a range of motion, different portions of the ligament are stressed more than other portions at each joint position as a consequence of simple geometry. In many ligaments, the collagen bundles are not strictly parallel; this represents an adaptation by the ligament to provide optimal strength at each angle of the joint.

Tendon

The composition of tendon is similar to that of ligament (see Table 8.1). Collagen comprises as much as 85% of tendon dry weight; of this, 95% is type I and 5% type III and/or type V (Cetta et al., 1982). Unlike ligament, elastin constitutes less than 3% of tendon dry weight. In most locations, the proteoglycan concentration is usually less than 2%; however tendons that curve around bony surfaces experience compressive forces, and the cells within them respond by producing more proteoglycans. Increasing proteoglycan concentrations give the tendon a more cartilaginous quality that decreases friction and enhances motion.

The structural hierarchy of tendon resembles that of ligament except that collagen fibers are arranged into discrete packets in the tendon called *fascicles*. As shown in Fig. 8.2, a *paratenon* surrounds each individual fascicle. Fascicles are bound together and supported by the *endotenon,* a loose connective tissue containing blood vessels, nerves, and lymphatics. The

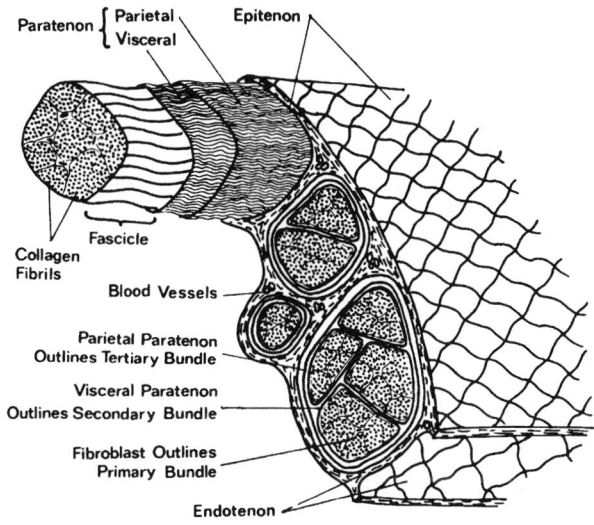

Paratenon { Parietal, Visceral }   Epitenon

Collagen Fibrils

Fascicle

Blood Vessels

Parietal Paratenon Outlines Tertiary Bundle

Visceral Paratenon Outlines Secondary Bundle

Fibroblast Outlines Primary Bundle

Endotenon

FIGURE 8.2. Internal structure of a tendon from a rat's tail. (Reproduced with permission from Rowe, The structure of rat tail tendon, *Connective Tissue Research*, Vol. 14:9–20, 1985, Gordon & Breach.)

endotenon is continuous with the *epitenon*, which encases the whole tendon. Long tendons such as the digital flexors are enclosed in a *synovial sheath* that lubricates the structure and enhances gliding. *Fibroblasts* are interspersed between collagen bundles and synthesize collagen for structural repair and maintenance. The fibroblasts in tendons are sometimes called *tenocytes*. On average, tendons contain fewer cells and are less metabolically active than ligaments.

Box 8.1
Technical Note: Gripping Slippery Tissues

The mechanical testing of tendons or ligaments requires that they be grasped at the ends for mounting in the testing apparatus; this is particularly difficult because these tissues are so slippery. Because tendon is viscoelastic, the specimen deforms with time when clamped and may slip at the interface when loads are applied. Methods and devices for gripping tendons include looping the tendon around a post with or without suture augmentation, use of grips with roughened or serrated surfaces, use of bone–muscle–tendon–bone preparations, and freezing the tendon to the clamps. We have found the latter two methods to work the best.

Bone–muscle–tendon–bone (or bone–ligament–bone) preparations, are an excellent choice if subfailure characteristics are to be determined. The bone affords a reliable grip site that can accommodate high loads. Much has been learned about ligamentous behavior using the rabbit medial collateral ligament (MCL) and methods developed by Woo et al. (1983); see upper figure. However, this method, too, has its limitations. When testing bone–ligament–bone preparations failure often occurs

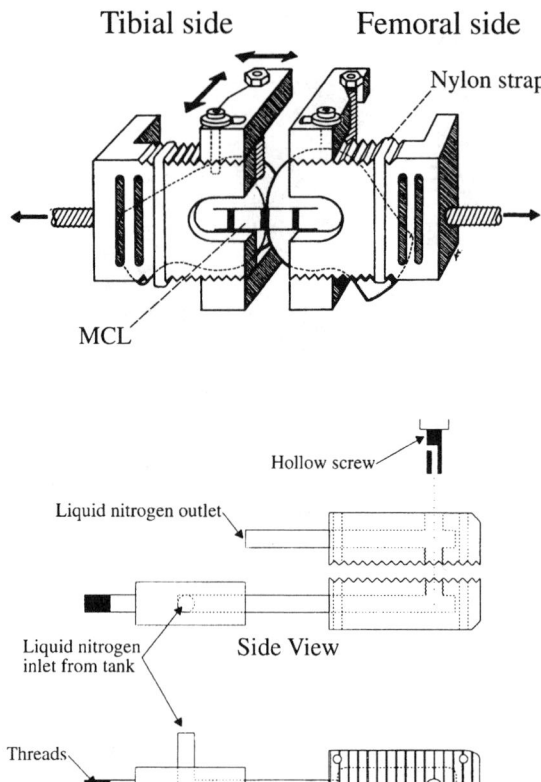

Upper figure: Grips for holding tibia and femur to test the medial collateral ligament (MCL). (Reproduced with permission from Woo et al., 1983.) *Lower figure:* Apparatus for gripping tendon or ligament tissue by freezing it within the clamp.

at the insertions rather than in the ligament. When testing muscle–tendon–bone preparations, the muscle fibers will fail at tendon strains greater than 4% or 5%, which precludes complete recordings of the tendon stress-strain relationship. In either case, appropriate orientation of the bony ends of the specimen is critical to achieve a physiologic application of stress.

Riemersma and Schamhardt (1982) were the first to use freeze clamps to grip these tissues during testing. In our modification of this method (lower figure), liquid nitrogen is pumped through hollow chambers in the clamp, causing the specimen to freeze in place. Because the mechanical properties of tendon and ligament are sensitive to temperature, it is important to have a long specimen so that the test region is well away from the clamps.

8.4 Mechanical Behavior

Quasistatic Tensile Properties

When loaded in the laboratory, ligaments and tendons produce character-istic load-deformation patterns that reflect their structural architecture. Materials such as steel and aluminum display linear stress-strain character-istics because of their crystalline nature. In these materials, the modulus of elasticity is constant throughout the elastic region and is representative of the atomic forces at work within their crystalline architecture. If the elastic limit is exceeded, crystalline materials will yield and eventually fail. In con-trast, noncrystalline materials display what is termed *rubbery elasticity*. Tensile tests of these materials produce upwardly concave load-displace-ment and stress-strain curves. In the noncrystalline materials, the elastic modulus is dependent on load magnitude because of intra- and intermole-cular forces and molecular cross-linking. Failure of these materials is often abrupt and can occur without a well-defined yield point. Tensile tests of tendon and ligament produce curves that display both crystalline and rub-bery characteristics, reflecting their complex structural hierarchy and mole-cular interactions.

A typical load-deformation (or stress-strain) curve for tendon or ligament (Fig. 8.3) can be divided into three regions. In the *toe region*, collagen crimps are removed by elongation. Initially minimal force is required for this elongation, but as the collagen fibers straighten, more and more force is required and the curve swings upward. The toe region typically ends at about 1.5% to 3.0% strain. It is a matter of some controversy whether the

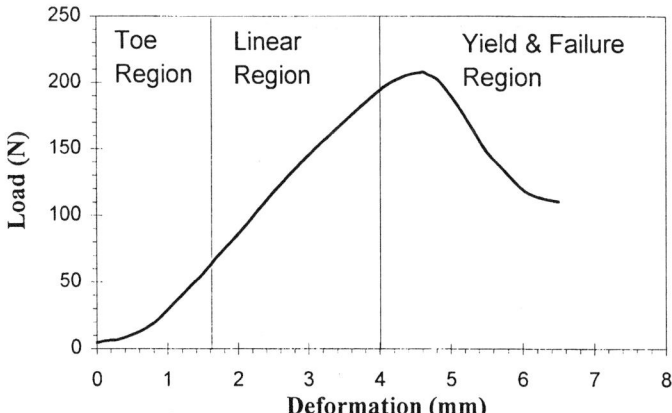

FIGURE 8.3. Typical load-deformation curve from a tendon or ligament showing the toe, linear, and yield and failure regions. The stress-strain curve would have a sim-ilar shape. (Reproduced with permission from Woo et al., 1994.)

toe region is physiologically or functionally important in vivo. It may actually be artifactual and the result of assuming that the normal resting length of the tendon or ligament in vivo occurs at zero load. Gillis et al. (1995) showed that in the forelimb of the standing horse the superficial flexor tendon has a strain of about 1.5% compared to the extracted tendon under zero load (Fig. 8.4). Therefore during standing, a resting position for the horse, the structure is prestrained beyond the toe region seen in laboratory tests of extracted tendons.

Further loading produces a *linear region* in which the molecular crosslinks of the collagen are stressed (see Fig. 8.3). The slope of this region of the load-deformation curve represents the stiffness of the tendon or ligament, and the slope of the stress-strain curve is a measure of the tissue's modulus of elasticity. In general the elastic modulus values reported for tendon and ligament range between 1.0 and 2.0 GPa; the modulus of highly specialized tissues (such as the *ligamentum nuchae*, described further in Section 8.7) can be substantially less.

In the final region of the curve, a reduction in slope marks the *yield point* or onset of cross-link or fiber damage. If loading is continued, the tendon or ligament will eventually fail. Tendons and ligaments can be strained to between 5% and 7% without damage. The maximum strain that a ligament or tendon can endure before failure is generally in the neighborhood of 12% to 15%; however, substantial damage to the collagen network can occur well before complete rupture. Ligaments with unusually high elastin content can be strained to 30% or more without damage. Ultimate tensile strengths for tendons and ligaments range from 50 to 150 MPa (for comparison, recall that the tensile strength of bone is about 150 MPa).

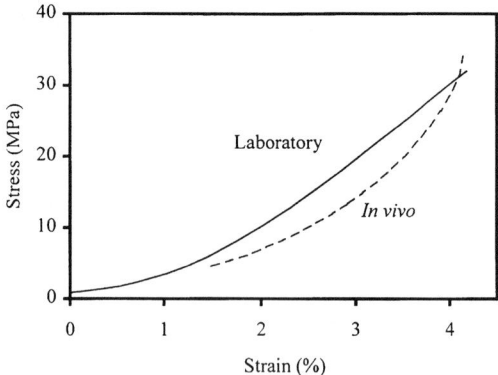

FIGURE 8.4. Different stress-strain curves for equine superficial flexor tendon are obtained when measured in vivo *(dashed line)* and on specimens removed from the animal and loaded in the laboratory *(solid line)*. (Reproduced with permission from Gillis et al., 1995, *American Journal of Veterinary Research*.)

Box 8.2 Technical Note
Measurement of Specimen Cross-Sectional Area

Derivation of stress-strain behavior from load-elongation data requires knowledge of specimen cross-sectional area (CSA) and resting length (L_0). In testing rigid materials these parameters are easily determined and controlled, but the viscoelastic nature of tendon and ligament makes accurate measurement of CSA and L_0 difficult. Errors in measurement of these variables can profoundly affect the values obtained for stress, strain, and elastic modulus.

A variety of techniques have been developed to establish the cross-sectional area of ligaments or tendons, including measurement of specimen weight or volume per unit length, physical measurement by area micrometer, shadow amplitude contour reconstruction, laser micrometer measurement, ultrasonographic measurement, and computer digitization of cross sections, photomicrographs, or inkblots. Significant systematic differences between these methodologies have been demonstrated by Ellis (1969) and by Woo et al. (1990).

If the density of the tissue is known, one can estimate the specimen's mean CSA by dividing its weight by its length and density. Alternatively, the specimen's volume can be determined by water displacement and divided by length to obtain mean cross-sectional area. These gravitational methods can be very reproducible but are generally useful only with long specimens of uniform dimensions. If a circular or elliptical cross section can be assumed, an alternative approach is to simply measure its diameter(s) with calipers at the site of interest and calculate the area. This technique produces much more variable results, however.

The area micrometer yields fairly reproducible values but can introduce substantial systematic error. The portion of the specimen where the CSA is to be determined is pushed into a rectangular slot of known width using a standardized force. Over time the specimen conforms to the geometry of the slot, after which its height is measured and the area calculated as height × slot width. Clearly, the result will depend on the force used to deform the ligament or tendon, the tissue's normal architecture, and the time between force application and height measurement. If these variables are carefully controlled, the method can produce excellent results.

Shadowing amplitude contour reconstruction, a fairly involved technique that requires a collimated light source, a detector, and a mechanical apparatus to rotate the sample, may give a more precise measure of specimen CSA; however, it is less reproducible than other techniques, and the apparatus is expensive to fabricate. Likewise, the laser micrometer described by Woo et al. (1990) is accurate but not readily available. Both these methods offer the advantage of nondestructive measurement on intact specimens, and they can also be used to measure CSA while the structure is loaded. Ultrasonography has recently been added to the list of cross-sectional area measurement techniques (Gillis et al., 1995). This methodology has been shown to be reliable and offers the unique advantage of in vivo measurement.

Computer digitization of histologic cross sections taken from the test specimen after testing is accurate and reproducible. Measurement of multiple sites and averaging the area can enhance this technique. This type of analysis cannot be used when testing ligament or tendon beyond the yield point, however.

Viscoelastic Properties

The mechanical behavior of ligament and tendon is *viscoelastic*; thus, their load-displacement and stress-strain relationships are **rate-** and **history dependent**. Testing parameters must always be carefully considered when establishing or comparing material properties such as ultimate strength, ultimate strain, and modulus of elasticity. Two representative stress strain curves for rat tail tendon are plotted in Fig. 8.5. The longer and shorter curves were produced using strain rates of 720%/s and 3.6%/s, respectively. As can be seen, the higher strain rate resulted in higher stress and strain at both failure and yield, and a slightly increased elastic modulus. Compare this behavior with that of bone in Fig. 4.12. Note that the ductility (energy absorption) of bone is decreased at higher strain rates, but that of tendon is increased.

Figure 8.6 demonstrates the effect of loading history. The mechanical behavior in cycle four is clearly different than that in cycle one. The shift

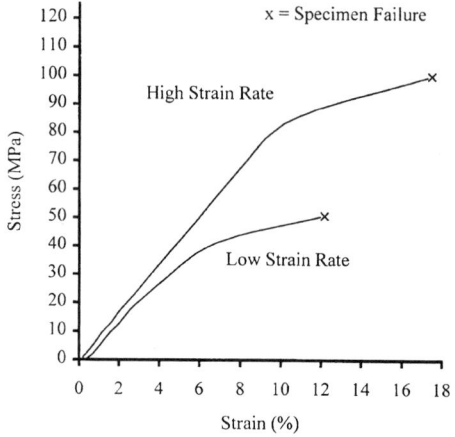

FIGURE 8.5. Effect of strain rate on stress-strain curve for rat tail tendon. (Reproduced with permission from Haut, 1983.)

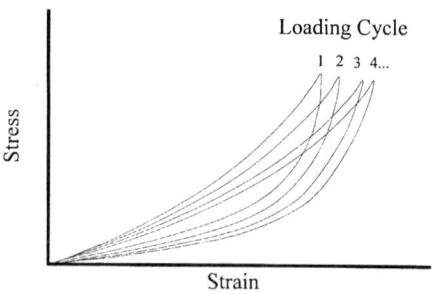

FIGURE 8.6. When a tendon is repeatedly loaded and unloaded, the stress-strain curve changes as shown here. A steady state develops after 10–20 cycles.

to the right reflects a decrease in tendon or ligament stiffness and may be accompanied by a decrease in ultimate strength. For most tendons and ligaments a steady state is achieved after 10 to 20 cycles of loading. However, if the peak load is increased on a subsequent cycle, the steady state is disrupted and additional cycles of loading will be required to reestablish stability. In life, tendons are often cyclicly loaded and therefore laboratory testing after conditioning cycles is considered more reflective of physiologic behavior. On the other hand, tendons and ligaments are also subjected to constant loads or deformations, and, like bone and cartilage, they exhibit *creep* under constant load and *stress relaxation* under constant deformation (Fig. 8.7).

Finally, tendon and ligament display *hysteresis,* another behavior caused, at least in part, by viscoelasticity. Hysteresis refers to loss of energy during a load-deformation cycle. If a purely elastic material is loaded in its elastic range and then immediately unloaded, the load-deformation curve for the loading phase is identical to the load-deformation curve for the unloading phase. The energy absorbed during loading and returned to the system during unloading will be the same, and the system will be 100% efficient. The loading-unloading profiles of ligaments and tendons are not identical, however. The area under the loading curve is greater than the area under the unloading curve, and the difference between the two represents the energy lost during the load cycle, which is referred to as hysteresis. The amount of hysteresis observed depends on the particular ligament or tendon and on the testing conditions. Tendons involved in gait tend to be efficient storers

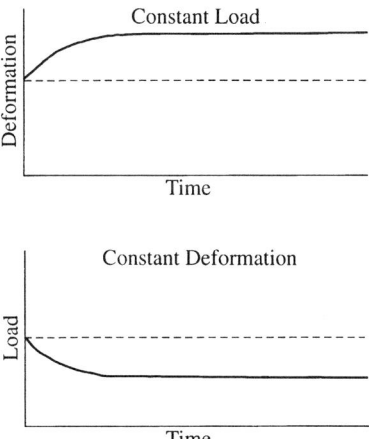

FIGURE 8.7. **A** When a constant load is applied to a tendon or ligament, its deformation gradually "creeps" to a stable value. **B** When a constant deformation is applied, the stress "relaxes" to a stable value. The times required for these processes are of the order of minutes or hours.

of elastic energy (i.e., have minimal hysteresis). For example, the digital flexor tendons of horses studied by Gillis et al. (1995) lost only about 5% of their strain energy during a loading-unloading cycle.

Where the lost energy goes is difficult to determine. Frictional forces at work within the sample cause some energy to be converted to heat, some of which flows to the surroundings while the remainder remains within the sample, raising its temperature and altering its mechanical properties. When the load is removed, the accumulated thermal energy slowly dissipates and the tendon or ligament may shorten further with time. However, *residual strain* may remain in the tissue for some time or indefinitely. Some of the lost strain energy has probably been used to break molecular cross-links or otherwise disrupt the molecular structure of the tissue.

Mathematical Modeling

When modeling musculotendon dynamics, many investigators have chosen to simulate tendon as a purely elastic component for simplicity's sake. Zajac (1989) assumed pure tendon elasticity in developing a generic musculotendon force-strain curve using two parameters, the *tendon slack length*, L_S, and the *peak isometric muscle force*, F_0^M. L_S is the length at which the tendon just begins to exhibit some resistance to lengthening. Tendon strain is defined as $\varepsilon = (L - L_S)/L_S$. Suppose that the force-strain curve for a tendon has the shape shown in Fig. 8.8. The peak isometric force is measured for the tendon's muscle and used to normalize the tendon force values (as shown on the right-hand axis). By dividing the force values by tendon cross-sectional area, the normalized force-strain curve, can be turned into a stress-strain curve which defines the response of a generic tendon. This relationship can then be scaled to obtain the force-length response of a musculotendon unit having a different area, F_0^M, and L_S.

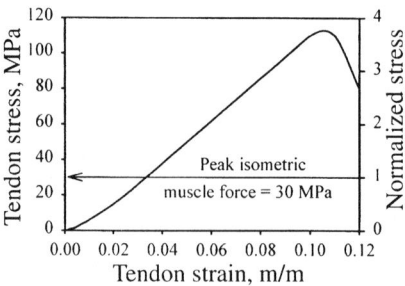

FIGURE 8.8. The stress-strain curve for tendon may be normalized by the peak isometric muscle force, F_0^M (expressed per unit cross-sectional area of the muscle), to obtain a generic tendon response shown by the scale on the right side of the graph. (After Zajac, 1989.)

Actually, of course, tendon tissue is viscoelastic and nonlinear. Lieber and associates (1992) developed an empirical model for muscle-tendon units by isolating and measuring the individual mechanical behavior of the frog semitendinosis muscle and its tendon and *aponeurosis* (the connection between the two; see following). They obtained the following regression equations for tendon and aponeurosis material behavior:

$$F_T = 10^{(\varepsilon + 0.633)/1.35} \tag{8.1}$$

$$F_A = 10^{(\varepsilon + 3.63)/5.66} \tag{8.2}$$

where ε is percent tendon or aponeurosis strain and F_T and F_A are the corresponding tensions in the tendon or aponeurosis, respectively, expressed as percent of the maximum tetanic muscle tension.

The viscoelastic properties of tendon or ligament are often represented using rheological models composed of two basic components: springs and dashpots (Fig. 8.9). The elastic component of their behavior is represented by the spring element, in which strain is linearly proportional to the applied stress. The dashpot represents the viscous behavior, in which the **strain rate** is linearly proportional to the applied stress. (The dashpot is a "leaky" piston in a cylinder; under load the fluid slowly flows past the piston, allowing it to move at a velocity proportional to the load.) In practice, combinations of elements are connected to simulate various behaviors. Two common models are the Kelvin solid and the three-parameter solid (Fig. 8.10). It is interesting to compare the behaviors of these models to those

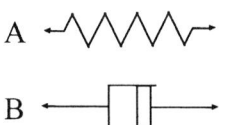

FIGURE 8.9. Symbols used for the two basic rheological model components: a spring (A) and a dashpot (B).

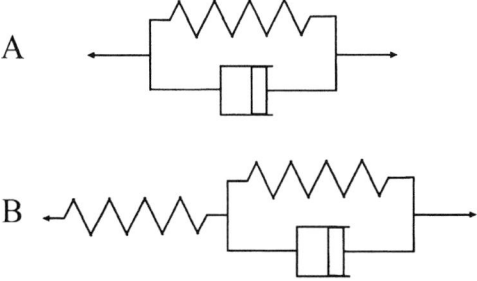

FIGURE 8.10. Two rheological models: a Kelvin solid (A) and a three-parameter solid (B).

seen in Fig. 8.7. For the Kelvin solid, the strain, ε, is the same in each element, and the total stress supported by the combination is

$$\sigma = E\varepsilon + \eta\dot{\varepsilon} \qquad (8.3)$$

where $\dot{\varepsilon}$ is strain rate, E is the elastic modulus of the spring, and η is the viscosity of the dashpot. In a creep test, one has a constant load, σ_0, and Eq. 8.3 can be integrated to obtain

$$\varepsilon = (\sigma_0/E) [1 - e^{-Et/\eta}] \qquad (8.4)$$

As t becomes very large, the strain asymptotically approaches σ_0/E. Initially, however, $t = 0$ and $\varepsilon = 0$; contrary to Fig. 8.7, there is no instantaneous elastic response in a creep test. When a constant strain, ε_0, is applied, the second term in Eq. 8.3 is zero, and the stress is constant at $E\varepsilon_0$. Thus, there is no relaxation effect for this model, either.

For the three-parameter model, a single spring is in series with a Kelvin solid. The equation for the series spring is $\sigma = E'\varepsilon'$ and that for the Kelvin element is $\sigma = E''\varepsilon'' + \eta\dot{\varepsilon}''$. Using the LaPlace transform method (Flugge, 1967), these equations may be combined to obtain

$$(E' + E'') \, \sigma + \eta\dot{\sigma} = E'E''\varepsilon + E'\eta\dot{\varepsilon} \qquad (8.5)$$

Using the LaPlace method, one may obtain for the creep solution

$$\varepsilon = (\sigma_0/E') [E^*(1 - e^{-\lambda t}) + e^{-\lambda t}] \qquad (8.6)$$

where $E^* = (E' + E'')/E''$ and $\lambda = E''/\eta$. In this case, at $t = 0$ there is an instantaneously produced strain σ_0/E', and the strain subsequently approaches $\sigma_0 E^*/E'$. For the relaxation solution, one obtains

$$\sigma = E'\varepsilon_0 [(1 - e^{-\lambda t})/E^* + e^{-\lambda t}] \qquad (8.7)$$

where ε_0 is the applied strain and $\gamma = (E' + E'')/\eta$. Again, there is a stress $E'\varepsilon_0$ at $t = 0$, and this subsequently relaxes asymptotically to $E'\varepsilon_0/E^*$. Thus, this model exhibits the general behavior observed in tendon and ligament. By fitting its equations to experimental data, values for E', E'', and η can be obtained. These results may be useful for predictive purposes, but the model is still phenomenological rather than mechanistic. That is, we have not learned much about how the viscoelastic behavior is caused by the tissue structure. What parts of the tendon have the attributes associated with E' and E''? Presumably η is associated with a flow of water, but where, exactly? These questions remain open.

A more elaborate nonlinear viscoelastic theory was described by Fung (1972). This theory mathematically describes the stress relaxation of collagenous tissues as the product of independent functions of time and strain:

$$\sigma [\varepsilon(t) \, ; \, t] = G(t)^* \, \sigma^e(\varepsilon) \qquad (8.8)$$

where σ^e is the *elastic response*, a function of strain (ε) only, and G is the *reduced relaxation function*, which is dependent on time (t) only. Versions of this theory have been successfully applied by other investigators as well (Haut and Little, 1972; Jenkins and Little, 1974; Woo, 1982). However, these models, too, are phenomenological rather than mechanistic.

Age and Mechanical Behavior

A number of animal studies have shown that the mechanical properties of tendons and ligaments change with age. During maturation, the size of the toe region of the stress-strain curve decreases and the elastic modulus and ultimate strength increase (Tipton et al., 1978; Woo et al., 1990). Evidence suggests that three processes are responsible for these changes in behavior. The first appears to be an increase in collagen content per unit area, as reflected by a steady decline in water content during maturation. The second is a gradual decrease in collagen crimp, which would account for the reduced toe region. The third is a gradual increase in collagen cross-links, which renders the molecules stiffer. Actually, these changes may be two sides of the same coin. The reduced crimping may just be a manifestation of the altered conformation of the collagen molecules caused by their additional cross-linking, and the straightening of the collagen may promote additional cross-linking. Increasing tensile strength and stiffness plateau after maturity and apparently deteriorate during old age (Noyes and Grood, 1976; Neumann et al., 1994).

Figure 8.11 demonstrates the increasing strength and modulus of rat tail tendon during maturation (Kastelic and Baer, 1980). Figure 8.12 shows how

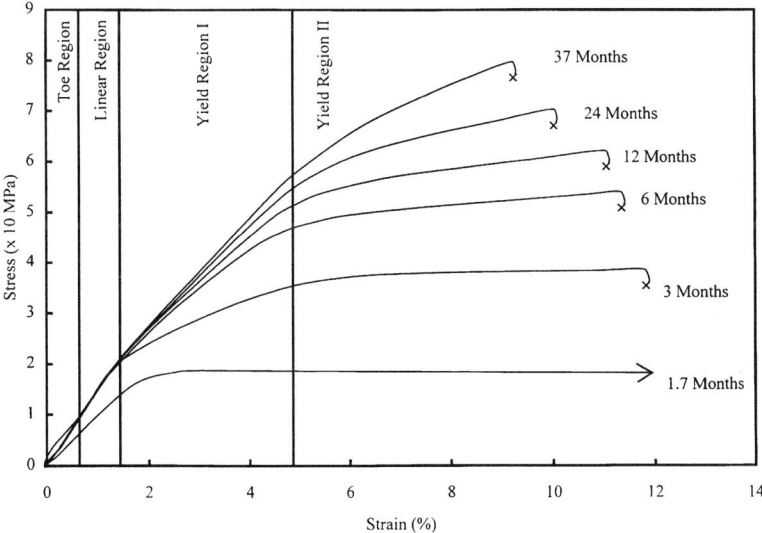

FIGURE 8.11. Effects of age on stress-strain curves for rat tail tendon. (Reproduced with permission from Kastelic, J, Deformation in tendon collagen, *Symp Soc Exp Biol*, Vol. 34, p. 397, 1980, Cambridge University Press.)

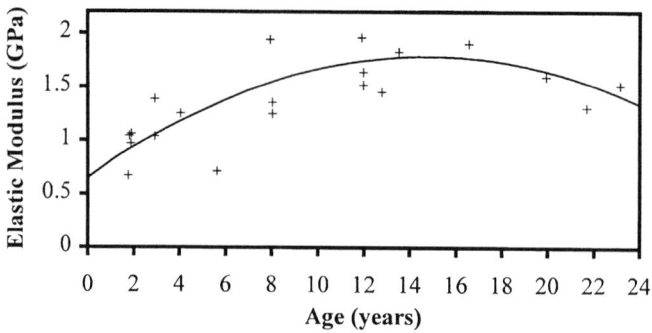

FIGURE 8.12. Effects of age on elastic modulus of equine flexor tendon. (Reproduced with permission from Gillis et al., 1995.)

the elastic modulus of equine flexor tendon changes with age (Gillis et al., 1995). Note that, in the horse, tendon stiffness gradually increases and then begins to decline in the older animals. Rats may simply not live long enough for this to happen to their tendons.

8.5 Mechanical Testing

Accurate and reproducible laboratory measurement of tendon and ligament mechanical properties is not an exercise for the faint of heart. Unlike metals, plastics, or even bone, tendons and ligaments cannot be machined into standardized testing specimens. To further compound matters, they are difficult to grip, perhaps even more sensitive to moisture and temperature than bone, and have complex internal structures that make data interpretation difficult.

Box 8.3 Technical Note
Measurement of Specimen Elongation

Methodologies to measure sample elongation can be grouped into two broad categories: measuring the displacement of the testing machine crosshead and measuring changes in a defined gauge length on the test specimen itself. The latter can be done either mechanically or optically.

Using crosshead displacement as a measure of sample elongation is the least preferred. Such experiments usually yield erroneously high strain values because of extraneous displacement (i.e., slipping) at the grips. Bone–ligament–bone preparations can reduce but not always eliminate this testing artifact. Because the error is systematic, methods that use machine displacement may be adequate for comparative experiments, but the error may increase nonlinearly with strain.

Measurement of elongation within the central region of the specimen, far from the grip sites, is much more likely to reflect the actual material properties of the

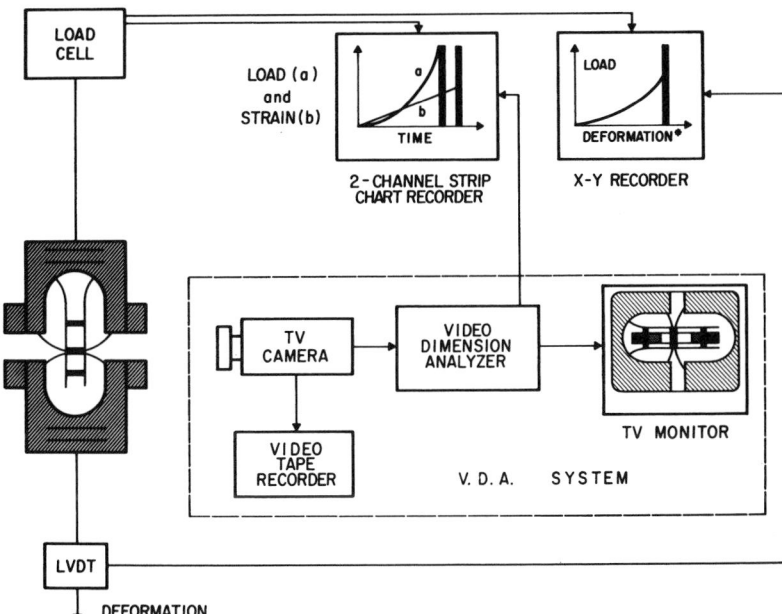

Video dimension analyzer (VDA) system used by Woo et al. (1983) to videograph-
ically record tendon deformations and plot load-deformation curves.

tendon or ligament. Mechanical techniques for local measurement of elongation
use *extensometers* supplied by the testing apparatus manufacturer or more generic linear displacement transducers. In either case the instrument is coupled to the
sample directly, which can present problems. Many tendons and ligaments have
complex fiber geometry or a helical fiber arrangement; pins inserted through the
body of these structures for instrument attachment often rotate and produce out-
of-plane motions that hamper axial elongation measurement. Using suture for
attachment can reduce this problem but may produce results that reflect the
behavior of the collagen fibers in the direct vicinity of the suture and not the tendon or ligament as a whole.

Perhaps the best methods of elongation measurement currently available are
optical. Cinematography or sequential still frame photographs have been used in
the past (Holmes et al., 1991), but video and computerized imaging techniques
(see figure) are easier and much more common (Woo et al., 1983). These techniques also have shortcomings, however. Optical techniques always measure surface strain; internal behavior may be somewhat different. The technique also
requires markers. Small beads glued to the surface may be misleading if they are
attached to different fibers, and stain or dye lines tend to disperse under load,
which makes interline distance difficult to determine. In addition, the recorded
behavior may be influenced by the surrounding fascia if the experiment is being
done in situ (Bay et al., 1997b). An alternative approach uses radiography and
radiopaque spheres placed within the substance of the tendon or ligament.
Ultrasonography can also be used (Gillis et al., 1995).

Determination of Resting Length

When you read in a research paper that a tendon or ligament was stretched 5%, it is important to also know how the unstretched length was designated. Viscoelastic materials, such as ligament and tendon, do not have an obvious gauge length, that is, resting length in the absence of load. The reason is that their length depends not only on current loads but also on past loading history. Several conventions are used to handle this problem. One is to apply a small load to the structure for a fixed period of time and to arbitrarily call the resulting length the gauge length. An alternative method is to extrapolate from the linear portion of the load-deformation curve to a point of zero load, and refer to the length intercept at that point as the gauge length.

A closely related problem is the determination of tendon or ligament lengths in vivo and in situ. For both ligaments and tendons, this depends on body position, joint angles, and muscle activity. For example, strains in the ulnar collateral ligament of the elbow joint are diminished when the joint is near full extension if the flexor muscles are active, which compresses the joint. In tendons, the relationships between body position, muscle activity, tendon compliance, and the architecture of the muscle-tendon construct are extremely complex. In addition, because of the viscoelastic features of these structures, cyclic loading alters resting length. Finally, we must keep in mind that there is considerable individual variability in the innate tightness of ligaments; some individuals have very lax ligaments and hypermobile joints, whereas others have very tight ligaments. Women generally have more compliant tendons and ligaments than men, and this difference can be further increased during pregnancy. The viscoelastic properties of the cervix and other tissues in the birth canal are greatly altered by hormonal factors when mammals give birth (see pp. 97–99 in Alexander, 1968), and this effect "spills over" into other connective tissues as well.

Architectural Implications

Although the collagen fibers in tendons typically are parallel, those in ligaments often are not. The pathways from origin to insertion of particular fibers are such that at any given angle of joint rotation certain fibers are stressed more than others. In testing a bone–ligament–bone preparation, orientation of the bones or the bone plugs to which the ligament is attached becomes critical. If only one particular joint angle is simulated, the characteristic behavior of the ligament at other angles of the joint will not be seen. Moreover, introduction of any extraneous twist or malalignment of the tissue may alter the measured mechanical properties. For testing to failure of either ligament or tendon, failure at the bony insertion always raises the concern that slight malpositioning of the bony block in the apparatus has resulted in nonphysiologic stresses at the bone-tendon or bone–ligament junction.

Because of ligament nonuniformity in relation to joint angles, elongation in loaded ligaments is often best mapped optically in two dimensions so that local variations in strain can be appreciated. Mapping in three dimensions would be ideal, but is technically unattainable at present.

Testing Environment

The treatment and storage of ligament and tendon specimens and the environment in which testing is actually conducted are important practical considerations. Because it is not always possible to test tissue immediately after death, the issue of storage by freezing is particularly crucial. Investigations as early as 1847 have studied the effects of various storage methods on the mechanical properties of tendons and ligaments, and although the data are somewhat conflicting, it would appear that freezing at −20°C or below has little effect on their mechanical behavior (Wertheim, 1847; Barad et al., 1982; Noyes and Grood, 1976). Later, Woo et al. (1986) evaluated the effects of freezing rabbit medial collateral ligaments for up to 3 months. The only parameter significantly changed by freezing was the area of hysteresis in the first few cycles; by cycle 10 this difference became insignificant. Cyclic stress relaxation, load-deformation characteristics, and energy-absorbing capacity were unchanged.

Carefully controlled studies have shown that storage of specimens at room temperature has time-dependent effects that appear to be amplified in small specimens. Davison (1989) noted that the breaking strength of rat tail tendon decreased with time. Because this could be prevented by immediately washing the harvested rat tail tendon with enzyme inhibitors and keeping them on ice, Davison concluded that the collagen molecules were being subjected to gradual enzymatic degradation. An acidic environment (pH below 6) can also induce changes in the collagen molecule and lead to decreased material strength. Figure 8.13 shows the breaking strengths of rat tail tendon after being soaked in buffers of varying pH for 30 min at 0°C.

Because tendon and ligament are viscoelastic and contain about 60% water in their native state, testing should be conducted in a physiologic solution (such as Ringers solution or normal saline; see Box 8.4) at body

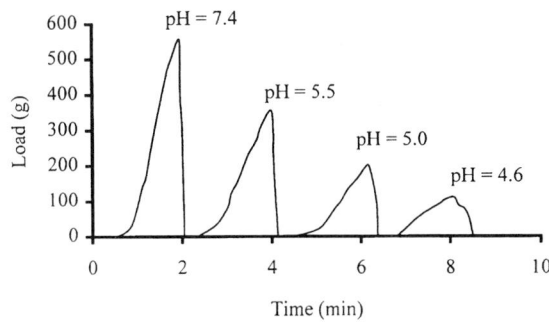

FIGURE 8.13. Rat tail tendon strength declines with diminishing pH. (Reproduced with permission from Davison, 1989.)

temperature. For large structures, such as equine tendons, water content can be maintained reasonably well by wrapping the sample in wet tissue paper; but for small specimens with larger surface-to-volume ratios, total immersion is required to prevent drying. Another approach is to coat the surfaces of the wet specimen with petroleum jelly.

Temperature dependence is another characteristic of viscoelasticity, and good experimental technique dictates that this parameter be controlled during laboratory testing of any skeletal soft tissue. Body temperature best simulates the in vivo situation. Substantial heat can be generated in cyclicly loaded tendon. In the laboratory situation, the amount of energy stored as strain energy or released to the environment as heat is dependent on the heat conduction characteristics and temperature of the test system. It is interesting to note that Wilson and Goodship (1994) measured greater than 44°C in the superficial flexor tendons of racehorses and postulated heat-induced cell death as a possible cause of tendon degeneration, a frequent problem in the equine athlete.

Box 8.4 Technical Note
Physiologic Testing Fluids

Mechanical testing of biologic structures and tissues should be conducted in environments that resemble the in vivo environment as closely as possible. Because biologic tissues are viscoelastic, with mechanical behavior dependent on fluid pressures and flow, the importance of tissue hydration cannot be overemphasized. This is even more true for soft tissues than for bone. To ensure complete hydration, skeletal tissues should be kept immersed, or at least saturated, in water or physiologic testing fluids throughout experimentation. It is also important that the fluid used have an appropriate chemistry because the mechanical behaviors of these tissues may depend on ionic forces. The fluids most frequently used are physiologic saline and Ringer's solution. Both are administered intravenously to rehydrate patients and expand their blood volume. The former is a 0.9% (g of solute/ml of solvent) solution of sodium chloride. This ionic concentration of salt is termed isotonic because its osmotic pressure is the same as that within the intracellular fluid of most cells. Cells remain intact and normally shaped when bathed in this fluid. Cells bathed in hypotonic solutions of lower ionic concentration take on water, swell, and eventually burst, releasing hydrolases and other potentially harmful enzymes into the extracellular matrix. Ringer's is a more elaborate isotonic electrolyte solution that closely matches the electrolyte concentration of the blood. It contains small amounts of potassium and calcium in addition to sodium and chloride. When testing or storing bone in such solutions for periods of several days, mineral ions may leach out, changing mechanical properties. Buffering the solution with calcium chloride can prevent this (Gustafson et al., 1995).

In addition to fluid balance and ionic concentration, the temperature of the fluid bath, its pH, and its enzymatic activity are parameters that can be manipulated in the laboratory. Testing conditions are important and should be closely examined when reviewing previous work. Discrepancies between experimental results may be explained by the different conditions under which the materials were evaluated. Obviously, conditions closely matched to the biologic environment are best able to reproduce the normal behavior of these tissues within the living body.

8.6 Functions at the Junctions

Suppose that you were given the daunting task of designing the interfaces between muscles and tendons and between tendons and bones. How would you design an interface between two soft tissues of very different tensile stiffnesses so that it could resist forces of hundreds of newtons? Or one between flexible and rigid tissues where the angle of pull would vary considerably? We now examine the solutions found in the skeleton.

The Myotendinous Junction

The transition from muscle to tendon occurs at the *aponeurosis*, a flat thin sheet of collagenous material that gradually narrows and thickens to form tendon. Individual muscle fibers attach to these tendinous extensions both proximally and distally. At each junction the force developed by the intracellular contractile apparatus of the fiber is transferred to the extracellular matrix of the tendon. The junction site is characterized by extensive membrane folding, which increases the surface area available for force transmission and thus decreases the junctional stress. The interdigitation also enables more effective load transmission because the interface is subjected primarily to shear stress rather than tensile stress. In fact, the angles at the junctions between muscle fibers and tendon collagen are usually close to zero, which greatly enhances junctional strength. Increases in the junctional angle may be associated with disuse atrophy and junction failure (Tidball, 1983).

Binding of the muscle fiber to the tendon collagen apparently occurs at discrete sites along the interface (Trotter et al., 1981; Tidball and Daniel, 1986). Subsarcolemmal "densities" have been identified within skeletal muscle cells at these sites. The cytoskeletal proteins *alpha-actinin, vinculin*, and *talin* and the adhesive proteins *integrin* and *vitronectin* have all been implicated in the load transmission process, but our understanding of myotendinous junctions is far from complete.

Several recent studies of musculotendinous preparations have shown that muscle strain injuries usually occur at or near the junction of muscle and its tendon of origin. Lieber and associates (1991), using frog semitendinosis muscle–tendon–bone preparations, demonstrated that during passive stretch the aponeurosis was significantly more compliant than both the tendon and the bone–tendon junction (Fig. 8.14). In another study using frog semitendinosis units, Tidball et al. (1993) reported large increases in failure load (30%) and energy absorption (112%) in stimulated (actively contracting) preparations as compared to nonstimulated controls. If the muscle-tendon unit is strained to failure with the muscle stimulated, most fibers fail at the myotendinous junction, whereas failure in nonstimulated muscle–tendon preparations occurs within the inactive fibers a short distance from the junction.

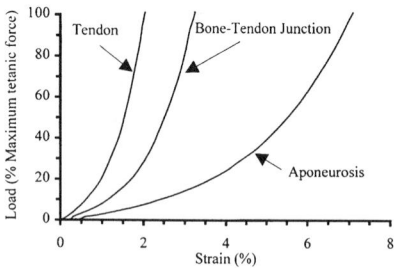

FIGURE 8.14. Variations in compliance of the three junctional components. (Reproduced with permission from Leiber et al., 1991.)

Insertions: Sites of Ligament-to-Bone and Tendon-to-Bone Attachment

Insertion sites of tendons and ligaments are anatomically and functionally diverse but can generally be classified as *direct* or *indirect*. Direct insertions often meet the adjoining bone at right angles, and microscopically these have four transitional zones. Zone 1 is represented by the normal collagen and fibroblasts of tendon or ligament; zone 2 is made of fibrocartilage and contains cells with a morphology similar to chondrocytes; zone 3 is mineralized fibrocartilage; and zone 4 is bone. Between zones 2 and 3 is a mineralization tidemark. The transition from tendon or ligament to bone is 300–700 μm across.

Collagen fibers at indirect insertion sites approach the bone at an acute angle and do not have a fibrocartilaginous zone of transition. However, there is a mineralization tidemark or demarcation between the tendon or ligament and the bone. At indirect insertions, more of the fibers blend with the periosteum, which is firmly attached to the underlying bone by *Sharpey's fibers*—collagen fibers originating in the periosteum and extending into peripheral lamellar bone. These fibers eventually become buried in bone matrix during new subperiosteal bone formation and form a radial collagen network within the underlying bone (Fig. 8.15).

Failures near insertion sites are common clinical injuries. Stouffer et al. (1985), in a study using human patellar tendon-bone units, concluded that bone-tendon strains were much larger than those measured within the substance of the tendon. This finding is consistent with the existence of stress concentrations at such interfaces resulting from the different mechanical properties of tendon and bone. However, the damage usually occurs in the underlying bone or in the tendon tissue adjacent to the junction rather than at the junction itself. Woo et al. (1986), using rabbit femur—medial collateral ligament—tibia complexes, showed that the specific failure location within these sites is age dependent. Young animals were more susceptible to avulsion of the ligament from the tibia, whereas older animals were more likely to fail by midsubstance tearing of the ligament (Table 8.2). In still older animals, the ligament was unlikely to fail either in its midsubstance or by avulsion from the tibia; instead the femur fractured. Generally speak-

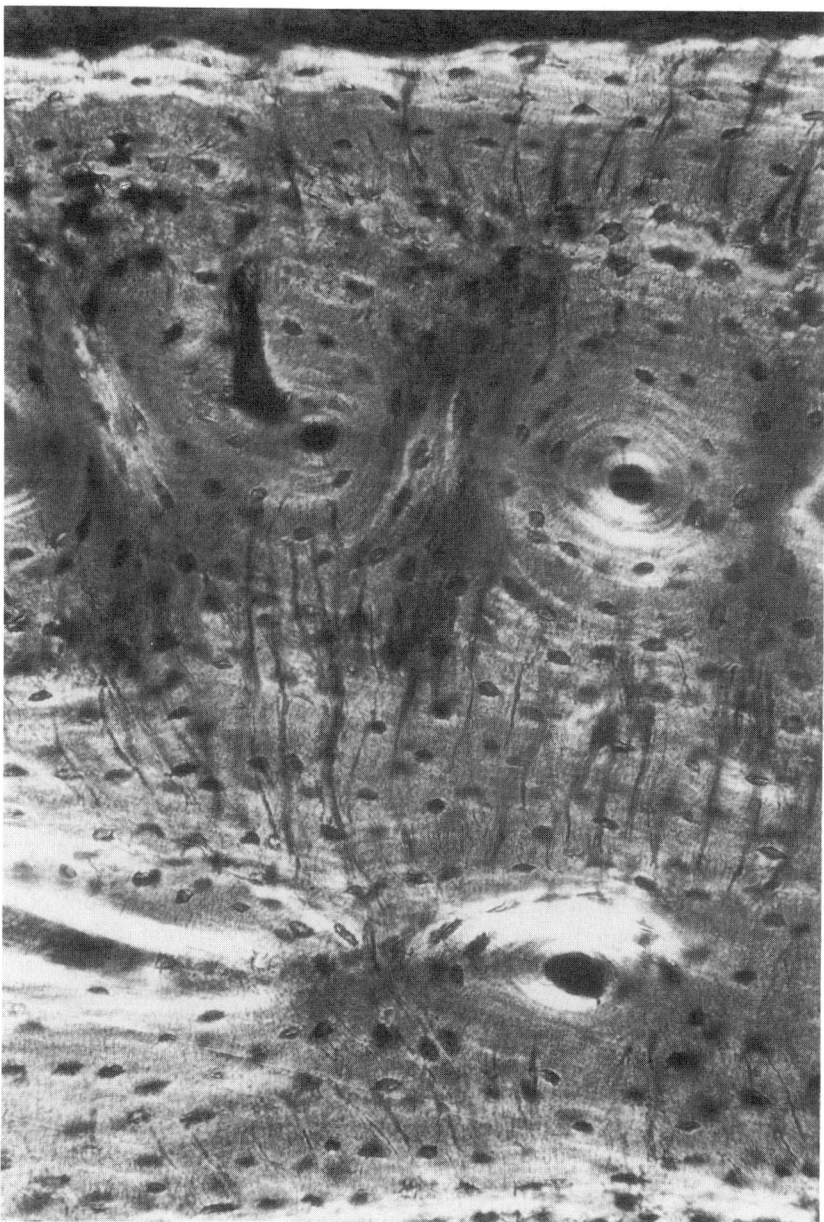

FIGURE 8.15. Sharpey's fibers appear as *dark lines* radiating inward from a tendo-
nous attachment at top of this photomicrograph of a canine radius. These fibers
have become embedded in the bone matrix as a result of ongoing periosteal apposition during growth. Note how they have also been interrupted by the formation
of secondary osteons. Basic fuchsin-stained, undemineralized cross section; field
width, approximately 0.5 mm.

TABLE 8.2. Failure modes of the medial collateral ligament as a function of age

	Age, months	No.	Failure modes, %		
			Tibial avulsion	Midsubstance	Femur fracture
Growing animals:	1.5	11	100	0	0
	4–5	20	100	0	0
Adult animals:	6–7	18	22	67	11
	12–15	16	0	56	44

From Woo et al., 1986a.

ing, the cross-linking of collagen within connective tissues increases with age; this may explain these experimental results.

It is often asserted, and some studies have confirmed, that failure occurs within the substance of the ligament at high strain rates, whereas at lower strain rates bone avulsion at the attachment site is the usual mode of failure. However, other studies have failed to confirm these findings. Several studies have shown that joint angle and loading direction have profound influences on failure mechanics at insertion sites. This experimental consideration is important when evaluating the biomechanical properties of tendon and ligament in the laboratory.

8.7 Functional Adaptation and Specialization

As previously mentioned, ligaments and tendons demonstrate tremendous variation in mechanical properties that apparently reflect the functional demands placed upon them. For example, the ligamentum flavum of the human lumbar spine functions to prestress the intervertebral disk and also keeps it from protruding dorsally. To function effectively over a large range of flexion-extension angles, the ligament must be highly extensible. This is indeed the case; Nachemson and Evans (1968) have shown that the ligament can withstand strains exceeding 60% without rupture. Figure 8.16 presents load-deformation curves for human anterior cruciate (top) and ligamentum flavum (bottom) ligaments. The ligamentum flavum has rubbery elasticity and a very low elastic modulus (0.1 GPa) compared to most other ligamentous structures. The ligamentum nuchae is another highly elastic, low-stiffness ligament that runs along the dorsal aspect of the spine. In the neck of a horse or cow, this ligament can store energy over prolonged periods of time during grazing and then use it to help raise the animal's head. At the other extreme, the elastic modulus of calcified turkey leg tendon approaches that of bone (10 GPa), presumably because it is under constant high stress from the continually flexed position of the turkey's lower limbs.

Specialized function is achieved by physiologic optimization of matrix composition, collagen fiber orientation, and myotendinous and insertional

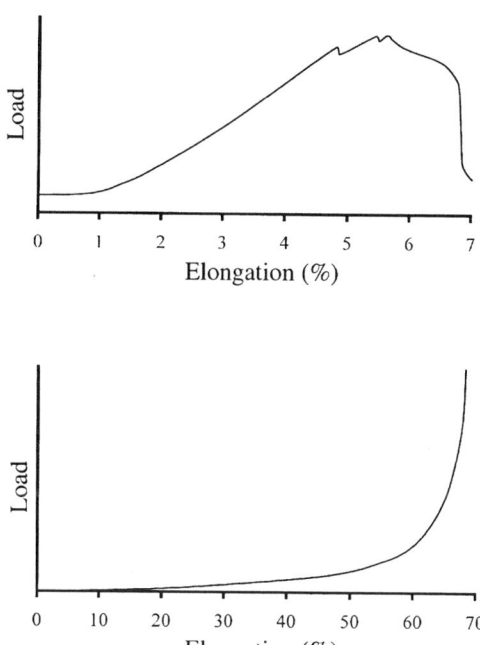

FIGURE 8.16. Load-deformation curves for human anterior cruciate *(top)* and liga-mentum flavum *(bottom)* ligaments. Note 10-fold difference in abcissa scales.

arrangements. Genetic coding undoubtedly controls skeletal tissue devel-opment; however, a growing body of evidence supports the notion that the expression of this information is affected by local mechanical conditions. Fibroblasts have been shown to be responsive to strain in vitro, and many studies have demonstrated enhanced healing in tendons subjected to active or passive motion. It has been postulated that, in addition to stimulating fibroblastic activity, mechanical factors may also play a role in regulating the extracellular assembly and orientation of collagen fibrils in the matrix.

The compositional influences on mechanical characteristics are dictated primarily by the relative concentrations of collagen, elastin, proteoglycan, and, occasionally, hydroxyapatite (mineral). The elastin content in the lig-amentum flavum described earlier approaches 70%, compared to only 10% in most ligaments, and even less in tendons. As is seen in Fig. 8.16, the ratio of collagen to elastin is a strong determinate of mechanical behavior. Calcification also correlates strongly with increases in structural stiffness, as shown by Leonard et al. (1976) in turkey tendon.

Determinants of Tendon Architecture

Tendon architecture is determined by considerations of stiffness as well as strength. In thinking about the various factors that govern tendon architecture,

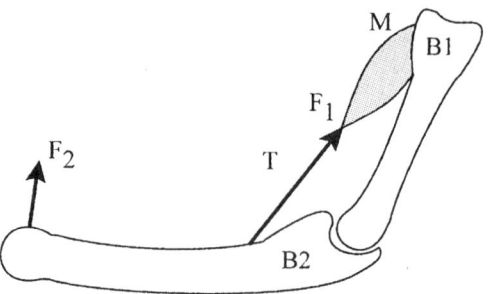

FIGURE 8.17. Sketch of a simple model for tendon mechanics. Bones B1 and B2 are articulated at a hinge joint; muscle M pulls on tendon T with to force F1, flexing the joint. The distal end of bone B2 produces force F2 against an object in the environment.

it is useful to consider a simple model of tendon action: a loop consisting of a bone, a hinge joint, a second bone, a muscle attached to the first bone, and the tendon of the muscle attached to the second bone so that contraction of the muscle flexes the joint (Fig. 8.17).

Contraction of the muscle exerts a force (F_1) on the tendon that causes rotation of the joint and in turn enables the effector end of the second bone to exert a force of some kind on the environment (F_2). The relationship of the two forces, F_1 and F_2, depends in part on the relative moment arms of the tendon insertion and the effector point (with respect to the joint center of rotation). The relationship of the two forces also depends on the compliance of the tendon as a structure, which is in turn determined by its elastic modulus, cross-sectional area, and length. If we assume for the moment that the tendon is perfectly elastic, then if it were twice as thick it would stretch only half as much at a given force, resulting in greater motion at the joint for a given level of muscle contraction. Similarly, if it were half as long, a given level of muscle motion would result in greater motion at the joint. From the perspective of an organism that must adapt efficiently, the variables which can be manipulated include muscle architecture, tendon cross-sectional area, tendon length, tendon moment arm, and moment arm of the force effector. From an evolutionary perspective, these variables may change together, with different evolutionary "problems" giving rise to different adaptive solutions. For example, if you have a long, compliant tendon, you could make it shorter or thicker (if greater forcefulness were sought), or you could increase fiber length (excursion) of the muscle if precision were more important, or you could increase both muscle fiber length and tendon stiffness while decreasing the tendon insertion moment arm if speed were the critical factor. In some cases, adaptation has led to tendons that are very robust compared to the size of their muscles. An extreme example is the massive suspensory ligament of the horse, which is not, strictly speaking, a ligament, but the tendon for a muscle so tiny that it appears as a reddish fuzz of rudimentary fibers. This tendon functions principally for energy storage during gait; it is loaded primarily by gravity rather than by its muscle. At the other extreme, some of the human finger flexor muscles have extremely long and compliant tendons; these are adapted to maximize precision of finger motion at the expense of force or elastic energy storage (which is not a factor in the fingers).

Sites of Tendon Compression

Proteoglycan is another component of tendon and ligament that increases under special conditions. High proteoglycan content in tendon or ligament produces a tissue called fibrocartilage. Localized areas of *fibrocartilage* are found in regions of tendon that wrap around and press against bone. Recall that proteoglycan is a major constituent of cartilage and is the molecule most responsible for this tissue's ability to withstand compression. Fibrocartilaginous tissues contain specialized cells that show a gradient of phenotypic expression between that of a fully differentiated chondrocyte and that of a fully differentiated fibroblast. These cells produce large amounts of proteoglycan and collagen. (The mix of different collagen types is also varied.) Cell shape is reflective of extracellular matrix composition and a *"fibrochondrocyte"* is more spindle shaped in areas of low proteoglycan (such as tendon or ligament) and more rounded in areas containing large amounts of proteoglycan (like cartilage).

An interesting mechanical theory of tissue differentiation based on localized stress and strain conditions has been proposed by Giori and coauthors (1993). This theory is related to the Carter–Wong theory of cartilage calcification discussed in Section 6.6. Here again, it is postulated that cells produce an extracellular matrix that varies according to the balance between distortional and hydrostatic stresses imposed on them. In this case, however, the theory addresses the balance between cartilage and tendon rather than the calcification of cartilage. A graphical representation of the theory is presented in Fig. 8.18. The horizontal axis represents the amount of compressive or tensile hydrostatic stress, and the vertical axis denotes the amount of distortional strain, at a point in the tendon. It is postulated that cartilaginous tissue is produced in an environment dominated by compressive hydrostatic stress and that fibrous tissue occurs in an environment dominated by tensile hydrostatic stress. Fibrocartilage, such as that seen in tendons that pass around and are forced against a bone, is produced in an environment with high compressive hydrostatic stress combined with high distortional strain. Fibrous tissue is maintained in a situation that combines high distortional strain with tensile hydrostatic stress.

To find out which of these situations obtains in various regions of a tendon wrapped around a bone, the authors developed a two-dimensional finite element model. They assumed tendon tissue to be a two-phase, linear elastic fiber-reinforced composite consisting of longitudinally oriented collagen fibers and an extrafibrillar matrix. Their model did not iteratively vary the tissue properties according to the local stress history, as the Carter–Wong models did, but simply studied the distribution of the stresses in a tendon wrapped around and forced against a cylindrical bone. Fig. 8.19 shows typical distributions of hydrostatic stress (left) and distortional strain (right) in the finite element model. The distortional strain is relatively high throughout the tendon and greatest along the surface contacting the bone. Thus, the tendon's mechanical situation is entirely in the upper half of Fig. 8.18. The hydrostatic stress is high and compressive (negative) where the tendon presses against the bone; it is slightly positive elsewhere. The region in contact with the bone is in the "fibrocartilage" zone because its hydrostatic stress is negative. The other portions of the tendon are considered to be in the "fibrous tissue maintenance" zone because their hydrostatic stresses are slightly positive. Therefore, the mechanical situation in the tendon and the distribution of tissue types correspond to the biologic effects postulated in Fig. 8.18.

Matyas et al. (1995) proposed a mechanical hypothesis for tissue differentiation at direct ligamentous insertions. (Recall that direct insertions contain a band of fibrocartilage.) They examined histologic sections of the femoral insertions of rabbit medial collateral ligaments and quantified cell shape as an index of tissue phenotype. They assumed that rounded cells produce a more cartilaginous material and spindle-shaped cells a more fibrous material. To perform their study, the investigators used automated

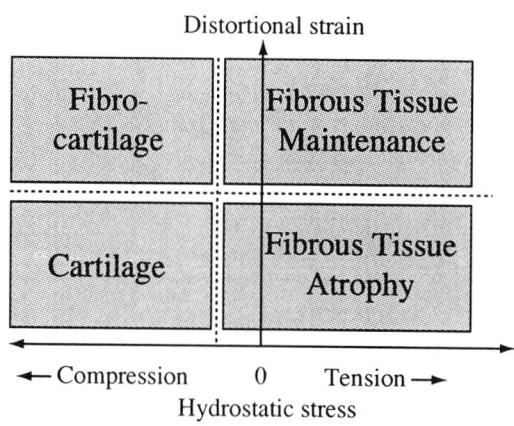

FIGURE 8.18. Hypothesis for the effects of stress and strain on tissues produced by cells in a tendon. The hydrostatic portion of the stress is plotted on the abcissa, with compression to the left and tension to the right. Distortional (or shear) strain is plotted on the ordinate. Depending on the balance between these hydrostatic and distortional influences, the cells are postulated to produce the kinds of tissue shown. (After Giori et al., 1993.)

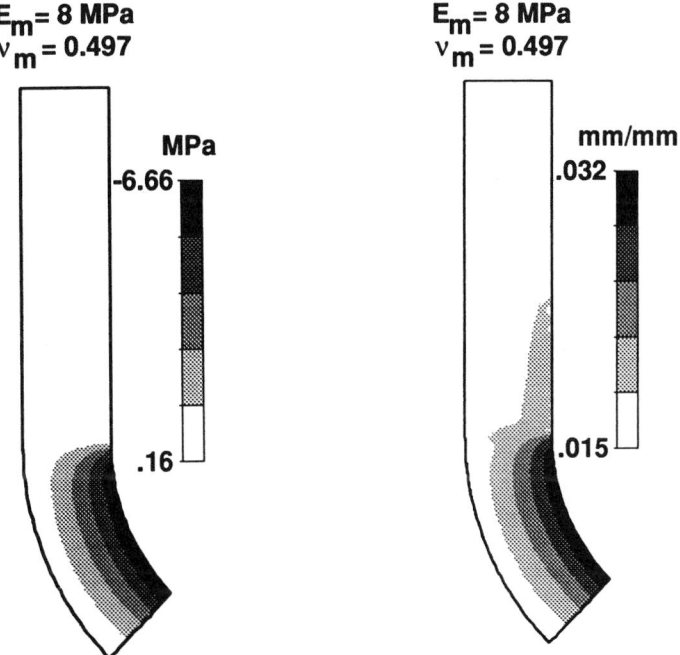

FIGURE 8.19. Hydrostatic stress *(left)* and distortional strain *(right)* in a finite element model of a tendon passing around a cylindrical bone. The elastic modulus and Poisson's ratio values used in the model are shown *above* and the scales of shading used to depict stress results to the *right* of each figure. The bone, which would be adjacent to the *dark region* in the tendon, is not shown, and only the upper half of the tendon appears because the model utilizes the symmetry of the situation. (Reproduced with permission from Giori et al., 1993.)

image processing and analysis techniques that differentiated the margins of dark-staining cell nuclei. A one-to-one correspondence of nuclear shape to cytoplasmic shape was assumed based on the authors' preliminary work. High-magnification images of histologic slides were captured with a CCD (charge-coupled device) camera and digitized. The software automatically measured the location, area, and perimeter of each nucleus and calculated a roundness coefficient for each cell as follows:

$$\text{Roundness} = (4\pi \times \text{area})/\text{perimeter}^2$$

Using this equation, a perfect circle would have a roundness coefficient of 1. Finite element techniques were used to characterize the local mechanical environments in and about the insertion site, and the results were correlated to the histologic maps of cell roundness. Corresponding gradients of cell shape and principal stress were noted in the fibrocartilaginous zones at insertion sites. Several mechanical properties correlated

TABLE 8.3. Mechanical correlations to cell shape

Mean nuclear roundness vs.:	Pearson correlation
Principal stresses	
Compression	−0.68
Tension	−0.53
Derivatives of principal stress	
Hydrostatic stress	−0.67
von Mises stress	−0.31
Strain energy density	−0.41
Principal strains	
ε_1	−0.40
ε_2	+0.69
ε_3	−0.20
Derivatives of principal strain	
Volumetric strain	−0.67
Distortional strain	−0.31

From Matyas et al., 1995

with tissue phenotype (cell roundness) (Table 8.3), providing further evidence that functional adaptation is controlled, at least in part, by the mechanical environment.

Flexors and Extensors

The long flexor and extensor tendons that contribute to finger motion in humans and quadrupedal gait in animals are also highly specialized structures. The flexors usually have different mechanical properties than the extensors. Shadwick (1990) showed that porcine flexor tendons were consistently stronger, stiffer, and more energy conserving than extensor tendons (Figure 8.20). Thus, flexor tendons were more effective biologic springs than the extensors, which makes good sense biologically in terms of reducing the amount of muscular energy, the metabolic cost of locomotion. As the body decelerates when the foot contacts the ground, kinetic and potential energy are converted to strain energy stored in the flexor tendons of the limb. Elastic recoil of the stretched tendons converts this momentarily stored energy back to kinetic and potential energy at push-off. When the tendons are more efficient at storing energy, less work is required for locomotion. Extensor tendons function differently and are not elongated during deceleration, thus efficient energy management is less crucial in these structures. Interestingly, Shadwick also showed that the two types of tendon possess identical mechanical characteristics at birth, providing further evidence that functional demand regulates structure and mechanical behavior.

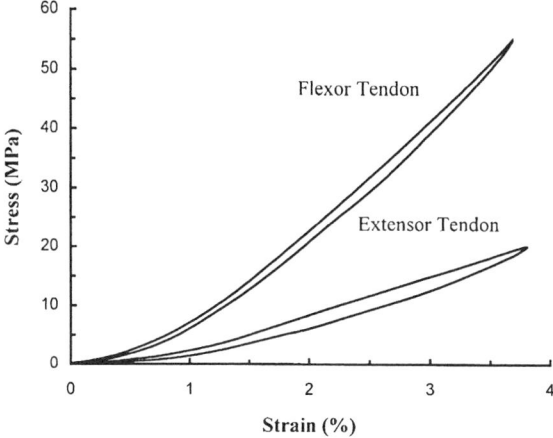

FIGURE 8.20. Comparison of load-deformation curves of flexor and extensor tendons from pigs. (Reproduced with permission from Shadwick, *J Appl Physiol,* 68:1033–1040, 1990, American Physiological Society.)

Tendon stiffness has been shown to increase with increasing load, largely because of the straightening of crimps in collagen fibers. Stiffness increases dramatically as the toe region of the load-deformation curve is surpassed (see Fig. 8.3); when crimp is removed, stiffness remains reasonably constant. Typically, relative muscle force at the point of transition is less than 25% of maximum. This behavior is amplified in human finger flexors because they are so long. In these structures, high compliance when muscle force is low serves to damp contractile variations, producing smooth and finely controlled motion at the fingers. When higher muscle force is applied, this damping effect is diminished, but high-force tasks usually do not require fine motor control. The greater tendon stiffness during high-load tasks (such as tight gripping) provides more efficient force transfer.

The synovial sheath that surrounds the long flexors and extensors is another important functional adaptation of these tendons. It produces and retains synovial fluid that bathes the tendon and facilitates smooth excursion with minimal friction. Without the synovial membrane these tendons would be subjected to substantial drag because of their length.

8.8 Pathology and Healing

Ligament

Ligament failure occurs by midsubstance tear or bony avulsion. Midsubstance failure is much more common in adults and avulsions are seen more frequently in children. *Sprains,* stretching the ligament beyond its elastic limit, elicit an inflammatory response that will eventually repair the compromised tissue. Ligament healing has been divided into the same three phases as fracture healing. Phase I (48–72 h), *acute inflammation,* is characterized by pain, swelling, redness, and warmth from vasodilation. During this phase cells are recruited and a hematoma forms. Phase II (up to 6 weeks post injury), *proliferation,* is characterized by increased numbers of fibroblasts, collagen synthesis, and scar tissue formation. Phase III (12 months or longer) involves *remodeling* of the scar. Collagen synthesis decreases, water content drops, and the extracellular matrix becomes more organized. (Usually this means a more longitudinal arrangement of fibers.) Scar contraction coincides with increasing tensile strength. The injured ligament may never regain full strength; primate studies indicate tensile strength to be 80% of the intact ligament at 1 year (Clancy et al., 1981).

Tendon

Like ligament, tendon can fail in midsubstance or at its bony insertion. Avulsion is less common for tendons than ligaments but does occur in young athletes. Tendon lacerations, particularly of the finger flexors and extensors, are common injuries. Overloading of tendon can cause damage to the collagen network, as can repetitive motion and abrasion. Diseases that compromise blood flow, such as diabetes, can cause ischemia-induced degeneration, particularly in areas with little collateral blood supply. Steroid injection has also been associated with tendon degeneration and rupture. Tendon healing follows the same pathway as described earlier for ligaments, but is more problematic. The length and quality of the repair process can vary considerably as the result of major differences in blood supply from site to site. In addition, *adhesions* after surgical repair are a significant clinical problem, particularly in the digital flexors/extensors. Scar formation extends into surrounding tissues, tacks the tendon down, and prevents sliding. Relatively new *passive motion protocols,* judiciously applied after surgery, can reduce this and improve clinical outcomes.

Immobilization, Exercise, and Passive Motion

We have learned that skeletal tissues, including bone, tendon, and ligament, respond in predictable ways to the mechanical demands placed upon them. Specifically, it has been demonstrated that the mass and strength of bone rise and fall according to mechanical demand. To some degree, tendons and ligaments appear to behave similarly, but the data are conflicting. Woo and associates (1981a) have shown that exercise increases the tensile strength of extensor tendons and flexor tendon insertions but not flexor tendons themselves. Figure 8.21 compares the mechanical behavior of digital **extensor** tendons extracted from exercised (40 km/week for 1 year) and nonexercised swine. The improved performance of the exercised tendons is obvious. The lack of similar strength changes in the **flexor** tendons

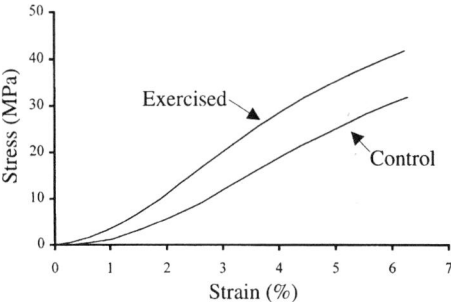

FIGURE 8.21. Extensor tendon load-deformation curves for exercised and control swine. (Reproduced with permission from Woo et al., 1981a.)

may be because tendons and ligaments that are regularly subjected to high loads by normal activities of daily living are already optimally designed. Perhaps the only way to increase the load-bearing capacity of these structures is to increase cross-sectional area, which, unlike bone, would be naturally self-limiting because of geometric and kinematic contraints. This reasoning may also explain why bone–tendon and bone–ligament complexes respond to increased demand by improving strength at the insertion site rather than by architectural modifications of the tendon or ligament itself (Woo et al., 1981a;Tipton et al., 1967). Conversely, Laros et al. (1971) and Noyes et al. (1974) demonstrated in dogs and primates, respectively, that the mechanical strength at insertion sites quickly diminishes in response to immobilization. In the primate study, recovery was still incomplete after 5 months of reconditioning exercise (Table 8.4). Woo and co-workers

TABLE 8.4. Mechanical properties of normal, immobilized, and reconditioned rhesus monkey anterior cruciate ligaments

Parameter	Normal ($n = 30$)	Immobilized ($n = 18$)	Five-month reconditioning ($n = 22$)	Twelve-month reconditioning ($n = 20$)
Maximum load (kg)	87.33	53.14	68.68	79.23
	(19.40)	(12.79)	(18.35)	(17.01)
Strain at maximum load (%)	46.66	43.88	38.53	42.91
	(8.63)	(9.08)	(7.80)	(6.22)
Energy at failure (cm-kg)	34.02	23.16	26.53	31.42
	(10.72)	(5.27)	(7.19)	(8.16)
Stiffness (kg/mm)	18.49	12.76	17.13	18.10
	(3.01)	(3.29)	(3.01)	(2.77)

SD in parentheses.
From Noyes et al., 1974a,b

(1987b) have also shown inferior mechanical properties in the rabbit medial collateral ligament after immobilization. All these studies indicate that the insertion sites may be more prone to disuse atrophy, and take longer to regain theire original strength, than the ligaments themselves.

Mechanically regulated tissue behavior can present difficult clinical problems for soft tissues as well as bone. Too little loading after surgical repair of a tendon or ligament leads to inadequate tissue remodeling, low tensile strength at the repair site, and diminished overall structural strength; too much loading will pull the repair apart. In addition, absence of motion can permit runaway scar formation, leading to adhesions and poor functional outcome after flexor tendon repair.

Active motion is perhaps the best postoperative treatment after ligament reconstruction or repair of partial tendon laceration. Vailas and associates (1981) demonstrated increased collagen synthesis and strength in rat medial collateral ligaments subjected to progressive exercise during healing as compared to nonexercised controls. In cases of complete tendon laceration, *passive motion* of the involved joint is a popular alternative that subjects the healing tendon to modest stress and lowers the risks of gap formation at the repair. Freehan and Beauchene (1990) and Takai et al. (1991) have shown that passive motion improves healing and increases wound strength as compared to immobilized controls. Recent studies have advocated new repair techniques that place more strands of suture across the laceration site. These new techniques increase the postoperative tensile strength of the repair so that active rather than passive motion rehabilitation can begin soon after surgery.

Flexor tendon adhesions and joint contractures continue to be problematic despite advances in surgical techniques and postoperative rehabilitation. In many instances *tenolysis*, surgically freeing the tendon from surrounding tissue, is required to achieve acceptable function. Patients not committed to extensive postoperative rehabilitation may never regain adequate function.

8.9 Surgical Repair

Surgical repair of damaged tendons and ligaments is often a difficult task, in large part because of the problems involved in replacing tissues with very specialized properties. Learning something about these surgical procedures helps to clarify the practical significance of the material presented in this chapter.

Anterior Cruciate Ligament

The anterior cruciate ligament (ACL) of the knee is a site of frequent injury, particularly in athletes. The ACL is an intraarticular ligament bathed in synovial fluid; it originates on the posteromedial aspect of the lateral femoral condyle and inserts centrally into the joint surface of the anterior tibia. The ligament restricts anterior translation of the tibia. Figure 8.22 shows that the

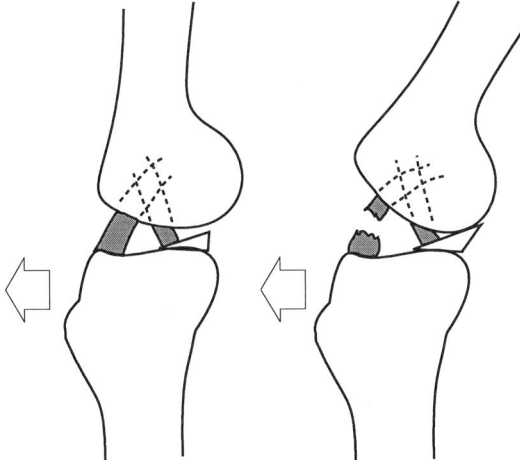

FIGURE 8.22. The anterior cruciate ligament (ACL) of the human knee restrains anterior translation of the tibia *(at bottom)* relative to the distal femur *(above)*.

ACL restrains translation of the tibia relative to the femoral condyles. The posterior horns of the menisci offer a second line of defense in case of ACL rupture, as does active contraction of the hamstrings muscle. Many patients with ACL rupture do well with conservative treatment, which usually consists of hamstring strengthening exercises and bracing, but 30% to 50% of all patients choose reconstruction; more than 100,000 such reconstructions are performed annually. No method of reconstruction is perfect, which in part accounts for the extremely large body of literature focused on the ACL and its repair. In the late 1970s and 1980s synthetic grafts were a popular alternative to autogenous (the patient's own tissue) graft materials, but it was soon realized that synthetic materials rapidly wear out and degrade in the environment of the joint. Fatigue damage was a common finding, emphasizing the need for continual maintenance and renewal of collagen fibers within the normal ligament.

Biologic grafts have proven to be much more effective because they are able to be repopulated with host cells (i.e., the patient's own cells) and subsequently remodeled into reasonable facsimiles of the original structure. *Allografts*, tissue taken from another human, were popular in the early 1980s but are used with less frequency today. Nonetheless, good results have been reported following allograft reconstruction using freeze-dried fascia lata or Achilles tendon. The most popular and successful techniques currently use *autografts*, the patient's own tissue, for ACL replacement. Table 8.5 compares the strengths of many of the autograft tissues that have been used for ACL reconstruction with that of the normal ACL. The central one-third of the inferior patellar tendon is a popular graft because it can be harvested as a bone–tendon–bone preparation, which can be rigidly attached at the repair site. Table 8.5 shows that the patellar tendon is also a very strong choice. It

TABLE 8.5. Failure loads of tissues used for anterior cruciate ligament (ACL) reconstruction

Tissue	Maximum load (N)	Percent of normal ACL
Normal anterior cruciate ligament ($n = 6$)	1725	100
Bone—patellar tendon—bone		
Central third ($n = 7$)	2900	168
Medial third ($n = 7$)	2734	159
Semitendinosis ($n = 11$)	1216	70
Gracilis ($n = 17$)	838	49
Distal iliotibial tract (18-mm width; $n = 10$)	769	44
Fascia lata (16-mm width; $n = 18$)	628	36
Retinaculum—patellar tendon		
Medial ($n = 7$)	371	21
Central ($n = 6$)	266	15
Lateral ($n = 7$)	249	14

From Noyes et al., 1984.

must be borne in mind, however, that any autograft will try to alter its mechanical properies to suit its new location and function. The question is whether it will be able to do so (in a joint initially inflamed by surgery) while it bears the loads applied by the recovering patient.

Supraspinatus Tendon (Rotator Cuff)

The *rotator cuff* of the shoulder joint is another site of injury frequently treated surgically. The rotator cuff comprises the supraspinatus, infraspinatus, subscapularis, and teres minor muscles. All these muscles originate on the scapula and insert on the proximal humerus close to the glenohumeral articulation. They assist in arm elevation and also stabilize the humeral head against the upward shear forces generated by deltoid contraction. The supraspinatus is the tendon most frequently affected, and tears of the rotator cuff are usually synonymous with tears or rupture of the supraspinatus tendon. The supraspinatus is the most superiorly oriented rotator cuff muscle; its tendon passes within a narrow space bordered superiorly by the acromion and coracoacromial ligament and inferiorly by the humeral head.

The etiology of supraspinatus damage is not well understood and more than one mechanism of injury is likely. Some investigators believe that *impingement* is the predominate cause of injury to the tendon. In this scenario the tendon incurs damage by being repetitively compressed against the undersurface of the acromion or coracoacromial ligament (Fig. 8.23). For this mechanism of injury to occur, either the humeral head must translate superiorly or the superior structures must invade the space of

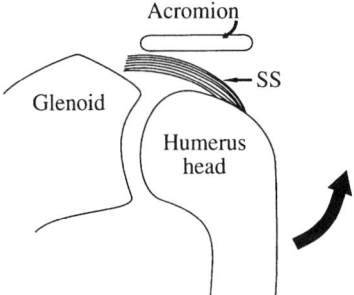

FIGURE 8.23. Cross section through the shoulder joint shows the precarious position of the supraspinatus tendon *(SS)*. Abduction of the *humerus* (as indicated by the *arrow*) may pinch the tendon against the *acromion* as the humeral head rotates.

supraspinatus travel. Either way, the tendon becomes chronically irritated and painful, making arm elevation difficult or impossible. Complete rupture is not uncommon. Cadaveric studies suggest that many people have non-symptomatic damage to the supraspinatus tendon.

Surgical techniques are designed to repair tendon damage and relieve or *decompress* the tendon. Surgery once consisted of complete acromion removal, but this compromised normal deltoid function and was cosmetically damaging. Over the years the procedure has become progressively less radical. The current treatment uses either open or arthroscopic techniques to remove the undersurface of the acromion and repair the tendon. Surgical intervention is not always successful and is indicated only after conservative treatment, consisting of rest and exercise designed to strengthen the inferior rotator muscles, has failed.

Flexor Tendons of the Hand

The flexor tendons are often cut in two by lacerations of the hand. Such injuries demand immediate surgical repair for successful restoration of function. Several suture techniques are available for reconnecting these tendons. Most techniques are aimed at minimizing the amount of suture within the repair site to maximize the area available for collagen fiber deposition and fusion (Fig. 8.24). New evidence indicates that suturing to provide strength sufficient to allow active flexion postoperatively may be a more important consideration because disuse during the postoperative recovery period leads to adhesions between the tendon and surrounding tissues. Autograft tendon from several different anatomical locations may be used to reconstruct missing segments of flexor tendon. Autografts taken from tendons within synovial sheaths, like those of the flexors they will replace, yield a better functional outcome than tendons taken from sites without an adjoining synovium.

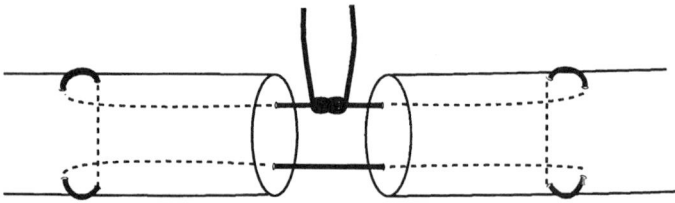

FIGURE 8.24. Typical suturing technique for repairing lacerated flexor tendons in the hand.

As has been previously noted, scar formation and tendon adhesion are complicating factors in flexor tendon repair. Experimental evaluations of flexor tendon repair should include tendon gliding and measurement of the resulting range of joint motion, in addition to the material properties of the tendon itself. Achieving superior tendon strength without proper function is useless. Unfortunately, these two factors tend to be inversely related, so that a repair technique and biologic response that yield a strong tendon may also cause significant scar deposition and adhesion to surrounding tissues.

8.10 Summary and Further Reading

Tendons transmit muscle forces to bones, storing and releasing energy in the process; ligaments are passive joint stabilizers. Both these structures provide proprioceptive information that is vital to all kinds of motor activity. Both are composed largely of type I collagen, with ligaments containing somewhat more elastin. Both tendon and ligament readily exhibit disuse atrophy, and their compositions and mechanical properties seem to vary in accordance with their function; for example, flexor tendons are generally stiffer than their corresponding extensor tendons. It is hypothesized that this is controlled by cellular responses to mechanical signals. Viscoelasticity is an important feature of the mechanical properties of these tissues, so that they are sensitive to strain rate and load history. They also creep, relax, and exhibit hysteresis. Mechanical testing is complicated by their soft, slippery, wet, and viscoelastic nature, making gripping, dimensional analysis, and strain measurement problematic. When lacerated or torn, ligaments and tendons will heal, but the healed tissue may never regain full strength and tendon motion is frequently constrained by adhesions to adjacent tissues.

An extensive review of all aspects of ligament and tendon tissue biology and mechanics may be found in the book *Injury and Repair of the Musculoskeletal Soft Tissues,* edited by Woo and Buckwalter (1988). Other excellent reviews include book chapters by Woo et al. (1985), Woo (1986), and Viidik (1996).

8.11 Exercises

8.1. Show that Zajac's peak isometric muscle force-normalized generic muscle model implicitly assumes that the cross-sectional areas of tendons are proportional to the physiologic cross-sectional areas of their muscles.

8.2. Equations 8.1 and 8.2 represent empirical force-strain relationships for the tendon and aponeurosis of a frog muscle-tendon unit. In these, force is expressed as percent of the maximum tetanic muscle force. Calculate the tendon and aponeurosis strains when the muscle is exerting this maximum tetanic force. Which tissue has the greater strain? In general, what is the relationship between the tendon and aponeurosis strains for this muscle-tendon unit?

8.3. Show that Eq. 8.4 is the solution to Eq. 8.3.

8.4. Use the LaPlace transform method to show that Eq. 8.6 is the creep solution to Eq. 8.5.

8.5. Your collaborator, a small animal surgeon, has recently taken an interest in the canine medial collateral ligament of the elbow and is exploring different surgical techniques for its reconstruction. As a first step you have decided to completely define the mechanical behavior of the ligament including its failure characteristics. Describe your approach, including the parameters that will be measured, your experimental methods and instrumentation, and problems which you anticipate.

8.6. You are testing anterior cruciate ligament (ACL) reconstructions with Dr. Kneely, an orthopaedic surgeon. You recorded the following load-deformation data for an experimental femur–ACL–tibia reconstruction in an animal model. Failure was by bony avulsion at the tibia. The ACL cross-sectional area was 150 mm^2 and the bone-to-bone resting length of the ligament was 3.5 cm. Calculate the important structural and material properties of the reconstruction.

Displacement, mm	Load, N	Displacement, mm	Load, N	Displacement, mm	Load, N
0.00	0	0.12	214	0.30	1908
0.01	0	0.14	379	0.32	1915
0.02	0	0.16	583	0.34	1940
0.03	0	0.18	796	0.36	2111
0.04	0	0.20	1000	0.37	2037
0.05	0	0.22	1202	0.38	1540
0.06	10	0.24	1406	0.39	1120
0.08	37	0.26	1611	0.40	821
0.10	96	0.28	1792	0.41	698

8.7. It was noted at the end of section 8.6 that tendon temperatures in race-horses may exceed 44°C. Suppose a typical tendon has an elastic modulus of 1 GPa and cycles between 0.5% and 2% strain during running. If the stress-strain relationship is linear in this range, what is the fluctuation in strain energy density as the tendon stress rises and falls? If that energy were expressed as heat, how much would it be in calories/ml of tendon/cycle? (The mechanical equivalent of heat is 4.19 joules/cal.) If the specific gravity and specific heat capacity of tendon tissue are similar to those of water (1 g/ml and 1 cal/g per °C, respectively), what percent of the strain energy must be converted to heat to raise the tissue temperature 1°C in 1000 load cycles, assuming adiabatic conditions (i.e., no heat leaves the site)? Is it plausible that tendons could be heated by loading during running?

8.8. In Chapter 1 we learned that the skeleton is arranged as a system of levers that amplify the small displacements produced by contracting muscles. Instruments known as micromanipulators use levers to demagnify finger motions. In this chapter you have read that the flexor tendons of the fingers themselves have characteristics that serve to increase fine motor control; that is, they demagnify somewhat the motions of the fingers which would otherwise be produced by the muscle contractions. Using the simple joint shown in Fig. 8.17 as a model, derive an equation to show how this can be achieved by making flexor tendons longer and more compliant. How would the tendon's length, diameter, and elastic modulus affect the resulting motion?

8.9. In Exercise 8.8, how would "pulleys" (loops of tissue to hold the tendon against the bones) near the joint contribute further to fine motor control? What other function would such pulleys have?

8.10. When a kangaroo switches from "pentapedal" (four limbs and tail) locomotion to hopping, its oxygen consumption drops, presumably because it then stores more energy in elastic tissues (Dawson and Taylor, 1973). One of these tissue "springs" in kangaroos and other animals is the Achilles tendon. A kangaroo's Achilles tendon was found to be 1.5 cm in diameter and 35 cm long. If both Achilles tendons had an elastic modulus of 1 GPa and were loaded to 2% strain, how much strain energy would they possess? Based strictly on energy considerations, how high could this amount of energy lift a 40-kg kangaroo?

Bibliography

CRC Handbook of Chemistry and Physics (1983). CRC Press, Boca Raton.

Encyclopaedia Britannica (1997). Encyclopaedica Britannica, Chicago.

Abbott, S., Trinkaus, E., and Burr, D.B. (1996). Dynamic bone remodeling in later Pleistocene fossil hominids. *American Journal of Physical Anthropology* **99**:585–601.

Abendschein, W., and Hyatt, G.W. (1970). Ultrasonics and selected physical properties of bone. *Clinical Orthopaedics and Related Research* **69**:294–301.

Adams, L.M. (1993). The anatomy of joints related to function. In *Mechanics of the Human Joints* (edited by V. Wright and E.L. Radin), 27–81. Dekker, New York.

Afoke, A., Hutton, W.C., and Byers, P.D. (1990). Pressure measurement in the human hip joint using Fujifilm. In *Methods in Cartilage Research* (edited by A. Maroudas and K. Kuettner), 281–287. Academic Press, London.

Agur, A.M.R. (1991). *Grant's Atlas of Anatomy*. Williams & Wilkins, Baltimore.

Ahlqvist, J., and Damsten, O. (1969). A modification of Kerley's method for the microscopic determination of age in human bone. *Journal of Forensic Sciences* **14**:205–213.

Albright, J.A., and Brand, R.A. (1987). *The Scientific Basis of Orthopaedics*. Appleton-Century-Crofts, New York.

Alexander, R.M. (1968). *Animal Mechanics*. University of Washington Press, Seattle.

Alexander, R.M., and Bennet-Clark, H.C. (1977). Storage of elastic strain energy in muscle and other tissues. *Nature* (London) **265**:114–117.

Alexander, R.M. (1988). *Elastic Mechanisms in Animal Movement*. Cambridge University Press, Cambridge, UK.

Alexander, R.M. (1996). *Optima for Animals*. Princeton University Press, Princeton, N.J.

Alto, A., and Pope, M.H. (1979). On the fracture toughness of equine metacarpi. *Journal of Biomechanics* **12**:415–421.

Amsel, S., Maniatis, A., Tavassoli, M., and Crosby, W.H. (1969). The significance of intramedullary cancellous bone in the repair of bone marrow tissue. *Anatomical Record* **164**:101–111.

Ascenzi, A., and Bonucci, E. (1964). The ultimate tensile strength of single osteons. *Acta Anatomica* **58**:160–183.

Ascenzi, A., and Bonucci, E. (1967). The tensile properties of single osteons. *Anatomical Record* **158**:375–386.

Ascenzi, A., and Bonucci, E. (1968). The compressive properties of single osteons. *Anatomical Record* **161**:377–391.

Ascenzi, A., and Bonucci, E. (1972). The shearing properties of single osteons. *Anatomical Record* **172**:499–510.

Ashman, R.B., Mayer, D.C., Cowin, S.C., and Van Buskirk, W.C. (1983). Elastic properties of human and canine femora. *Transactions of the Orthopaedic Research Society* **8**:127.

Ashman, R.B., Cowin, S.C., Van Buskirk, W.C., and Rice, J.C. (1984). A continuous wave technique for the measurement of the elastic properties of cortical bone. *Journal of Biomechanics* **17**:349–361.

Ashman, R.B., Van Buskirk, W.C., Cowin, S.C., Sandborn, P.M., Wells, M.K., and Rice, J.C. (1985a). The mechanical properties of immature osteopetrotic bone. *Calcified Tissue International* **37**:73-76.

Ashman, R.B., Rosinia, G., Cowin, S.C., and Fontenot, M.G. (1985b). The bone tissue of the canine mandible is elastically isotropic. *Journal of Biomechanics* **18**:717–721.

Ashman, R.B., and Van Buskirk, W.C. (1987). The elastic properties of a human mandible. *Advances in Dental Research* **1(1)**:64–67.

Ashman, R.B., and Rho, J.Y. (1988). Elastic modulus of trabecular bone material. *Journal of Biomechanics* **21(3)**:177–181.

Atwater, A.E. (1990). Gender differences in distance running. In *Biomechanics of Distance Running* (edited by P. R. Cavanaugh). Human Kinetics Books, Champaign.

Bab, I., Gazit, D., Massarawa, A., and Sela, J. (1985). Removal of the tibial marrow induces increased formation of bone and cartilage in rat mandibular condyle. *Calcified Tissue International* **37**:551–555.

Balazs, E.A., and Gibbs, D.A. (1970). The rheological properties and biological function of hyaluronic acid. In *Chemistry and Molecular Biology of the Intercellular Matrix* (edited by E. A. Balazs). Academic Press, New York.

Barad, S., Cabaud, H.E., and Rodrigo, J.J. (1982). Effects of storage at −80°C. as compared to 4°C. on the strength of rhesus monkey anteriocruciate ligaments. *Transactions of the Orthopaedic Research Society* **7**:378.

Barbenel, J.C. (1972). The biomechanics of the temporomandibular joint: a theoretical study. *Journal of Biomechanics* **5**:251–256.

Bargren, J.H., Bassett, C.A.L., and Gjelsvik, A. (1974). Mechanical properties of hydrated cortical bone. *Journal of Biomechanics* **7**:239–245.

Barnes, G.R.G., and Pinder, D.N. (1974). In vivo tension and bone strain measurement and correlation. *Journal of Biomechanics* **7**:35-42.

Barnett, C.H., and Cobbold, A.F. (1962). Lubrication within living joints. *Journal of Bone and Joint Surgery* **44B**:662–674.

Baron, R., Tross, R., and Vignery, A. (1984). Evidence of sequential remodeling in rat trabecular bone morphology, dynamic histomorphometry, and changes during skeletal maturation. *Anatomical Record* **208**:137–145.

Barou, O., Lafage-Proust, M.H., Palle, S., Vico, L., and Alexandre, C. (1996). Effects of immobilization on preosteoblast proliferation assessed histomorphometrically using BRDU uptake on epon-embedded rat bone (abstract). *Bone* (New York) **19**:131S.

Bartel, D.L., Schryver, H.F., Lowe, J.E., and Parker, R.A. (1978). Locomotion in the horse: a procedure for computing the internal forces of the digit. *American Journal of Veterinary Research* **39**:1721–1733.

Bassett, C.A.L. (1965). Electrical effects in bone. *Scientific American* **213**:18–25.

Bassett, C.A.L., Mitchell, S.N., and Gaston, S.R. (1981). Treatment of ununited tibial diaphyseal fractures with pulsing electromagnetic fields. *Journal of Bone and Joint Surgery* **63A**:511–523.

Bateson, G. (1963). The role of somatic change in evolution. *Evolution* **17**:529–539.

Bateson, G. (1987). *Steps to an Ecology of Mind*. Jason Aronson, Northvale, NJ.

Bay, B.K., Hamel, A.J., Olsen, S.A., and Sharkey, N.A. (1997a). Statically equivalent load and support conditions produce different hip joint contact pressures and periacetabular strains. *Journal of Biomechanics* **30**:193–196.

Bay, B.K., Sharkey, N.A., and Szabo, R.M. (1997b). Biomechanical behavior of the median nerve in and about the carpal tunnel. *Journal of Hand Surgery* **22A**:621–627.

Beaupre, G.S., Orr, T.E., and Carter, D.R. (1990a). An approach for time-dependent bone modeling and remodeling—theoretical development. *Journal of Orthopaedic Research* **8**:651-661.

Beaupre, G.S., Orr, T.E., and Carter, D.R. (1990b). An approach for time-dependent bone modeling and remodeling—application: a preliminary remodeling simulation. *Journal of Orthopaedic Research* **8**:662–670.

Beaupre, G.S., and Carter, D.R. (1992). Finite element analysis in biomechanics. In *Biomechanics—Structures and Systems* (edited by A. A. Biewener), 149–174. Oxford University Press, Oxford.

Beck, T.J., Ruff, C.B., and Bissessur, K. (1993). Age-related changes in female femoral neck geometry: implications for bone strength. *Calcified Tissue International* **53**:S41–S46.

Behiri, J.C., and Bonfield, W. (1984). Fracture mechanics of bone—the effects of density, specimen thickness and crack velocity on longitudinal fracture. *Journal of Biomechanics* **17**:25–34.

Behrens, J.C., Walker, P.S., and Shoji, H. (1974). Variations in strength and structure of cancellous bone at the knee. *Journal of Biomechanics* **7**:201–207.

Bell, C. (1827). *Animal Mechanics, or Proofs of Design in the Animal Frame.* Morrill Wyman, Cambridge, MA.

Benninghoff, A. (1925). Form und bau der gelenhknorpel in ihren bezeiehungen zur funktion. *Erste Mittalung Zeitschrift fur Anatomie und Entwicklungsgeschichte* **76**:43.

Bensusan, J.S., Davy, D.T., Heiple, K.G., and Verdin, P.J. (1983). Tensile, compressive, and torsional testing of cancellous bone. *Transactions of the Orthopaedic Research Society* **8**:132.

Bentolila, V., Hillam, R.A., Skerry, T.M., Boyce, T.M., Fyhrie, D.P., and Schaffler, M.B. (1997). Activation of intracortical remodeling in adult rat long bones by fatigue loading. *Transactions of the Orthopaedic Research Society* **22**:578.

Bergmann, G., Graichen, F., Siraky, J., Jendrzynski, H., and Rohlmann, A. (1988). Multichannel strain gauge telemetry for orthopaedic implants. *Journal of Biomechanics* **21**:169–176.

Bergmann, G., Graichen, F., and Rohlmann, A. (1993). Hip joint loading during walking and running measured in two patients. *Journal of Biomechanics* **26**:969–990.

Bertram, J.E., and Biewener, A.A. (1988). Bone curvature: sacrificing strength for load predictability? *Journal of Theoretical Biology* 131:75–92.

Bertram, J. E. A., and Swartz, S. M. (1991). The 'Law of Bone Transformation': a case of crying Wolff? *Biological Reviews* **66**:245–273.

Biewener, A.A., Thomason, J., Goodship, A., and Lanyon, L.E. (1983). Bone stress in the horse forelimb during locomotion at different gaits: A comparison of two experimental methods. *Journal of Biomechanics* **16**:565–576.

Bilezikian, J.P., Raisz, L.G., and Rodan, G.A. (1996). *Principles of Bone Biology.* Academic Press, New York.

Bonfield, W., and Li, C.H. (1967). Anistropy of non-elastic flow in bone. *Journal of Applied Physiology* **38**:2450.

Bonfield, W., and Clark, E.A. (1973). Elastic deformation of compact bone. *Journal of Materials Science* **8**:1590–1594.

Bonfield, W., and Datta, P.K. (1976). Fracture toughness of compact bone. *Journal of Biomechanics* **9**:131–134.

Bonfield, W., and Tully, A.E. (1982). Ultrasonic analysis of the Young's modulus of cortical bone. *Journal of Biomedical Engineering* **4**:23–27.

Borelli, G.A. (1989). *On the Movement of Animals* (1989 translation by P. Maquet). Springer-Verlag, Berlin.

Borsa, P.A., Lephart, S.M., Irrgang, J.J., Saffran, M.R., and Fu, F.H. (1997). The effects of joint position and direction of joint motion on proprioceptive sensibility in anterior cruciate ligament-deficient athletes. *American Journal of Sports Medicine* **25**:336–340.

Borysenko, M., and Beringer, T. (1984). *Functional Histology*. Little, Brown, Boston.

Bourgery, J.M. (1832). Traite Complet de l'Anatomie de l'Homme. I. Osteologie. Paris.

Bourne, G.H. (1972-6). *The Biochemistry and Physiology of Bone, Vols. I–IV.* Academic Press, New York.

Boussinesq, J. (1885). *Applications des potentiels a l'etude de l'equilibre et du mouvement des solides elastique.* Gauthier-Villars, Paris.

Bouvier, M., and Ubelaker, D.H. (1977). A comparison of two methods for the microscopic determination of age at death. *American Journal of Physical Anthropology* **46**:391–394.

Boyde, A. (1972). Scanning electron microscope studies of bone. In *The Biochemistry and Physiology of Bone* (edited by G.H. Bourne). Academic Press, New York.

Brighton, C.T., and Pollack, S.R. (1985). Treatment of recalcitrant non-union with a capacitively coupled electrical field. A preliminary report. *J Bone Joint Surg [Am]* **67**:577–85.

Brighton, C.T., Friedlaender, G.E., and Lane, J.M. (1994). *Bone Formation and Repair.* American Academy of Orthopaedic Surgeons, Rosemont, IL.

Brown, S.D., Biddulph, R.B., and Wilcox, P.D. (1964). A strength-porosity relation involving different pore geometry and orientation. *American Ceramic Society Journal* **47**:320.

Brown, T.D., and Ferguson, A.B. (1980). Mechanical property distributions in the cancellous bone of the human proximal femur. *Acta Orthopaedica Scandinavica* **51**:429–437.

Buckwalter, J.A., and Mow, V.C. (1992). Cartilage repair in osteoarthritis. In *Osteoarthritis: Diagnosis and Medical/Surgical Management* (edited by R.W. Moskowitz, D.S. Howell, V.M. Goldberg, et al.), 71–107. Saunders, Philadelphia.

Burny, F., Donkerwolcke, M., Bourgois, R., Domb, M., and Saric, O. (1984). Twenty years experience in fracture healing measurements with strain gauges. *Orthopedics* **7**:1823–1829.

Burr, D.B., Gerven, D.P.V., and Gustav, B.L. (1977). Sexual dimorphism and mechanics of the human hip: a multivariate assessment. *American Journal of Physical Anthropology* **47**:273–278.

Burr, D.B., Martin, R.B., Schaffler, M.B., and Radin, E.L. (1985). Bone remodeling in response to in vivo fatigue microdamage. *Journal of Biomechanics* **18**:189–200.

Burr, D.B., Schaffler, M.B., and Fredrickson, R.G. (1988). Composition of the cement line and its possible mechanical role as a local interface in human compact bone. *Journal of Biomechanics* **21**:939–945.

Burr, D.B., Schaffler, M.B., Yang, K.H., Wu, D.D., Lukoschek, M., Kandzari, D., Sivaneri, N., Blaha, J.D., and Radin, E.L. (1989a). The effects of altered strain environment on bone tissue kinetics. *Bone* (New York) **10**:215–221.

Burr, D. B., Schaffler, M. B., Yang, K. H., Lukoschek, M., Sivaneri, N., Blaha, J. D., and Radin, E. L. (1989b). Skeletal change in response to altered strain environments: is woven bone a response to elevated strain? *Bone* (New York) **10**:223–233.

Burr, D.B., and Stafford, T. (1990). Validity of the bulk-staining technique to separate artifactual from in vivo bone microdamage. *Clinical Orthopaedics and Related Research* **260**:305–308.

Burr, D.B., Ruff, C.B., and Thompson, D.D. (1990). Patterns of skeletal histologic change through time: comparison of an archaic native American population with modern populations. *Anatomical Record* **226**:307–313.

Burr, D.B. (1992). Estimated intracortical bone turnover in the femur of growing macaques: implications for their use as models in skeletal pathology. *Anatomical Record* **232**:180–189.

Burr, D.B., Milgrom, C., Fyhrie, D., Forwood, M., Nyska, M., Finestone, A., Hoshaw, S., Saiag, E., and Simkin, A. (1996). In vivo measurement of human tibial strains during vigorous activity. *Bone* (New York) **18**:405–10.

Burr, D.B. (1997). Bone, exercise, and stress fractures. *Exercise and Sport Sciences Reviews* **25**:171–194.

Burr, D.B., Forwood, M.K., Fyhrie, D.P., Martin, R.B., Schaffler, M.S., and Turner, C.H. (1997). Bone microdamage and skeletal fragility in osteoporotic and stress fractures. *Journal of Bone and Mineral Research* **12**:6–15.

Burstein, A.H., and Frankel, V.H. (1968). The viscoelastic properties of some biological materials. *Annals of the New York Academy of Sciences* **146**:158–165.

Burstein, A.H., and Frankel, V.H. (1971). A standard test for laboratory animal bone. *Journal of Biomechanics* **4**:155-158.

Burstein, A.H., Currey, J.D., Frankel, V.H., and Reilly, D.T. (1972). The ultimate properties of bone tissue: the effects of yielding. *Journal of Biomechanics* **5**:35–44.

Burstein, A.H., Zika, J.M., Heiple, K.G., and Klein, L. (1975). Contribution of collagen and mineral to the elastic-plastic properties of bone. *Journal of Bone and Joint Surgery* **57A**:956–961.

Caler, W.E., and Carter, D.R. (1989). Bone creep-fatigue damage accumulation. *Journal of Biomechanics* **22**:625–635.

Cameron, A. (1981). *Basic Lubrication Theory.* Halsted Press, New York.

Carter, D.R., Hayes, W.C., and Schurman, D.J. (1976). Fatigue life of compact bone. II. Effects of microstructure and density. *Journal of Biomechanics* **9**:211–218.

Carter, D.R., and Hayes, W.C. (1976). Fatigue life of compact bone. I. Effects of stress amplitude, temperature and density. *Journal of Biomechanics* **9**:27–34.

Carter, D.R., and Hayes, W.C. (1977a). The compressive behavior of bone as a two-phase porous structure. *Journal of Bone and Joint Surgery* **59A**:954–962.

Carter, D.R., and Hayes, W.C. (1977b). Compact bone fatigue damage. I. Residual strength and stiffness. *Journal of Biomechanics* **10**:325–337.

Carter, D.R., and Hayes, W.C. (1977c). Compact bone fatigue damage: a microscopic examination. *Clinical Orthopaedics and Related Research* **127**:265–274.

Carter, D.R., and Spengler, D.M. (1978). Mechanical properties and composition of cortical bone. *Clinical Orthopaedics and Related Research* **135**:192–217.

Carter, D.R., Schwab, G.H., and Spengler, D.M. (1980). Tensile fracture of cancellous bone. *Acta Orthopaedica Scandinavica* **51**:733–741.

Carter, D.R., Caler, W.E., Spengler, D.M., and Frankel, V.H. (1981). Uniaxial fatigue of human cortical bone. The influence of tissue physical characteristics. *Journal of Biomechanics* **14**:461–470.

Carter, D.R., and Caler, W.E. (1983). Cycle-dependent and time-dependent bone fracture with repeated loading. *Journal of Biomechanical Engineering* **105**:166–170.

Carter, D.R., and Caler, W.E. (1985). A cumulative damage model for bone fracture. *Journal of Orthopaedic Research* **3**:84–90.

Carter, D.R. (1987). Mechanical loading history and skeletal biology. *Journal of Biomechanics* **20**:1095–1109.

Carter, D.R., Orr, T.E., Fyhrie, D.P., and Schurman, D.J. (1987a). Influences of mechanical stress on prenatal and postnatal skeletal development. *Clinical Orthopaedics and Related Research* **219**:237–250.

Carter, D.R., Fyhrie, D.P., and Whalen, R.T. (1987b). Trabecular bone density and loading history: regulation of connective tissue biology by mechanical energy. *Journal of Biomechanics* **20**:785–794.

Carter, D.R., Blenman, P.R., and Beaupre, G.S. (1988). Correlations between mechanical stress history and tissue differentiation in initial fracture healing. *Journal of Orthopaedic Research* **6**:736–748.

Carter, D.R., and Wong, M. (1988). Mechanical stresses and endochondral ossification in the chondroepiphysis. *Journal of Orthopaedic Research* **6**:148–154.

Carter, D.R., Orr, T.E., and Fyhrie, D.P. (1989). Relationships between loading history and femoral cancellous bone architecture. *Journal of Biomechanics* **22**:231–244.

Cetta, G., Tenni, R., and Zanaboni, G. (1982). Biochemical and morphological modifications in rabbit Achilles tendon during maturation and aging. *Journal of Biochemistry* (Tokyo) **204**:61–67.

Chamay, A., and Tschantz, P. (1972). Mechanical influences in bone remodeling. Experimental research on Wolff's law. *Journal of Biomechanics* **5**:173–180.

Charnley, J. (1960). The lubrication of animal joints in relation to surgical reconstruction by arthroplasty. *Annals of the Rheumatic Diseases* **19**:10–19.

Charnley, J. (1973). Arthroplasty of the hip: a new operation (reprinted from *Lancet*, pp. 1129–1132, 1961). *Clinical Orthopaedics and Related Research* **95**:4–8.

Chen, H.L., and Gundjian, A.A. (1974). Determination of the bone-crystallites distribution function by x-ray diffraction. *Medical and Biological Engineering* **12**:531–535.

Choi, K., Kuhn, J.L., Ciarelli, M.J., and Goldstein, S.A. (1990). The elastic moduli of human subschodral, trabecular, and cortical bone tissue and the size-dependency of cortical bone modulus. *Journal of Biomechanics* **23**:1103–1114.

Clancy, W.G., Narechania, R.G., and Rosenberg, T.D. (1981). Anterior and posterior cruciate ligament reconstruction in rhesus monkeys: a histological, microangiographic, and biomechanical analysis. *Journal of Bone and Joint Surgery* **63A**:1270–1284.

Coccia, P.F., Krivit, W., Cervenka, J., Clawson, C.C., Kersey, J.H., Kim, T.H., Nesbit, M.E., and Ramsay, N.K.C. (1980). Successful bone marrow transplantation for infantile malignant osteopetrosis. *New England Journal of Medicine* **302**:701–708.

Cochran, G.B.V. (1972). Implantation of strain gages on bone in vivo. *Journal of Biomechanics* **5**:119–123.

Cochran, G.B.V. (1974). A method for direct recording of electromechanical data from skeletal bone in living animals. *Journal of Biomechanics* **7**:563–565.

Cohen, J., and Harris, W.H. (1958). The three-dimensional anatomy of haversian systems. *Journal of Bone and Joint Surgery* **40A**:419–434.

Collins, J.J. (1995). The redundant nature of locomotor optimization laws. *Journal of Biomechanics* **28**:251–267.

Collier, R.J., and Donarski, R. (1987). Non-invasive method of measuring resonant frequency of a human tibia in vivo. *Journal of Biomedical Engineering* **9**:329–331.

Cooper, R.R., Milgram, J.W., and Robinson, R.A. (1966). Morphology of the osteon. *Journal of Bone and Joint Surgery* **48A**:1239–1271.

Corondan, G., and Haworth, W.L. (1986). A fractographic study of human long bone. *Journal of Biomechanics* **19**:207–218.

Cowin, S.C., and Hegedus, D.H. (1976). Bone remodeling. I: A theory of adaptive elasticity. *Journal of Elasticity* **6**:313–325.

Cowin, S.C., and Nachlinger, R.R. (1978). Bone remodeling. III: Uniqueness and stability in adaptive elasticity theory. *Journal of Elasticity* **8**:285–295.

Cowin, S.C., and Van Buskirk, W.C. (1978). Internal bone remodeling induced by a medullary pin. *Journal of Biomechanics* **11**:269–275.

Cowin, S.C., and Firoozbakhsh, K. (1981). Bone remodeling of diaphysial surfaces under constant load: theoretical predictions. *Journal of Biomechanics* **7**:471–484.

Cowin, S.C. (1985). The relationship between the elasticity tensor and the fabric tensor. *Mechanics of Materials* **4**:137–147.

Cowin, S.C., Hart, R.T., Balser, J.R., and Kohn, D.H. (1985). Functional adaptation in long bones: establishing in vivo values for surface remodeling rate coefficients. *Journal of Biomechanics* **18(9)**:665–684.

Cowin, S.C. (1986). Wolff's law of trabecular architecture at remodeling equilibrium. *Journal of Biomechanical Engineering* **108**:83–88.

Cowin, S.C. (1989). *Bone Mechanics*. CRC Press, Boca Raton.

Cowin, S.C., and Mehrabadi, M.M. (1989). Identification of the elastic symmetry of bone and other materials. *Journal of Biomechanics* **22**:503–515.

Cowin, S.C., Weinbaum, S., and Zeng, Y. (1995). A case for bone canaliculi as the anatomic site of strain generated potentials. *Journal of Biomechanics* **28**:1281–1297.

Cowin, S.C. (1997). The false premise of Wolff's law. *Forma* 12:247-262.

Crowninshield, R.D., and Pope, M.H. (1974). The response of compact bone in tension at various strain rates. *Annals of Biomedical Engineering* **2**:217–225.

Cruz-Coke, R. (1978). A genetic description of high-altitude populations. In *The Biology of High Altitude Peoples* (edited by P.T. Baker), 47–63. Cambridge University Press, Cambridge.

Cumming, D.C. (1996). Exercise-associated amenorrhea, low bone density, and estrogen replacement therapy. *Archives of Internal Medicine* **156**:2193–2195.

Currey, J.D. (1959). Differences in the tensile strength of bone of different histological types. *Journal of Anatomy* **98**:87–95.

Currey, J.D. (1964). Three analogies to explain the mechanical properties of bone. *Biorheology* **2**:1–10.

Currey, J.D. (1969a). The mechanical consequences of variation in the mineral content of bone. *Journal of Biomechanics* **2**:1–11.

Currey, J.D. (1969b). The relationship between the stiffness and the mineral content of bone. *Journal of Biomechanics* **2**:477–480.

Currey, J.D., and Butler, G. (1975). The mechanical properties of bone tissue in children. *Journal of Bone and Joint Surgery* **57A**:810–814.

Currey, J.D. (1981). What is bone for? Property-function relationships in bone. In *Mechanical Properties of Bone* (edited by S.C. Cowin), 13–26. American Society of Mechanical Engineers, New York.

Currey, J.D. (1984). *The Mechanical Adaptations of Bones*. University Press, Princeton.

Currey, J.D. (1986). Effects of porosity and mineral content on the Young's modulus of bone. *Proceedings, European Society of Biomechanics* **5**:104.

Currey, J.D. (1988). The effect of porosity and mineral content on the Young's modulus of elasticity of compact bone. *Journal of Biomechanics* **21**:131–139.

Dannucci, G.A., Martin, R.B., and Patterson-Buckendahl, P. (1987). Ovariectomy and trabecular bone remodeling in the dog. *Calcified Tissue International* **40**:194–199.

Davison, P.F. (1989). The contribution of labile crosslinks to the tensile behavior of tendons. *Connective Tissue Research* **18**:293–305.

Davy, D.T., Kotzar, G.M., Brown, R.H., Heiple, K.G., Goldberg, V.M., Heiple, J.G., Berrile, J., and Burstein, A.H. (1988). Telemetric force measurements across the hip after total arthroplasty. *Journal of Bone and Joint Surgery* **70A**:45–50.

Dawson, T.J., and Taylor, C.R. (1973). Energetic cost of locomotion in kangaroos. *Nature* (London) **246**:313–314.

Devas, M.B. (1975). *Stress Fractures*. Churchill Livingstone, London.

Dintenfass, D. (1963a). Lubrication in synovial joints: a theoretical analysis. *Journal of Bone and Joint Surgery* **45A**:1241–1256.

Dintenfass, D. (1963b). Lubrication in synovial joints. *Nature* (London) **197**:496–497.

Dowson, D. (1979). *History of Tribology*. Longman, London.

Dowson, D., and Jin, Z.M. (1986). Micro-elastohydrodynamic lubrication of synovial joints. *Engineering in Medicine* **15**:63–65.

Dumbleton, J.H. (1981). *Tribology of Natural and Artificial Joints*. Elsevier, Amsterdam.

Dyson, E.D., and Whitehouse, W.J. (1968). Composition of trabecular bone in children and its relation to radiation dosimetry. *Nature* (London) **217**:576–578.

Ellis, D.G. (1969). Cross-sectional area measurements for tendon specimens: a comparison of several methods. *Journal of Biomechanics* **2**:175–186.

English, T.A., and Kilvington, T.M. (1979). In vivo records of hip loads using a femoral implant with telemetric output. *Journal of Biomedical Engineering* **1**:111–115.

Enlow, D.H., and Brown, S.O. (1956). A comparative histological study of fossil and recent bone tissues. Part I. Introduction, methods, fish and amphibian bone tissues. *Texas Journal of Science* **8**:405–443.

Enlow, D.H., and Brown, S.O. (1957). A comparative histological study of fossil and recent bone tissues. Part II. Reptilian and bird bone tissues. *Texas Journal of Science* **9**:186–214.

Enlow, D.H., and Brown, S.O. (1958). A comparative histological study of fossil and recent bone tissues. Part III. Mammalian bone tissues. General discussion. *Texas Journal of Science* **10**:187–230.

Eriksson, C. (1974). Streaming potentials and other water dependent effects in mineralized tissues. *Annuals of the New York Academy of Science* **238**:321–338.

Evans, F.G. (1957). *Stress and Strain in Bones*. Thomas, Springfield.

Evans, F.G., and Bang, S. (1967). Differences and relationships between the physical properties and microscopic structure of human femoral, tibial and fibular cortical bone. *American Journal of Anatomy* **120**:79–88.

Evans, F.G., and Vincentelli, R. (1969). Relation of collagen fiber orientation to some mechanical properties of human cortical bone. *Journal of Biomechanics* **2**:63–71.

Evans, F.G., and Riolo, M.L. (1970). Relations between the fatigue life and histology of adult human cortical bone. *Journal of Bone and Joint Surgery* **52A**:1579–1586.

Evans, F.G., and Vincentelli, R. (1974). Relations of the compressive properties of human cortical bone to histological structure and calcification. *Journal of Biomechanics* **7**:1–10.

Fawcett, D.W. (1994). *A Textbook of Histology*. Chapman & Hall, New York.

Finkel, V. (1985). *Portrait of a Crack* (translated from the Russian by Y. Nadler). Mir Publications, Moscow.

Firoozbakhsh, K., and Cowin, S.C. (1980). Devolution of inhomogeneities in bone structure: Predictions of adaptive elasticity theory. *Journal of Biomechanical Engineering* **102**:287–293.

Firoozbakhsh, K., and Cowin, S.C. (1981). An analytical model of Pauwel's functional adaptation mechanism for bone. *Journal of Biomechanical Engineering* **103**:246–252.

Flugge, W. (1967). *Viscoelasticity*. Blaisdell, Waltham, MA.

Ford, C.M., and Keaveny, T.M. (1996). The dependence of shear failure properties of trabecular bone on apparent density and trabecular orientation. *Journal of Biomechanics* **29**:1309–1317.

Forwood, M.R., and Turner, C.H. (1994). The response of rat tibiae to incremental bouts of mechanical loading: a quantum concept for bone formation. *Bone* (New York) **15**:603–609.

Frank, C., Woo, S., Andriacchi, T., Brand, R., Oakes, B., Dahners, L., DeHaven, K., Lewis, J., and Sabiston, P. (1988). Normal ligament: structure, function, and composition. In *Injury and Repair of the Musculoskeletal Soft Tissues* (edited by S.L.-Y. Woo and J.A. Buckwalter), 45–128. American Academy of Orthopaedic Surgeons, Park Ridge, IL.

Frankel, V.H., and Burstein, A.H. (1970). *Orthopaedic Biomechanics*. Lea & Febiger, Philadelphia.

Frasca, P. (1981). Scanning electron microscopy studies of ground substance in the cement lines, resting lines, hypercalcified rings and reversal lines of human cortical bone. *Acta Anatomica* **109**:115–121.

Freehan, L.M., and Beauchene, J.G. (1990). Early tensile properties of healing chicken flexor tendons: early controlled passive motion versus postoperative immobilization. *Journal of Hand Surgery* **15**:63–68.

Freeman, M.A.R., and Wyhe, B.D. (1966). Articular contributions to limb muscle reflexes: the effects of partial neurectomy of the knee-joint on postural reflexes. *British Journal of Surgery* **53**:61–69.

Freund, L.B. (1990). *Dynamic Fracture Mechanics*. Cambridge University Press, Cambridge.

Friedenberg, Z.B., Andrews, E.T., Smolenski, B.I., Pearl, B.W., and Brighton, C.T. (1970). Bone reaction to varying amounts of direct current. *Surgery, Gynecology, & Obstetrics* **131**:894–899.

Frost, H.M. (1958). Preparation of thin undecalcified bone sections by a rapid manual method. *Stain Technology* **33**:272.

Frost, H.M. (1960a). Presence of microscopic cracks in vivo in bone. *Henry Ford Hospital Medical Bulletin* **8**:25-35.

Frost, H.M. (1960b). Observations on osteoid seams: the existence of a resting state. *Henry Ford Hospital Medical Bulletin* **8**:220–224.

Frost, H.M. (1960c). Measurement of bone formation in a 57-year-old man by means of tetracycline. *Henry Ford Hospital Medical Bulletin* **8**:212–219.

Frost, H.M. (1960d). Measurement of osteocytes per unit volume and volume components of osteocytes and canaliculae in man. *Henry Ford Hospital Medical Bulletin* **8**:208–211.

Frost, H.M., Roth, H., and Villanueva, A.R. (1961). Physical characteristics of bone. Part III. A semimicro measurement of unit shear stress. *Henry Ford Hospital Medical Journal* **9**:157–162.

Frost, H.M. (1964a). *The Laws of Bone Structure*. Thomas, Springfield.

Frost, H.M. (1964b). *Mathematical Elements of Lamellar Bone Remodelling*. Thomas, Springfield.

Frost, H.M. (1969). Tetracycline-based histological analysis of bone remodeling. *Calcified Tissue Research* **3**:211–237.

Frost, H.M. (1973a). *Orthopaedic Biomechanics*. Thomas, Springfield.

Frost, H.M. (1973b). *Bone Remodeling and Its Relationship to Metabolic Bone Disease*. Thomas, Springfield.

Frost, H.M. (1983). Bone histomorphometry: correction of the labeling "escape error". In *Bone Histomorphometry: Techniques and Interpretation* (edited by R.R. Recker). CRC Press, Boca Raton.

Frost, H.M. (1985). The pathomechanics of osteoporoses. *Clinical Orthopaedics and Related Research* **200**:198–225.

Frost, H.M. (1986). *Intermediary Organization of the Skeleton*. CRC Press, Boca Raton.

Frost, H.M. (1987a). Secondary osteon populations: an algorithm for estimating the missing osteons. *Yearbook of Physical Anthropology* **30**:239–254.

Frost, H. M. (1987b). Bone "mass" and the "mechanostat": a proposal. The Anatomical Record 219:1-9.

Frost, H.M. (1990). Structural adaptations to mechanical usage (SATMU): 3. The hyaline cartilage modeling problem. *Anatomical Record* **226**:423–432.

Frost, H.M. (1994). Perspectives: a vital biomechanical model of synovial joint design. *Anatomical Record* **240**:1–18.

Fukada, E., and Yasuda, I. (1957). On the piezoelectric effect of bone. *Journal of the Physical Society of Japan* **10**:1158–1169.

Fung, Y.C. (1969). *A First Course in Continuum Mechanics*. Prentice Hall, Englewood Cliffs, NJ.

Fung, Y.C.B. (1972). Stress-strain-history relations of soft tissues in simple elongation. In *Biomechanics: Its Foundations and Objectives* (edited by Y.C.B. Fung, N. Perrone, and M. Anliker). Prentice-Hall, Englewood Cliffs.

Fung, Y.C. (1977). *A First Course in Continuum Mechanics*. Prentice-Hall, Englewood Cliffs.

Galileo, G. (1638). *Discorsi e dimonstrazioni matematiche, intorno a due nuove scienze attentanti alla meccanica ed a muovementi locali*. University of Wisconsin Press, Madison.

Gannon, R.E., Harris, G.M., and Vasilos, T. (1965). Effect of porosity on mechanical thermal, and dielectric properties of fused silica. *American Ceramic Society Bulletin* **44**:460.

Gerlanc, M., Haddad, D., Hyatt, G.W., Lanloh, H.T., and Hilaire, P. (1975). Ultrasonic study of normal and fractured bone. *Clinical Orthopaedics and Related Research* **111**:175–180.

Gibson, L.J. (1985). The mechanical behavior of cancellous bone. *Journal of Biomechanics* **18**:317–328.

Gibson, V.A., Stover, S.M., Martin, R.B., Gibeling, J.C., Willits, N.H., Gustafson, M.B., and Griffin, L.V. (1996). Fatigue behavior of the equine third metacarpus: mechanical property analysis. *Journal of Orthopaedic Research* **13**:861–868.

Gillis, C., Sharkey, N.A., Stover, S.M., Pool, R.R., and Meagher, D.M. (1995). The effect of maturation and aging on the material and ultrasonographic properties of equine superficial digital flexor tendon. *American Journal of Veterinary Research* **56**:1345–1350.

Giori, N.J., Beaupre, G.S., and Carter, D.R. (1993). Cellular shape and pressure may mediate mechanical control of tissue composition in tendons. *Journal of Orthopaedic Research* **11**:581–591.

Giraud-Guille, M.M. (1988). Twisted plywood architecture of collagen fibrils on human compact bone osteons. *Calcified Tissue International* **42**:167–180.

Goldstein, S.A., Wilson, D.L., Sonstegard, D.A., and Matthews, L.S. (1983). The mechanical properties of human tibial trabecular bone as a function of metaphyseal location. *Journal of Biomechanics* **16**:865–969.

Goldstein, S.A. (1987). The mechanical properties of trabecular bone: dependence on anatomic location and function. *Journal of Biomechanics* **20**:1055–1061.

Gong, J.K., Arnold, J.S., and Cohn, S.H. (1964). Composition of trabecular and cortical bone. *Anatomical Record* **149**:325–332.

Goodship, A.E., Lanyon, L.E., and McFie, H. (1979). Functional adaption of bone to increased stress. *Journal of Bone and Joint Surgery* **61A**:539–546.

Gottesman, T., and Hashin, Z. (1980). Analysis of viscoelastic behavior of bones on the basis of microstructure. *Journal of Biomechanics* **13**:89–96.

Gray, R.J., and Korbacher, G.K. (1974). Compressive fatigue behaviour of bovine compact bone. *Journal of Biomechanics* **7**:287–292.

Griffin, L.V., Gibeling, J.C., Martin, R.B., Gibson, V.A., and Stover, S.M. (1997). A model of flexural fatigue damage accumulation for cortical bone. *Journal of Orthopaedic Research* **15**:607–614.

Griffith, A.A. (1920). The phenomenon of rupture and flow in solids. *Philosophical Transactions of the Royal Society of London* **A221**:163–198.

Gu, W.Y., Lai, W.M., and Mow, V.C. (1993). Transport of fluid and ions through a porous-permeable charged-hydrated tissue, and streaming potential data on normal bovine articular cartilage. *Journal of Biomechanics* **26**:709–723.

Gunness, M., and Hock, J.M. (1993). Anabolic effect of parathyroid hormone on cancellous and cortical bone histology. *Bone* (New York) **14**:277–281.

Gustafson, M.B., Gibson, V.A., Stover, S.M., and Martin, R.B. (1995). Requirement of calcium buffering to maintain equine cannon bone stiffness in saline. *Transactions of the Orthopaedic Research Society* **20**:297.

Hall, B.K. (1990-2). *Bone, Vols. 1–7*. Telford Press, Caldwell, N.J.

Harrigan, T.P., and Mann, R.W. (1984). Characterization of microstructural anisotropy in orthotropic materials using a second rank tensor. *Journal of Materials Science* **19**:761–767.

Harrigan, T.P., Jasty, M., Mann, R.W., and Harris, W.H. (1988). Limitations of the continuum assumption in cancellous bone. *Journal of Biomechanics* **21**:269–275.

Harrigan, T.P., and Hamilton, J.J. (1994). Necessary and sufficient conditions for global stability and uniqueness in finite element simulations of adaptive bone remodeling. *International Journal of Solids and Structures* **31**:97–107.

Harris, W.H. (1960). A microscopic method of determining rates of bone growth. *Nature* (London) **188**:1038–1039.

Hart, R.T., Davy, D.T., and Heiple, K.G. (1984a). Mathematical modeling and numerical solutions for functionally dependent bone remodeling. *Calcified Tissue International* **36**:S104–S109.

Hart, R.T., Davy, D.T., and Heiple, K.G. (1984b). A computational method for stress analysis of adaptive elastic materials with a view toward applications in strain-induced bone remodeling. *Journal of Biomechanical Engineering* **106**:342–350.

Hart, R.T. (1989). The finite element method. In *Bone Mechanics* (edited by S.C. Cowin), 53–74. CRC Press, Boca Raton.

Hart, R.T., and Davy, D.T. (1989). Theories of bone modeling and remodeling. In *Bone Mechanics* (edited by S.C. Cowin), 253–277. CRC Press, Boca Raton.

Hasegawa, K., Turner, C.H., Chen, J., and Burr, D.B. (1995). Effect of disc lesion on microdamage accumulation in lumbar vertebrae under cyclic compression loading. *Clinical Orthopedics* **311**:190–198.

Hashin, Z., and Shtrikman, S. (1963). A variation approach to the theory of the elastic behavior of multiphase materials. *Journal of Mechanical and Physical Solids* **11**:127.

Hattner, R., and Frost, H.M. (1963). Mean skeletal age: its calculation and theoretical effects on skeletal tracer physiology and on the physical characteristics of bone. *Henry Ford Hospital Medical Bulletin* **11**:201–216.

Haut, R.C., and Little, R.W. (1972). A constitutive equation for collagen. *Journal of Biomechanics* **5**:423–430.

Haut, R.C. (1983). Age-dependent influence of strain rate on the tensile failure of rat-tail tendon. *Journal of Biomechanical Engineering* **110**:296–299.

Havers, C. (1691). *Osteologia Nova, or Some New Observations of the Bones, and the Parts Belonging to Them, with the Manner of Their Accretion, and Nutrition, Communicated to the Royal Society in Several Discourses.* Samuel Smith, London.

Hayes, W.C., Keer, L.M., Herrmann, G., and Mockros, L.F. (1972). A mathematical analysis for indentation tests of articular cartilage. *Journal of Biomechanics* **5**:541–551.

Hayes, W.C., and Snyder, B. (1981). Toward a quantitative formulation of Wolff's law in trabecular bone. In *Mechanical Properties of Bone* (edited by S.C. Cowin). 43–68. American Society of Mechanical Engineers, New York.

Hegedus, D.H., and Cowin, S.C. (1976). Bone remodeling II. Small strain adaptive elasticity. *Journal of Elasticity* **6**:337–352.

Hert, J., Kucera, P., Vavra, M., and Volenik, V. (1965). Comparison of the mechanical properties of both the primary and Haversian bone tissue. *Acta Anatomica* **61**:412–423.

Higginson, G.R., and Snaith, J.E. (1979). The mechanical stiffness of articular cartilage in confined oscillating compression. *Engineering in Medicine* **8**:11–14.

Hill, R. (1952). The elastic behavior of a crystalline aggregate. *Proceedings of the Physical Society of London* **A65**:349.

Hill, R. (1963). Elastic properties of reinforced solids: some theoretical principles. *Journal of Mechanical and Physical Solids* **11**:357.

Hillam, R.A., and Skerry, T.M. (1995). Inhibition of bone resorption and stimulation of formation by mechanical loading of the modeling rat ulna in vivo. *Journal of Bone and Mineral Research* **10**:683–689.

Hills, B.A. (1989). Oligolamellar lubrication of joints by surface active phospholipid. *Journal of Rheumatology* **16**:82–91.

Hirsch, T.J. (1962). Modulus of elasticity of concrete affected by elastic moduli of cement paste matrix and aggregate. *Proceedings of American Concrete Society* **59**:427.

Holmes, G., Sharkey, N.A., Syftestad, G., and Reiser, K. (1991). Changes in collagen crosslinks, fibril diameter, and mechanical strength in aging chicken tendons. *Transactions of the Orthopaedic Research Society* **16**:620.

Holtrop, M.E. (1990). Light and electron microscopic structure of bone-forming cells. In *Bone. Vol. 1. The Osteoblast and Osteocyte* (edited by B.K. Hall), 1–39. Telford Press, Caldwell, N.J.

Huiskes, R., Weinans, H., Grootenboer, H J., Dalstra, M., Fudala, B., and Sloof, T. J. (1987). Adaptive bone remodeling theory applied to prosthetic-design analysis. *Journal of Biomechanics* **20**:1135–1150.

Hulsen, C. (1898). Specifisches Gewicht, Elasticitat und Festigkeit des Knochengewebes. *Bulletin of Laboratory Biology St. Petersburg* **1**:7–37.

Hunter, W. (1743). Of the structure and diseases of articulating cartilages. *Philosophical Transactions of the Royal Society of London* **43**:514.

Hvid, I., and Hansen, S.L. (1985). Trabecular bone strength patterns at the proximal tibial epiphysis. *Journal of Orthopaedic Research* **3**:464–472.

Hvid, I. (1988). Mechanical strength of trabecular bone at the knee. *Danish Medical Bulletin* **35**:345–365.

Hylander, W.L. (1981). Patterns of stress and strain in the macaque mandible. In *Craniofacial Biology* (edited by D.S. Carlson). Center for Human Growth & Development, Ann Arbor.

Inman, V.T. (1947). Functional aspects of the abductor muscles of the hip. *Journal of Bone and Joint Surgery* **29A**:607–619.

Irwin, G.R. (1957). Analysis of stresses and strains near the end of a crack traversing a plate. *Journal of Applied Mechanics* **24**:361–364.

Jacobs, C.R. (1994) *Numerical simulation of bone adaptation to mechanical loading.* Doctoral dissertation, Stanford University, Palo Alto, CA.

Jansen, M. (1920). *On Bone Formation: Its Relation to Tension and Pressure.* Longmans, London.

Jaworski, Z.F., and Lok, E. (1972). The rate of osteoclastic bone erosion in Haversian remodeling sites of adult dog's rib. *Calcified Tissue Research* **10**:103–112.

Jaworski, J.F.G., and Hooper, C. (1980). Study of cell kinetics within evolving secondary haversian systems. *Journal of Anatomy* **131**:91–102.

Jaworski, Z.F.G., Liskova-Kiar, M., and Uhtoff, H.K. (1980). Effect of long-term immobilization on the pattern of bone loss in older dogs. *Journal of Bone and Joint Surgery* **62B**:104–110.

Jaworski, Z.F.G., Duck, B., and Sekaly, G. (1981). Kinetics of osteoclasts and their nuclei in evolving secondary haversian systems. *Journal of Anatomy* **133**:397–405.

Jaworski, Z.F.G., and Uhthoff, H.K. (1986). Reversability of nontraumatic disuse osteoporosis during its active phase. *Bone* (New York) **7**:431–439.

Jeanty, P., Rodesch, F., Delbeke, D., and Dumont, J.E. (1984). Estimation of gestational age from measurements of fetal long bones. *Journal of Ultrasound Medicine* **3**:75–79.

Jee, W.S.S., and Li, X.J. (1990). Adaptation of cancellous bone to overloading in the adult rat: a single photon absorptiometry and histomorphometric study. *Anatomical Record* **227**:418–426.

Jee, W.S.S., Li, X.J., and Schaffler, M.B. (1991). Adaptation of diaphyseal structure with aging and increased mechanical usage in the adult rat: a histomorphometrical and biomechanical study. *Anatomical Record* **230**:332–338.

Jenkins, R.B., and Little, R.W. (1974). A constitutive equation for parallel-fibered elastic tissue. *Journal of Biomechanics* **7**:397–402.

Johnson, L.C., Stradford, H.T., Geis, R.W., Dineen, J.R., and Kerley, E. (1963). Histogenesis of stress fractures. *Journal of Bone and Joint Surgery* **45A**:1542.

Johnson, L.C. (1964). Morphologic analysis in pathology: the kinetics of disease and general biology in bone. In *Bone Biodynamics* (edited by H.M. Frost). Little, Brown, Boston.

Johnson, L.C. (1966). The kinetics of skeletal remodeling. *Birth Defects Original Article Series* **2**:66–142.

Johnson, M., and Katz, J.L. (1984). Some new developments in the rheology of bone. *Biorheology* (Supplement) **1**:169–174.

Jones, B.H., Harris, J.M., Vinh, T.N., and Rubin, C. (1989). Exercise-induced stress fractures and stress reactions in bone: epidemiology, etiology, and classification. *Exercise and Sport Sciences Reviews* **17**:379–421.

Jones, E.S. (1934). Joint lubrication. *Lancet* **228**:1426–1427.

Jones, E.S. (1936). Joint lubrication. *Lancet* **230**:1043–1044.

Kaplan, S.J., Hayes, W.C., Stone, J.L., and Beaupre, G.S. (1985). Tensile strength of bovine trabecular bone. *Journal of Biomechanics* **18**:723–727.

Kastelic, J., and Baer, E. (1980). Deformation in tendon and collagen. In *The Mechanical Properties of Biological Materials* (edited by J.F.V. Vincent and J.D. Currey), 397–435. Cambridge University Press, Cambridge.

Katz, J.L. (1971). Hard tissue as a composite material. I. Bounds on the elastic behavior. *Journal of Biomechanics* **4**:455–473.

Katz, J.L., Yoon, H.S., Lipson, S., Maharidge, R., Meunier, A., and Christel, P. (1984). The effects of remodeling on the elastic properties of bone. *Calcified Tissue International* **36**:S31–S36.

Keaveny, T.M., Guo, X.E., Wachtel, E.F., McMahon, T.A., and Hayes, W.C. (1994). Trabecular bone exhibits fully linear elastic behavior and yields at low strain. *Journal of Biomechanics* **27**:1127–1136.

Kelin, M., and Frost, H.M. (1964). Evidence of periodic changes in the rate of formation of individual osteons in human bone. *Henry Ford Hospital Medical Bulletin* **12**:565–572.

Kendall, K. (1975). Control of cracks by interfaces in composites. *Proceedings of the Royal Society of London* **341**:409–428.

Kerley, E.R. (1965). The microscopic determination of age in human bone. *American Journal of Physical Anthropology* **23**:149–164.

Kerley, E.R., and Ubelaker, D.H. (1978). Revision in the microscopic method of estimating age at death in human cortical bone. *American Journal of Physical Anthropology* **49**:545–546.

Kimmel, D.B. (1985). A computer simulation of the mature skeleton. *Bone* (New York) **6**:369–372.

King, A.I., and Evans, F.G. (1967). Analysis of fatigue strength of human compact bone by the Weibull method. In *Digest of the 7th International Conference on Medical and Biological Engineering* (edited by B. Jacobson). The Royal Academy of Engineering Sciences, Stockholm.

Klesges, R.C., Ward, K.D., Shelton, M.L., Applegate, W.B., Cantler, E.D., Palmieri, G.M., Harmon, K., and Davis, J. (1996). Changes in bone mineral content in male athletes. Mechanisms of action and intervention effects. *Journal of the American Medical Association* **276**:226–230.

Knets, I.V. (1978). Mechanics of biological tissues. A review. *Polymer Mechanics [translation of Mekhanika Polimerov]* **13**:434–440.

Koch, J.C. (1917). The laws of bone architecture. *American Journal of Anatomy* **21**:17–293.

Krajcinovic, D., Trafimow, J., and Sumarac, D. (1987). Simple constitutive model for cortical bone. *Journal of Biomechanics* 20:779-784.

Krings, M., Stone, A., Schmitz, R.W., Krainitzki, H., Stoneking, M., and Paabo, S. (1997). Neandertal DNA sequences and the origin of modern humans. *Cell* **90**:19–30.

Krumme, H., Guenay, M., Nagel, H.-H., and Delling, G. (1984). Computerized simulation of bone remodeling: graphic demonstration of dynamic processes. *Metabolic Bone Disease and Related Research* **5**:253–257.

Kuhn, J.L., Goldstein, S.A., Ciarelli, M.J., and Mathews, L.S. (1989a). The limitations of canine trabecular bone as a model for human: A biomechanical study. *Journal of Biomechanics* **22**:95–107.

Kuhn, J.L., Goldstein, S.A., Choi, K.W., London, M., Feldkamp, L.A., and Mathews, L.S. (1989b). Comparison of the trabecular and cortical tissue moduli from human iliac crests. *Journal of Orthopaedic Research* **7**:876–884.

Lai, W.M., Hou, J.S., and Mow, V.C. (1991). A triphasic theory for the swelling and deformation behaviors of articular cartilage. *Journal of Biomechanical Engineering* **113**:245–258.

Laird, G.W., and Kingsbury, H.B. (1973). Complex viscoelastic moduli of bovine bone. *Journal of Biomechanics* **6(1)**:59–67.

Lakes, R.S., and Katz, J.L. (1974). Interrelationships among the viscoelastic functions for anisotropic solids: application to calcified tissues and related systems. *Journal of Biomechanics* **7**:259–270.

Lakes, R., and Saha, S. (1979). Cement line motion in bone. *Science* **204**:501–503.

Lakes, R.S., and Katz, J.L. (1979a). Viscoelastic properties of wet cortical bone. III. A nonlinear constitutive equation. *Journal of Biomechanics* **12**:689–698.

Lakes, R.S., and Katz, J.L. (1979b). Viscoelestic properties of wet cortical bone. II. Relaxation mechanisms. *Journal of Biomechanics* **12**:679–687.

Lakes, R.S., Katz, J.L., and Sternstein, S.S. (1979). Viscoelastic properties of wet cortical bone. I. Torsional and biaxial studies. *Journal of Biomechanics* **12**:657–678.

Lakes, R., and Saha, S. (1980). Long-term torsional creep in compact bone. *Journal of Biomechanical Engineering* **102**:178–180.

Lakes, R.S., Yoon, H.S., and Katz, J.L. (1986). Ultrasonic wave propagation and attenuation in wet bone. *Journal of Biomedical Engineering* **8**:143–148.

Landis, W.J. (1995). The strength of a calcified tissue depends in part on the molecular structure and organization of its constituent mineral crystals in their organic matrix. *Bone* (New York) **16**:533–544.

Lanyon, L.E., and Smith, R.N. (1969). Measurements of bone strain in the walking animal. *Research in Veterinary Science* **10**:93–94.

Lanyon, L.E., and Smith, R.N. (1970). Bone strain in the tibia during normal quadrupedal locomotion. *Acta Orthopaedica Scandinavica* **41**:238–248.

Lanyon, L.E. (1971). Strain in sheep lumbar vertebrae recorded during life. *Acta Orthopaedica Scandinavica* **42**:102–112.

Lanyon, L.E. (1972). In vivo bone strain recorded from thoracic vertebrae of sheep. *Journal of Biomechanics* **5**:277–281.

Lanyon, L.E. (1973). Analysis of surface bone in the calcaneus of sheep during normal locomotion. *Journal of Biomechanics* **6**:41–49.

Lanyon, L.E. (1974). Experimental support for the trajectorial theory of bone structure. *Journal of Bone and Joint Surgery* **56B**:160–166.

Lanyon, L.E., Hampson, W.G.J., Goodship, A.E., and Shah, J.S. (1975). Bone deformation recorded in vivo from strain gauges attached to the human tibial shaft. *Acta Orthopaedica Scandinavica* **46**:256–268.

Lanyon, L.E., and Baggott, D.G. (1976). Mechanical function as an influence on the structure and form of bone. *Journal of Bone and Joint Surgery* **58B**:436–443.

Lanyon, L.E., and Bourn, S. (1979). The influence of mechanical function on the development and remodeling of the tibia. *Journal of Bone and Joint Surgery* **62A**:263–273.

Lanyon, L.E. (1981). Osteoporosis and mechanically related bone modeling. In *Osteoporosis. Proceedings of the International Symposium, Jerusalem Osteoporosis Center* (edited by J. Menczel, G.C. Robin, M. Makin, and R. Steinberg). Wiley, Chichester.

Lanyon, L.E., Goodship, A.E., Pye, C.J., and MacFie, J.H. (1982). Mechanically adaptive bone remodeling. *Journal of Biomechanics* **15**:141–154.

Lanyon, L.E., and Rubin, C.T. (1984). Static vs. dynamic loads as an influence on bone remodeling. *Journal of Biomechanics* **17**:897–905.

Lanyon, L.E. (1984). Functional strain as a determinant of bone remodeling. *Calcified Tissue International* **36**:S56–S61.

Lanyon, L.E., Rubin, C.T., and Baust, G. (1986). Modulation of bone loss during calcium insufficiency by controlled dynamic loading. *Calcified Tissue International* **38**:209–216.

Lanyon, L.E. (1993). The importance of mechanical adaptation in controlling bone architecture and averting bone fracture. In *Architecture et Resistance Mecanique Osseuses* (edited by C. Marcelli and J.L. Sebert), 58–67. Masson, Paris.

Laporte, Y., and Lloyd, D.P.C. (1952). Nature and significance of reflex connections established by large afferent fibers of muscle origin. *American Journal of Physiology* **169**:609–621.

Laros, G.S., Tipton, C.M., and Cooper, R.R. (1971). Influence of physical activity on ligament insertions in the knees of dogs. *Journal of Bone and Joint Surgery* **53A**:275–286.

Lee, E.R., Lamplugh, L., Shepard, N.L., and Mort, J.S. (1995). The septoclast, a cathepsin B-rich cell involved in the resorption of growth plate cartilage. *Journal of Histochemistry and Cytochemistry* **43**:525–536.

Lekhnitskii, S.G. (1963). *Theory of Elasticity of an Anisotropic Body* (English translation). Holden-Day, San Francisco.

Leonard, F., Moscovitz, P., Hodge, J.W., and Adams, J.P. (1976). Age-related Ca-Mg content and strength in turkey tendon. *Calcified Tissue Research* **19**:331–336.

Leutenegger, W. (1972). Newborn size and pelvic dimensions of *Australopithecus*. *Nature* (London) **240**:568–569.

Lewis, P.R., and McCutchen, C.W. (1959). Experimental evidence for weeping lubrication in animal joints. *Nature* (London) **184**:1285–1285.

Li, X.J., Jee, W.S.S., Chow, S.Y., and Woodbury, D.M. (1990). Adaptation of cancellous bone to aging and immobilization: a single photon absorptiometry and histomorphometric study. *Anatomical Record* **227**:12–24.

Lieber, R.L., Leonard, M.E., Brown, C.G., and Trestik, C.L. (1991). Frog semitendinosus tendon load-strain and stress-strain properties during passive loading. *American Journal of Physiology* **261**:C86–C92.

Lieber, R.L., Brown, C.G., and Trestik, C.L. (1992). Model of muscle-tendon interaction during frog semitendinosus fixed-end contractions. *Journal of Biomechanics* **25**:421–428.

Linde, F., Hvid, I., and Madsen, F. (1992). The effect of specimen geometry on the mechanical behavior of trabecular bone specimens. *Journal of Biomechanics* **25**:359–368.

Linn, F.C. (1967). Lubrication of animal joints: I. The arthrotrip-someter. *Journal of Bone and Joint Surgery* **49A**:1079–1098.

Linn, F.C. (1968). Lubrication of animal joints: II. The mechanism. *Journal of Biomechanics* **1**:193–205.

Lipson, S.F., and Katz, J.L. (1984). The relationship between elastic properties and microstructure of bovine cortical bone. *Journal of Biomechanics* **17**:231–240.

Little, T., Freeman, M., and Swanson, A. (1969). Experience on friction in the human joint. In *Lubrication and Wear in Joints* (edited by V. Wright), 110–114. Sector Publishing, London.

Lovejoy, C.O., Heiple, K.G., and Burstein, A.H. (1973). The gait of australopithecus. *American Journal of Physical Anthropology* **38**:757–780.

MacDonald, P.B., Hedden, D., Pacin, O., and Sutherland, K. (1996). Proprioception in anterior cruciate ligament-deficient and reconstructed knees. *American Journal of Sports Medicine* **24**:774–778.

MacKenzie, J.K. (1950). The elastic constants of a solid containing spherical holes. *Proceedings of the Physical Society* **B63**:2–11.

Malcom, L.L. (1976) *An experimental investigation of the frictional and deformational responses of articular cartilage interfaces to static and dynamic loading.* Doctoral dissertation, University of California at San Diego.

Manson, J.D., and Waters, N.E. (1965). Observations on the rate of maturation of the cat osteon. *Journal of Anatomy, London* **99**:539–549.

Mansour, J.M., and Mow, V.C. (1976). The permeability of articular cartilage under compressive strain and high pressures. *Journal of Bone and Joint Surgery* **58A**:509–516.

Marcus, R., Feldman, D., and Kelsey, J. (1996). *Osteoporosis.* Academic Press, New York.

Margel-Robertson, D., and Smith, D.C. (1978). Compressive strength of mandibular bone as a function of microstructure and strain rate. *Journal of Biomechanics* **11**:455–471.

Maroudas, A. (1967). Hyaluronic acid films. *Proceedings, Institute of Mechanical Engineering, London* **181**:122–124.

Martin, R.B. (1972). The effects of geometric feedback in the development of osteoporosis. *Journal of Biomechanics* **4**:447–455.

Martin, R.B., Clark, R.N., and Advani, S. (1974). An electro-mechanical basis for osteonal mechanics. In *Advances in Bioengineering* (edited by C.T. Brighton and S. Goldstein). American Society of Mechanical Engineers, New York.

Martin, R.B., and Atkinson, P.J. (1977). Age and sex-related changes in the structure and strength of the human femoral shaft. *Journal of Biomechanics* **10**:223–231.

Martin, R.B., Pickett, J.C., and Zinaich, S. (1980). Studies of skeletal remodeling in aging men. *Clinical Orthopaedics and Related Research* **149**:268–282.

Martin, R.B., and Burr, D.B. (1982). A hypothetical mechanism for the stimulation of osteonal remodeling by fatigue damage. *Journal of Biomechanics* **15**:137–139.

Martin, R.B. (1984). Porosity and specific surface of bone. *Critical Reviews in Biomedical Engineering* **10**:179–222.

Martin, R. B. (1985). The usefulness of mathematical models for bone remodeling. Yearbook of Physical Anthropology 28:227-236.

Martin, R.B. (1986). Interpretation of fluorochrome labeling results in dynamic remodeling experiments. *Transactions, Orthopaedic Research Society* **11**:247.

Martin, R.B. (1987). Osteonal remodeling in response to screw implantation in canine femora. *Journal of Orthopaedic Research* **5**:445–452.

Martin, R.B., Dannucci, G.A., and Hood, S.J. (1987). Bone apposition rate differences in osteonal and trabecular bone. *Transactions of the Orthopaedic Research Society* **12**:178.

Martin, R.B. (1989). Label escape theory revisited: the effects of resting periods and section thickness. *Bone* (New York) **10**:255–264.

Martin, R.B., and Burr, D.B. (1989). *The Structure, Function, and Adaptation of Compact Bone*. Raven Press, New York.

Martin, R.B., and Ishida, J. (1989). The relative effects of collagen fiber orientation, porosity, density, and mineralization on bone strength. *Journal of Biomechanics* **22**:419–426.

Martin, R.B., Chow, B.D., and Lucas, P.A. (1990). Bone marrow fat content in relation to bone remodeling and serum chemistry in intact and ovariectomized dogs. *Calcified Tissue International* **46**:189–194.

Martin, R.B. (1991). On the significance of remodeling space and activation rate changes in bone remodeling. *Bone* (New York) **12**:391–400.

Martin, R.B., and Boardman, D.L. (1993). The effects of collagen fiber orientation, porosity, density, and mineralization on bovine cortical bone bending properties. *Journal of Biomechanics* **26**:1047–1054.

Martin, R.B. (1995). A mathematical model for fatigue damage repair and stress fracture in osteonal bone. *Journal of Orthopaedic Research* **13**:309–316.

Martin, R.B., Lau, S.T., Mathews, P.V., Gibson, V.A., and Stover, S.M. (1996). Collagen fiber organization is related to mechanical properties and remodeling in equine bone. A comparison of two methods. *Journal of Biomechanics* **29**:1515–1522.

Matyas, J.R., Anton, M.G., Shrive, N.G., and Frank, C.B. (1995). Stress governs tissue phenotype at the femoral insertion of the rabbit MCL. *Journal of Biomechanics* **28**:147–157.

Maughan, R.J., Watson, J.S., and Weir, J. (1983). Strength and cross-sectional area of human skeletal muscle. *Journal of Physiology* (London) **338**:37–49.

Mazess, R.B., and Cameron, J.R. (1972). Direct readout of bone mineral content using radionuclide absorptiometry. *International Journal of Applied Radiation and Isotopes* **23**:471–479.

McCutchen, C.W. (1962). The frictional properties of animal joints. *Wear* **5**:1–17.

McElhaney, J.H. (1966). Dynamic response of bone and muscle tissue. *Journal of Applied Physiology* **21**:1231–1236.

McElhaney, J., Alem, N., and Roberts, V. (1970). A porous block model for cancellous bones. *ASME Paper No. 70-WA/BHF-2*, 1–9. American Society of Mechanical Engineers, New York.

McHenry, H. (1994). Tempo and mode in human evolution. *Proceedings of the National Academy of Sciences* **91**:6780–6786.

McLeish, R.D., and Charnley, J. (1970). Abduction forces in the one-legged stance. *Journal of Biomechanics* **3**:191–209.

Melvin, J.W., and Evans, F.G. (1973). Crack propagation in bone. In *Biomaterials Symposium, AMD2* (edited by Y.C. Fung and J.A. Brighton). American Society of Mechanical Engineers, New York.

Milch, R.A., Rall, D.P., Tobie, J.E., Albrecht, J.M., and Trivers, G. (1958). Florescence of tetracycline antibiotics in bone. *Journal of Bone and Joint Surgery* **40A**:897–910.

Miller, S.C., and Jee, W.S.S. (1992). Bone lining cells. In *Bone, Vol. 4* (edited by B. K. Hall), 1-19. CRC Press, Boca Raton.

Minaire, P., Edouard, C., Arlot, M., and Meunier, P. (1984). Marrow changes in paraplegic patients. *Calcified Tissue International* **36**:338–340.

Monro, A. (1776). *The Anatomy of the Human Bones, Nerves and Lacteal Sac and Duct*. (Published in Dublin.)

Mori, S., and Burr, D.B. (1993). Increased intracortical remodeling following fatigue damage. *Bone* (New York) **14**:103–109.

Mori, S., Harruff, R., Ambrosius, W., and Burr, D.B. (1997). Trabecular bone volume and microdamage accumulation in the femoral heads of women with and without femoral neck fractures. *Bone* (New York) **21**:521–526.

Morris, C.B. (1948). The measurement of the strength of muscle relative to the cross section. *Research Quarterly of the American Association of Health, Physical Education, and Recreation* **19**:295–303.

Mosekilde, L., and Mosekilde, L. (1990). Sex differences in age-related changes in vertebral body size, density and biomechanical competence in normal individuals. *Bone* (New York) **11**:67–73.

Mow, V.C., Kuei, S.C., Lai, W.M., and Armstrong, C.G. (1980). Biphasic creep and stress relaxation of articular cartilage in compression. Theory and experiment. *Journal of Biomechanical Engineering* **102**:73–84.

Mow, V.C., Kwan, M.K., Lai, W.M., and Holmes, M.H. (1986). A finite deformation theory for nonlinearly permeable soft hydrated biological tissues. In *Frontiers in Biomechanics* (edited by G.W. Schmid-Schonbein, S.L.-Y. Woo, and B.W. Zweifach), 153–179. Springer-Verlag, New York.

Mow, V.C., and Mak, A.F. (1987). Lubrication of diarthroidal joints. In *Handbook of Bioengineering* (edited by R. Skalak and S. Chien), 5.1–5.34. McGraw-Hill, New York.

Mow, V.C., Ratcliffe, A., and Woo, S.L.-Y. (1990). *Biomechanics of Diarthroidal Joints*. Springer-Verlag, New York.

Mow, V.C., Zhu, W., and Ratcliffe, A. (1991). Structure and function of articular cartilage and meniscus. In *Basic Orthopaedic Biomechanics* (edited by V.C. Mow and W.C. Hayes), 143–198. Raven Press, New York.

Mow, V.C., and Soslowsky, L.J. (1991). Friction, lubrication and wear of diarthroidal joints. In *Basic Orthopaedic Biomechanics* (edited by V.C. Mow and W.C. Hayes), 245–292. Raven Press, New York.

Mow, V.C., Ratcliffe, A., Chern, K.Y., and Kelly, M.A. (1992). Structure and function relationships of the menisci of the knee. In *Knee Meniscus: Basic and Clinical Foundations* (edited by V.C. Mow, S.P. Arnoczky, and D.W. Jackson), 37–57. Raven Press, New York.

Mow, V.C., and Hayes, W.H. (1997). *Basic Orthopaedic Biomechanics.* Lippincott-Raven, Philadelphia.

Moyle, D.D., Welborn, J.W., and Cooke, F.W. (1978). Work to fracture of canine femoral bone. *Journal of Biomechanics* **11**:435–440.

Moyle, D.D., and Bowden, R.W. (1984). Fracture of human femoral bone. *Journal of Biomechanics* **17**:203–213.

Mullender, M.G., Huiskes, R., and Weinans, H. (1994). A physiologic approach to the simulation of bone remodeling as a self-organizational control process. *Journal of Biomechanics* **27**:1389–1394.

Mullender, M.G., and Huiskes, R. (1995). A proposal for the regulatory mechanism of Wolff's Law. *Journal of Orthopaedic Research* **13**:503–512.

Mullender, M.G., Meer, D.D. v. d., Huiskes, R., and Lips, P. (1996). Osteocyte density changes in aging and osteoporosis. *Bone* (New York) **18**:109–113.

Nachemson, A.L., and Evans, J.H. (1968). Some mechanical properties of the third human lumbar interlaminar ligament (ligamentum flavum). *Journal of Biomechanics* **1**:211–220.

Nash, C.D. (1966). *Fatigue of a Self-Healing Structure: A Generalized Theory of Fatigue Failure. ASME Publication #66-WA/BHF-3,* American Society of Mechanical Engineers, New York.

Neil, J.L., Demos, T.C., Stone, J.L., and Hayes, W.C. (1983). Tensile and compressive properties of vertebral trabecular bone. *Transactions of the Orthopaedic Research Society* **8**:344.

Neuberger, A., and Slack, H.G.B. (1953). The metabolism of collagen from liver, bone, skin, and tendon in the normal rat. *Journal of Biochemistry* (Tokyo) **53**:47–53.

Neumann, P., Ekstrom, L.A., Kellar, T.S., Perry, L., and Hansson, T.H. (1994). Aging, vertebral density, and disk degeneration alter the stress-strain characteristics of the human anterior longitudinal ligament. *Journal of Orthopaedic Research* **12**:103–112.

Neville, A.C. (1984). Cuticle: organization. In *Biology of the Integument. I. Invertebrates* (edited by J. Bereiter-Hahn, A.G. Matoltsy, and K.S. Richards), 611–625. Springer-Verlag, Berlin.

Nordin, B.E.C. (1973). *Metabolic Bone and Stone Disease.* Williams & Wilkins, Baltimore.

Nordin, M., and Frankel, V.H. (1989). *Basic Biomechanics of the Musculoskeletal System.* Lea & Febiger, Philadelphia.

Norman, T.L., Vashishth, D., and Burr, D.B. (1992). Effect of groove on bone fracture toughness. *Journal of Biomechanics* **25**:1489–1492.

Norman, T.L., Vashishth, D., and Burr, D.B. (1995). Fracture toughness of human bone under tension. *Journal of Biomechanics* **28**:309–320.

Norman, T.L., and Wang, Z. (1997). Microdamage of human cortical bone: incidence and morphology in long bones. *Bone* (New York) **20**:375–379.

Norrdin, R.W., Phemister, R.D., Jaenke, R.S., and LoPresi, C.A. (1977). Density and composition of trabecular and cortical bone in perinatally irradiated beagles with chronic renal failure. *Calcified Tissue Research* **24**:99–104.

Norrdin, R.W., Histand, M.B., Sheahan, H.J., and Carpenter, T.R. (1990). Effects of corticosteroids on mechanical strength of intervertebral joints and vertebrae in dogs. *Clinical Orthopaedics and Related Research* **259**:268–276.

Noyes, F.R., Delucas, J.L., and Torvik, P.J. (1974). Biomechanics of anterior cruciate ligament failure: an analysis of strain-rate sensitivity and mechanisms of failure in primates. *Journal of Bone and Joint Surgery* **56**:236–253.

Noyes, F.R., Torvik, P.J., Hyde, W.B., and DeLucas, J L. (1974b). Biomechanics of ligament failure. II: An analysis of immobilization, exercise, and reconditioning effects in primates. *Journal of Bone and Joint Surgery* **56A**:1406–1418.

Noyes, F R., and Grood, E.S. (1976). The strength of the anterior cruciate ligament in humans and rhesus monkeys. Age-related and species-related changes. *Journal of Bone and Joint Surgery* **58**:1074–1082.

Noyes, F.R., Butler, D.L., Grood, E.S., Zernicke, R.F., and Hefzy, M.S. (1984). Biomechanical analysis of human ligament grafts used in knee-ligamant repairs and reconstructions. *Journal of Bone and Joint Surgery* **66**:344–352.

Nunamaker, D.M., Butterweck, D.M., and Provost, M.T. (1990). Fatigue fractures in thoroughbred racehorses: Relationships with age, peak bone strain, and training. *Journal of Orthopaedic Research* **8**:604–611.

Obeid, E.M.H., Adams, M.A., and Newman, J.H. (1994). Mechanical properties of articular cartilage in knees with unicompartmental osteoarthritis. *Journal of Bone and Joint Surgery* **76B**:315–319.

Oden, Z.M., Hart, R.T., Forwood, M.R., and Burr, D.B. (1995). A priori prediction of functional adaptation in canine radii using a cell-based mechanistic model. *Transactions of the Orthopaedic Research Society* **20**:296.

Odgaard, A., and Weinans, H. (1995). *Bone Structure and Remodeling.* World Scientific, Singapore.

Olivo, O.M. (1937). Rispondenza della funzione meccanica varia degli osteoni con la loro diversa minuta architettura. *Bollenttino della Societa Italiana Biologia Sperimentale* **12**:400–401.

Olivo, O.M., Maj, G., and Toajari, E. (1937). Sul significato della minuta struttura del tessuto osseo compatto. *Bollenttino della Societa Italiana Biologia Sperimentale* **109**:369–394.

Osborn, J.W. (1996). Features of human jaw design which maximize the bite force. *Journal of Biomechanics* **29**:589–595.

Owan, I., Burr, D.B., Turner, C.H., Qui, J., Tu, Y., Onyia, J.E., and Duncan, R.L. (1997). Mechanotransduction in bone: osteoblasts are more responsive to fluid forces than mechanical strain. *American Journal of Physiology* **273**:C810–C815.

Owen, M. (1980). The origin of bone cells in the post-natal organism. *Arthritis and Rheumatism* **23**:1073–1079.

Ozkaya, N., and Nordin, M. (1991). *Fundamentals of Biomechanics.* Van Nostrand Reinhold, New York.

Palmer, I. (1958). Pathophysiology of the medial ligament of the knee joint. *Acta Chirurgica Scandinavica* **115**:312–318.

Panjabi, M.M., White, A.A., and Southwick, W.O. (1973). Mechanical properties of bone as a function of rate of deformation. *Journal of Bone and Joint Surgery* **55A**:322–330.

Parfitt, A.M. (1983). The physiologic and clinical significance of bone histomorphometric data. In *Bone Histomorphometry Techniques and Interpretation* (edited by R.R. Recker), 143–223. CRC Press, Boca Raton.

Parfitt, A.M. (1987). Bone and plasma calcium homeostasis. *Bone* (New York) **8**:S1–S8.

Parfitt, A.M., Drezner, M.K., Glorieux, F.H., Kanis, J.A., Malluche, H., Meunier, P.J., Ott, S.M., and Recker, R.R. (1987). Bone histomorphometry: standardization of nomenclature, symbols, and units. *Journal of Bone and Mineral Research* **2(6)**:595–610.

Paris, P.C. (1964). The fracture mechanics approach to fatigue. In *Fatigue—An Interdisciplinary Approach. Proceedings, Tenth Sagamore Army Materials Research Conference.* Syracuse University Press, Syracuse, NY.

Park, H.C., and Lakes, R.S. (1986). Cosserat micromechanics of human bone: Strain redistribution by a hydration sensitive constituent. *Journal of Biomechanics* **19**:385–397.

Pattin, C.A., Caler, W.E., and Carter, D.R. (1996). Cyclic mechanical property degradation during fatigue loading of cortical bone. *Journal of Biomechanics* **29**:69–80.

Pauwels, F. (1980). *Biomechanics of the Locomotor Apparatus: Contributions on the Functional Anatomy of the Locomotor Apparatus.* Springer-Verlag, Berlin.

Peterson, R.E. (1974). *Stress Concentration Factors.* Wiley, New York.

Petrie, S., Collins, J., Solomonow, M., Wink, C., and Chuinard, R. (1997). Mechanoreceptors in the palmar wrist ligaments. *Journal of Bone and Joint Surgery* **79B**:494–496.

Petrtyl, M., Hert, J., and Fiala, P. (1996). Spatial organization in the haversian bone in man. *Journal of Biomechanics* **29**:161–170.

Piekarski, K. (1970). Fracture of bone. *Journal of Applied Physics* **41**:215–223.

Piekarski, K. (1973a). Analysis of bone as a composite material. *Journal of Biomechanics* **11**:557–565.

Piekarski, K. (1973b). Analysis of bone as a composite aggregate. *International Journal of Engineering Science* **11**:557.

Pilkey, W.D. (1994). *Formulas for Stress, Strain, and Structural Matrices.* Wiley, New York.

Polig, E., and Jee, W.S.S. (1987). Bone age and remodeling: A mathematical treatise. *Calcified Tissue International* **41**:130–136.

Polig, E., and Jee, W.S.S. (1990). A model of osteon closure in cortical bone. *Calcified Tissue International* **47**:261–269.

Pope, M.H., and Murphy, M.C. (1974). Fracture energy of bone in a shear mode. *Medical and Biological Engineering* **12**:763–767.

Portigliatti-Barbos, M., Bianco, P., and Ascenzi, A. (1983). Distribution of osteonic and interstitial components in the human femoral shaft with reference to structure, calcification and mechanical properties. *Acta Anatomica* **115**:178–186.

Portigliatti-Barbos, M., Bianco, P., Ascenzi, A., and Boyde, A. (1984). Collagen orientation in compact bone: II. Distribution of lamellae in the whole of the human femoral shaft with reference to its mechanical properties. *Metabolic Bone Disease and Related Research* **5**:309–315.

Prendergast, P.J., and Taylor, D. (1994). Prediction of bone adaptation using damage accumulation. *Journal of Biomechanics* **27**:1067–1076.

Radin, E.L., Paul, I.L., and Lowy, M. (1970). A comparison of the dynamic force transmitting properties of subchondral bone and articular cartilage. *Journal of Bone and Joint Surgery* **52A**:444–456.

Rahn, B.A., Gallinaro, P., and Schenck, R.K. (1972). Compression interfragmentaire et surcharge locale de l'os. In *Interarthrite de l'Epaule. Osteogenese et Compression* (edited by A. Boitzy). Huber, Bern.

Ramaekers, J.G.M. (1977). The dynamic shear modulus of bone in dependence on the form. *Acta Morphologica Neerlando-Scandinavica* **15**:185–201.

Ramaekers, J.G.M. (1978). The rheological behavior of cortical bone and cartilage of Bos taurus. The importance of maximal damping capacity in collagen. *Acta Morphologica Neerlando-Scandinavica* **16**:55–67.

Ramaekers, J.G.M. (1979a). The mechanical properties of mallard humeral bone in dependence on the form. *Acta Morphologica Neerlando-Scandinavica* **17**:173–180.

Ramaekers, J.G.M. (1979b). The dynamic shear modulus and damping of compact bovine metacarpal bone in dependence on the topography along the bone shaft. *Netherlands Journal of Zoology* **29**:151–165.

Ramaekers, J.G.M. (1979c). The rheological behavior of skeletal material originating from several classes of vertebrates. *Netherlands Journal of Zoology* **29**:166–176.

Raux, P., Townsend, P.R., Miegel, R., Rose, R.M., and Radin, E.L. (1975). Trabecular architecture of the human patella. *Journal of Biomechanics* **8**:1–7.

Recker, R.R. (1983). *Bone Histomorphometry: Techniques and Interpretation.* CRC Press, Boca Raton.

Reddi, A.H. and Anderson, W.A. (1976). Collagenous bone matrix-induced endochondral ossification and hemopoiesis. *Journal of Cell Biology* **69**:557–572.

Reeve, J. (1986). A stochastic analysis of iliac trabecular bone dynamics. *Clinical Orthopaedics and Related Research* **213**:264–278.

Reifsnider, K.L. (1990). Damage and damage mechanics. In *Fatigue of Composite Materials* (edited by K.L. Reifsnider), 11–77. Elsevier, New York.

Reilly, D.T., and Burstein, A.H. (1974). The mechanical properties of cortical bone. *Journal of Bone and Joint Surgery* **56A**:1001–1021.

Reilly, D.T., Burstein, A.H., and Frankel, V.H. (1974). The elastic modulus for bone. *Journal of Biomechanics* **7**:271–275.

Reilly, D.T., and Burstein, A.H. (1975). The elastic and ultimate properties of compact bone tissue. *Journal of Biomechanics* **8**:393–405.

Reiter, T.J., and Rammerstorfer, F.G. (1993). Simulation of natural adaptation of bone material and application in optimum composite design. In *Optimal Design with Advanced Materials* (edited by P. Pedersen), 25–36. Elsevier, Amsterdam.

Reynolds, O. (1886). On the theory of lubrication and its application to Mr. Beauchamp Towers' experiments. *Philosophical Transactions of the Royal Society. London (Biology)* **177**:157.

Rho, J.-Y., Ashman, R.B., and Turner, C.H. (1993). Young's modulus of trabecular and cortical bone material: ultrasonic and microtensile specimens. *Journal of Biomechanics* **26**:111–119.

Rice, J.C., Cowin, S.C., and Bowman, J.A. (1988). On the dependency of the elasticity and strength of cancellous bone on apparent density. *Journal of Biomechanics* **21**:155–168.

Riemersma, D.J., and Schamhardt, H.C. (1982). The cryo-jaw, a clamp designed for in vitro rheology studies of horse digital flexor tendons. *Journal of Biomechanics* **15**:619–620.

Riemersma, D.J., Schamhardt, H.C., Hartman, W., and Lammertink, J.L M.A. (1988). Kinetics and kinematics of the equine hind limb: in vivo tendon loads and force plate measurements in ponies. *American Journal of Veterinary Research* **49**:1344–1352.

Roberts, W.E., Mozsary, P.G., and Klingler, E. (1982). Nuclear size as a cell-kinetic marker for osteoblast differentiation. *American Journal of Anatomy* **165**:373–384.

Roberts, W.E., Morey-Holton, E., and Gonsalves, M.R. (1984). Sensitivity of bone cell populations to weightlessness and simulated weightlessness. In *The Gravity Relevance in Bone Mineralisation Processes*, European Space Agency, Paris.

Roberts, W.E., Wood, H.B., and Burk, D.T. (1987). Vascularly oriented differentiation gradient of osteoblast precursor cells in rat periodontal ligament. *Journal of Periodontal Research* **22**:461–467.

Robinson, R.A., and Elliott, S.R. (1957). The water content of bone. I. The mass of water, inorganic crystals, organic matrix, and "CO2 space" components in a unit volume of dog bone. *Journal of Bone and Joint Surgery* **39A**:167–188.

Rodan, G.A. (1992). Introduction to bone biology. *Bone* (New York) **13**:S3–S6.

Roesler, H. (1981). Some historical remarks on the theory of cancellous bone structue (Wolff's law). In *Mechanical Properties of Bone* (edited by S.C. Cowin), 27–42. American Society of Mechanical Engineers, New York.

Roesler, H. (1987). The history of some fundamental concepts in bone biomechanics. *Journal of Biomechanics* **20**:1025–1034.

Rohl, L., Larsen, E., Linde, F., Odgaard, A., and Jorgensen, J. (1991). Tensile and compressive properties of cancellous bone. *Journal of Biomechanics* **24**:1143–1149.

Roux, W. (1881). *Der zuchtende Kampf der Teile, oder die 'Teilauslese' im Organismus (Theorie der 'funktionellen Anpassung')*. Wilhelm Engelmann, Leipzig.

Rowe, R.W.D. (1985). The structure of rat tail tendon. *Connective Tissue Research* **14**:9–20.

Rubin, C.T., and Lanyon, L.E. (1982). Limb mechanics as a function of speed and gait: a study of functional strains in the radius and tibia of horse and dog. *Journal of Experimental Biology* **101**:187–211.

Rubin, C.T. (1984). Skeletal strain and the functional significance of bone architecture. *Calcified Tissue International* **36**:S11–S18.

Rubin, C.T., and Lanyon, L.E. (1984). Regulation of bone formation by applied dynamic loads. *Journal of Bone and Joint Surgery* **66A**:397–402.

Rubin, C.T., and Lanyon, L.E. (1985). Regulation of bone mass by mechanical strain magnitude. *Calcified Tissue International* **37**:411–417.

Ruff, C.B., and Hayes, W.C. (1988). Sex differences in age-related remodeling of the femur and tibia. *Journal of Orthopaedic Research* **6**:886–896.

Ruff, C.B. (1995). Biomechanics of the hip and birth in early *Homo. American Journal of Physical Anthropology* **98**:527–574.

Rutishauser, E., and Majno, G. (1950). Lesions osseuses par sur charge dans le squelette normal et pathologique. *Bulletin Schweizerischen Akademie der Medizinischen Wissenschaften* **6**:333–341.

Rydell, N. (1965). Forces in the hip joint. Part II. Intravital measurements. In *Biomechanics and Related Bioengineering Topics* (edited by R.M. Kenedi), 351–357. Pergamon Press, Oxford.

Rydell, N. (1966). Forces acting in the femoral head prosthesis. *Acta Orthopaedica Scandinavica Supplement* **88**:1–46.

Saha, S. (1977). Longitudinal shear properties of human compact bone and its constituents, and the associated failure mechanisms. *Journal of Materials Science* **12**:1798–1806.

Schaffler, M.B., and Burr, D.B. (1984). Primate cortical bone microstructure: relationship to locomotion. *American Journal of Physical Anthropology* **65**:191–197.

Schaffler, M.B. (1985) Stiffness and Fatigue of Compact Bone at Physiological Strains and Strain Rates. Doctoral dissertation, West Virginia University, Morgantown, WV.

Schaffler, M.B., and Burr, D.B. (1988). Stiffness of compact bone: effects of porosity and density. *Journal of Biomechanics* **21**:13–16.

Schaffler, M.B., Radin, E.L., and Burr, D.B. (1989). Mechanical and morphological effects of strain rate on fatigue of compact bone. *Bone* (New York) **10**:207–214.

Schaffler, M.B. (1990). Immobilization induced bone loss: quantitative histological studies of cortical bone resorption. *Transactions of the Orthopaedic Research Society* **15**:187.

Schaffler, M.B., Choi, K., and Milgrom, C. (1995). Aging and matrix microdamage accumulation in human compact bone. *Bone* (New York) **17**:521–525.

Schinagl, R.M., Ting, M.K., Price, J.H., and Sah, R.L. (1996). Video microscopy to quantitate the inhomogeneous equilibrium strain within articular cartilage during confined compression. *Annals of Biomedical Engineering* **24**:500–512.

Schock, C.C., Noyes, F.R., and Villanueva, A.R. (1972). Measurement of haversian bone remodeling by means of tetracycline labeling in rib of rhesus monkeys. *Henry Ford Hospital Medical Bulletin* **20**:131–144.

Schutte, M.J., Dabezies, E.J., Zimny, M.L., and Happel, L.T. (1987). Neural anatomy of the anterior cruciate ligament. *Journal of Bone and Joint Surgery* **69A**:243–247.

Scully, T.J., and Besterman, G. (1982). Stress fracture—a preventable training injury. *Military Medicine* **147**:285–287.

Sedillot, M.C. (1865). De l'influence des causes mecaniques sur la forme et le developpement des os; moulage de ces organes par des matieres solidifiables injectees dans leur gaine periostee. *Comptes Rendus de l'Academie des Sciences* **60**:97–100.

Sedillot, M.C. (1869). Des modification que subissent les membres reseques pendant leur periode de developpement, et en particulier du siege et des degres du raccourvissement oberve a la suite de la resection coxo-femorale. *Comptes Rendus de l'Academie des Sciences* **68**:1444–1446.

Sedlin, E.D. (1965). A rheological model for cortical bone. *Acta Orthopaedica Scandinavica* **83**:1–78.

Shadwick, R.E. (1990). Elastic energy storage in tendons: mechanical differences related to function and age. *Journal of Applied Physiology* **68**:1033–1040.

Sharkey, N.A., Ferris, L., Smith, T.S., and Mathews, D.K. (1995). Strain and loading of the second metatarsal during heel-lift. *Journal of Bone and Joint Surgery* **77A**:1050–1057.

Simkin, A., and Robin, G. (1974). Fracture formation in differing collagen fiber pattern of compact bone. *Journal of Biomechanics* **7**:183–188.

Simon, S.R. (1994). *Orthopaedic Basic Science.* American Academy of Orthopaedic Surgeons, Rosemont, IL.

Sissons, H.A. (1955). Experimental study of the effect of radiation on bone growth. In *Progress in Radiobiology* (edited by J.S. Mitchell, B.F. Holmes, and C.L. Smith), 436–452. Oliver and Boyd, Edinburgh.

Smith, J.W., and Walmsley, R. (1959). Factors affecting the elasticity of bone. *Journal of Anatomy* **93**:503–522.

Smith, R.W., and Keiper, D.A. (1965). Dynamic measurement of viscoelastic properties of bone. *American Journal of Medical Electronics* October–December:156–160.

Smith, K.K., and Hylander, W.L. (1985). Strain gauge measurement of mesokinetic movement in the lizard *Varanus exanthematicus*. *Journal of Experimental Biology* **114**:53–70.

Smith, T.S., Martin, R.B., Hubbard, M., and Bay, B.K. (1997). Surface remodeling of trabecular bone using a tissue-level model. *Journal of Orthopaedic Research* **15**:593–600.

Sneddon, I.N. (1965). The relation between load and penetration in the axisymmetric Boussinesq problem for a punch of arbitrary profile. *International Journal of Engineering Science* **3**:47–57.

Snow, G.R., Cook, M.A., and Anderson, C.V. (1984). Oophorectomy and cortical bone remodeling in the beagle. *Calcified Tissue International* **36**:586–590.

Sokoloff, L. (1966). Elasticity of aging cartilage. *Federation Proceedings* **25**:1089–1095.

Stech, E.L. (1967). A descriptive model of lamellar bone anisotropy. In *Biomechanics Monographs* (edited by E.F. Byars), 236–245. American Society of Mechanical Engineers, New York.

Steindler, A. (1955). *Kinesiology of the Human Body under Normal and Pathological Conditions.* Charles C. Thomas, Springfield, IL.

Stouffer, D.C., Butler, D.L., and Hosney, D. (1985). The relationship between crimp pattern and mechanical response of human patellar tendon-bone units. *Journal of Biomechanical Engineering* **107**:158–165.

Stout, S.D., and Gehlert, S.J. (1980). The relative accuracy and reliability of of histologic aging methods. *Forensic Science International* **15**:181–190.

Stout, S.D., and Paine, R.R. (1994). Brief communication: bone remodeling rates: a test of an algorithm for estimating missing osteons. *American Journal of Physical Anthropology* **93**:123–129.

Stover, S.M., Pool, R.R., Martin, R.B., and Morgan, J.P. (1992). Histologic features of the dorsal cortex of the third metacarpal bone mid-diaphysis during postnatal growth in thoroughbred horses. *Journal of Anatomy* **181**:455–469.

Stover, S.M., Martin, R.B., Gibson, V.A., Gibeling, J., and Griffin, L. (1995). Osteonal pullout increases fatigue life of cortical bone. *Transactions of the Orthopaedic Research Society* **20**:129.

Strang, G., and Kohn, R.V. (1986). Optimal design in elasticity and plasticity. *International Journal for Numerical Methods in Engineering* **22**:183–188.

Swann, D.A., Slayter, H.S., and Silver, F.H. (1981). The molecular structure of lubricating glycoprotein-I, the boundary lubricant for articular cartilage. *Journal of Biological Chemistry* **256**:5921–5925.

Takai, S., Woo, S.L., Horibe, S., Tung, D.K., and Gelberman, R.H. (1991). The effects of frequency and duration of controlled passive mobilization on tendon healing. *Journal of Orthopaedic Research* **9**:705–713.

Tam, C.S., Harrison, J.E., Heersche, J.N.M., Jones, G., Wilson, D.R., Parsons, J.A., and Murray, T.M. (1980). Short term variation in the rate of apposition of mineralized bone matrix in small animals. In *Bone Histomorphometry. Third International Workshop* (edited by W.S.S. Jee and A.M. Parfitt). Societe Nouvelle Publications Medicales et Dentaires, Paris.

Tappen, N.C. (1977). Three dimensional studies of resorption spaces and developing osteons. *American Journal of Anatomy* **149**:301–332.

Tavassoli, M., and Yoffey, J.M. (1983). *Bone Marrow Structure and Function.* Liss, Inc., New York.

Thomas, T.R., Sayles, R.S., and Haslock, I. (1980). Human joint performance and the roughness of articular cartilage. *Journal of Biomechanical Engineering* **102**:50–56.

Thomason, J.J. (1991). Functional interpretation of locomotory adaptations during equid evolution. In *Biomechanics in Evolution* (edited by J.M.V. Rayner and R.J. Wootton), 213–227. Cambridge University Press, Cambridge, England.

Thompson, D.W. (1942). *On Growth and Form.* Cambridge University Press, London (1992 republication by Dover Publications, New York.)

Thompson, D.D. (1979). The core technique in the determination of age at death in skeletons. *Journal of Forensic Sciences* **44**:902–915.

Tidball, J.G. (1983). The geometry of actin filament-membrane associations can modify adhesive strenth of the myotendinosis junction. *Cell Motility* **3**:439–447.

Tidball, J.G., and Daniel, T.L. (1986). Myotendinous junctions of tonic muscle cells: structure and loading. *Cell and Tissue Research* **245**:315–322.

Tidball, J.G., Salem, G., and Zernicke, R. (1993). Site and mechanical conditions for failure of skeletal muscle in experimental strain injuries. *Journal of Applied Physiology* **74**:1280–1286.

Timoshenko, S.P. (1953). *History of Strength of Materials.* McGraw-Hill, New York.

Tipton, C.M., Schild, R.J., and Tomanek, R.J. (1967). Influence of physical activity on the strength of knee ligaments in rats. *American Journal of Physiology* **212**:783–787.

Tipton, C.M., Matthes, R.D., and Martin, R.K. (1978). Influence of age and sex on the strength of bone-ligament junctions in knee joints of rats. *Journal of Bone and Joint Surgery* **60A**:230–234.

Torrance, A.G., Mosley, J.R., Suswillo, R.F.L., and Lanyon, L.E. (1994). Noninvasive loading of the rat ulna in vivo induces a strain-related modeling response uncomplicated by trauma or periosteal pressure. *Calcified Tissue International* **54**:241–247.

Townsend, P.R., Raux, P., Rose, R.M., Miegel, R.E., and Radin, E.L. (1975). The distribution and anisotropy of the stiffness of cancellous bone in the human patella. *Journal of Biomechanics* **8**:363–367.

Trinkaus, E. (1983). Functional aspects of Neandertal pedal remains. *Foot and Ankle* **3**:377–390.

Trotter, J.A., Corbett, K., and Avner, B.P. (1981). Structure and function of the murine muscle-tendon junction. *Anatomical Record* **201**:293–302.

Tschantz, P., and Rutishauser, E. (1967). La surcharge mechanique de l'os vivant. Annales d'Anatomie Pathologique **12**:223–248.

Turner, A.S., Mills, E.J., and Gabel, A.A. (1975). In vivo measurement of bone strain in horse. *American Journal of Veterinary Research* **36**:1573–1579.

Turner, C.H., and Cowin, S.C. (1987). Dependence of elastic constants of an anisotropic porous material upon porosity and fabric. *Journal of Materials Science* **22**:3178–3184.

Turner, C.H. (1989). Yield behavior of bovine cancellous bone. *Journal of Biomechanical Engineering* **111**:256–260.

Turner, C.H., Cowin, S.C., Rho, J.Y., Ashman, R.B., and Rice, J.C. (1990). The fabric dependence of the orthotropic elastic constants of cancellous bone. *Journal of Biomechanics* **23**:549–561.

Turner, C.H., and Burr, D.B. (1993). Basic biomechanical measurements of bone: a tutorial. *Bone* (New York) **14**:595–608.

Turner, C.H., Forwood, M., Rho, J.-Y., and Yoshikawa, T. (1993a). Mechanical loading thresholds for lamellar and woven bone formation. *Journal of Bone and Mineral Research* **9**:87–97.

Turner, C.H., Boivin, G., and Meunier, P.J. (1993b). A mathematical model for fluoride uptake by the skeleton. *Calcified Tissue International* **52**:130–138.

Turner, C.H., Forwood, M., Rho, J.-Y., and Yoshikawa, T. (1994). Mechanical loading thresholds for lamellar and woven bone formation. *Journal of Bone and Mineral Research* **9**:87–97.

Turner, C.H., Chandran, A., and Pidaparti, R.M.V. (1995). Anisotropy of cortical bone and its constituents. *Transactions of the Orthopaedic Research Society* **20**:155.

Turner, C.H., and Burr, D.B. (1997). Orientation of collagen in osteonal bone. *Calcified Tissue International* **60**:90.

Ubelaker, D.H. (1989). *Human Skeletal Remains*. Smithsonian Institution Press, Washington, D.C.

Uhthoff, H.K., and Jaworski, Z.F.G. (1978). Bone loss in response to long-term immobilization. *Journal of Bone and Joint Surgery* **60B**:420–429.

Uhthoff, H.K., and Jaworski, Z.F.G. (1985). Periosteal stress-induced reactions resembling stress fractures. *Clinical Orthopaedics and Related Research* **199**:284–291.

Underwood, E.E. (1970). *Quantitative Stereology*. Addison-Wesley Publishing Co., Reading, MA.

Unsworth, A., Dowson, D., and Wright, V. (1975). The frictional behavior of human synovial joints: I. Natural joints. *Journal of Lubrication Technology* **97**:360–376.

Unsworth, A. (1981). Cartilage and synovial fluid. In *An Introduction to the Biomechanics of Joints and Joint Replacement* (edited by D. Dowson and V. Wright), 107–114. Mechanical Engineering, London.

Vahey, J.W., Lewis, J.L., and Vanderby, R. (1987). Elastic moduli, yield stress, and ultimate stress of cancellous bone in the canine proximal femur. *Journal of Biomechanics* **20**:29–33.

Vailas, A.C., Tipton, C.M., Mattes, R.D., and Gart, M. (1981). Physical activity and its influence on the repair process of medial collateral ligaments. *Connective Tissue Research* **9**:25–31.

Van Buskirk, W.C., Cowin, S.C., and Ward, R.N. (1981). Ultrasonic measurement of orthotropic elastic constants of bovine femoral bone. *Journal of Biomechanical Engineering* **103**:67–72.

van der Meulen, M.C.H., Beaupre, G.S., and Carter, D.R. (1993). Mechanobiological influences in long bone cross-sectional growth. *Bone* (New York) **14**:635–642.

van der Meulen, M.C.H., and Carter, D.R. (1995). Developmental mechanics determine long bone allometry. *Journal of Theoretical Biology* **172**:323–327.

Vangsness, C T., Ennis, M., Taylor, J.G., and Atkinson, R. (1995). Neural anatomy of the glenohumeral ligaments, labrum, and subacromial bursa. *Arthroscopy* **11**:180–184.

Viidik, A. (1996). Tendons and ligaments. In *Extracellular Matrix* (edited by W.D. Comper), 303–327. Harwood Academic, Melbourne.

Villanueva, A.R., and Frost, H.M. (1970). Evaluation of factors affecting the tissue-level haversian bone formation rate in man. *Journal of Dental Research* **49**:836–846.

Vincent, J. (1955). *Recherches sur la Constitution de l'Os Adult.* Arscia, Brussels.

Vincentelli, R., and Evans, F.G. (1971). Relations among mechanical properties, collagen fibers, and calcification in adult human cortical bone. *Journal of Biomechanics* **4**:193–201.

Vincentelli, R., and Grigorov, M. (1985). The effect of Haversian remodeling on the tensile properties of human cortical bone. *Journal of Biomechanics* **18**:201–207.

Vose, G.P., and Kubala, A.L. (1959). Bone strength—its relationship to x-ray-determined ash content. *Human Biology* 31:261-270.

Wainwright, S.A., Biggs, W.D., Currey, J.D., and Gosline, J.M. (1976). *Mechanical Design in Organisms.* Halsted Press, New York.

Walker, P.S., Dowson, D., Longfield, M.D., and Wright, V. (1968). Boosted lubrication in synovial joints by fluid entrapment and enrichment. *Annals of the Rheumatic Diseases* **27**:512–520.

Walker, R.A., Lovejoy, C.O., and Meindl, R.S. (1994). Histomorphological and geometric properties of human femoral cortex in individuals over 50: implications for histomorphological determination of age-at-death. *American Journal of Human Biology* **6**:659–667.

Walmsley, R., and Smith, J.W. (1957). Variation in bone structure and the value of Young's modulus. *Journal of Anatomy London* **91**:603.

Ward, F.O. (1838). *Outlines of Human Osteology.* Published in London, England.

Weinans, H., Huiskes, R., and Grootenboer, H.J. (1992). The behavior of adaptive bone-remodeling simulation models. *Journal of Biomechanics* **25**:1425–1441.

Weinbaum, S., Cowin, S.C., and Zeng, Y. (1994). A model for the excitation of osteocytes by mechanical loading-induced bone fluid shear stresses. *Journal of Biomechanics* **27**:339–360.

Weinhold, P.S., Gilbery, J.A., and Woodard, J.C. (1994). The significance of transient changes in trabecular bone remodeling activation. *Bone* (New York) **15**:577–584.

Wenzel, T.E., Schaffler, M.B., and Fyhrie, D P. (1996). In vivo trabecular microcracks in human vertebral bone. *Bone* (New York) **19**:89–95.

Wermel, J. (1934). Untersuchungen uber die kinetogenese und ihre bedeutung in der onto - und phylogenetischen entwicklung (Experimente und vergleichungen an wirbeltierextremitaten). Allgemeine einleitung, veranderungen der lange der knochen. *Morphologica Jahrbuch* **74**:143–169.

Wermel, J. (1935a). Untersuchungen uber die kinetogenese und ihre bedeutung in der onto - und phylogenetischen entwicklung (Experimente und vergleichungen an wirbeltierextremitaten). Veranderungen der dicke und der masses der knochen. *Morphologica Jahrbuch* **75**:92–127.

Wermel, J. (1935b). Untersuchungen uber die kinetogenese und ihre bedeutung in der onto - und phylogenetischen entwicklung (Experimente und vergleichungen an wirbeltierextremitaten). Veranderungender widerstandsfahigkeit der knochen. *Morphologica Jahrbuch* **75**:128–149.

Wermel, J. (1935c). Untersuchungen uber die kinetogenese und ihre bedeutung in der onto - und phylogenetischen entwicklung (Experimente und vergleichungen an wirbeltierextremitaten). Lageveranderungender skelettelemente und damit verbundene veranderungen der knochen - und gelenkform. *Morphologica Jahrbuch* **75**:180-288.

Wermel, J. (1935d). Untersuchungen uber die kinetogenese und ihre bedeutung in der onto - und phylogenetischen entwicklung (Experimente und vergleichungen an wirbeltierextremitaten). Regeneration der knochen und gelenke, sowoe newbuldung der letzteren. *Morphologica Jahrbuch* **75**:445–451.

Wertheim, M.G. (1847). Memoirs sur l'elasticite et la cohesion des principaux tissues des corps humain. *Anales de Quimica y Physica* **21**:385–414.

Whalen, R.T., Carter, D.R., and Steele, C.R. (1988). Influence of physical activity on the regulation of bone density. *Journal of Biomechanics* **21**:825–837.

White, A.A., Panjabi, M.M., and Southwick, W.O. (1977). The four biomechanical stages of fracture repair. *Journal of Bone and Joint Surgery* **59A**:188–192.

Whitehouse, W.J. (1974a). A stereological method for calculating internal surface areas in structures which have become anisotropic as the result of linear expansions or contractions. *Journal of Microscopy* (Oxford) **101**:169–180.

Whitehouse, W.J. (1974b). The quantitative morphology of anisotropic trabecular bone. *Journal of Microscopy* (Oxford) **101(2)**:153–168.

Whitehouse, W.J., and Dyson, E.D. (1974). Scanning electron microscope studies of trabecular bone in the proximal end of the human femur. *Journal of Anatomy* **118**:417–423.

Williams, W.S., and Breger, L. (1974). Analysis of stress distribution and piezoelectric response in cantilever bending of bone and tendon. *Annals of the New York Academy of Sciences* **238**:121–129.

Williams, M., and Lissner, H.R. (1977). *Biomechanics of Human Motion*. Saunders, Philadelphia.

Williams, J.L., and Lewis, J.L. (1982). Properties and an anisotropic model of cancellous bone from the proximal tibial epiphysis. *Journal of Biomechanical Engineering* **104**:50–56.

Williams, P.L. (1995). *Gray's Anatomy*. Churchill Livingston, New York.

Wilson, A.M., and Goodship, A.E. (1994). Exercise-induced hyperthermia as a possible mechanism for tendon degeneration. *Journal of Biomechanics* **27**:899–905.

Winter, D.A. (1990). *Biomechanics and Motor Control of Human Movement*. Wiley, New York.

Wolff, J. (1869). *Uber die bedeutung der architektur der spongiosa. Zentralblatt fur die Medizinische Wissenschaft VI. Jahrgang* S. 223–234.

Wolff, J. (1870). Uber die innere architektur der knochen und ihre bedeutung fur die frage vom knochenwachstum. *Virchows Archiv fuer Pathologische Anatomie und Physiologie* **50**:389–453.

Wolff, J. (1892). *The Law of Bone Remodeling*. (Translation of Wolff's *Das Gesetz der Transformation der Knochen* by P. Maquet and R. Furlong.) Springer-Verlag, Berlin.

Wong, M., and Carter, D.R. (1988). Mechanical stress and morphogenetic endochondral ossification of the sternum. *Journal of Bone and Joint Surgery* **70A(7)**:992–1000.

Woo, S.L.-Y., Akeson, W.H., and Jemmott, G.F. (1976). Measurements of nonhomogeneous, directional mechanical properties of articular cartilage in tension. *Journal of Biomechanics* **9**:785–791.

Woo, S.L.Y., Gomez, M.A., Amiel, D., Ritter, M.A., Gelberman, R.H., and Akeson, W.H. (1981a). The effects of exercise on the biomechanical and biochemical properties of swine digital flexor tendons. *Journal of Biomechanical Engineering* **103**:51–61.

Woo, S.L., Kuei, S.C., Amiel, D., Gomez, M.A., Hayes, W.C., White, F.C., and Akeson, W.H. (1981b). The effect of prolonged physical training on the properties of long bone: a study of Wolff's law. *Journal of Bone and Joint Surgery* **63A**:780–787.

Woo, S.L.-Y. (1982). Mechanical properties of tendons and ligaments: quasistatic and nonlinear properties. *Biorheology* **19**:385–396.

Woo, S.L.-Y., Gomez, M.A., Seguchi, Y., Endo, C.M., and Akeson, W.A. (1983). Measurement of mechanical properties of ligament substance from a bone-ligament-bone preparation. *Journal of Orthopaedic Research* **1**:22–29.

Woo, S.L.-Y., Gomez, M.A., and Akeson, W.H. (1985). Mechanical behaviors of soft tissues: measurements, modifications, injuries, and treatment. In *The Biomechanics of Trauma* (edited by A.M. Nahum and J. Melvin), 109–133. Appleton-Century-Crofts, Norwalk.

Woo, S.L.Y., Orlando, C.A., Gomez, M.A., Frank, C.B., and Akeson, W.H. (1986). Tensile properties of the medial collateral ligament as a function of age. *Journal of Orthopaedic Research* **4**:133–141.

Woo, S.L.-Y. (1986). Biomechanics of tendons and ligaments. In *Frontiers in Biomechanics* (edited by G.W. Schmid-Schonbein, S.L.-Y. Woo, and B.W. Zweifach), 180–195. Springer-Verlag, New York.

Woo, S.L.-Y., Mow, V.C., and Lai, W.M. (1987a). Biomechanical properties of articular cartilage. In *Handbook of Bioengineering* (edited by R. Skalak and S. Chien), 4.1–4.44. McGraw-Hill, New York.

Woo, S.L., Gomez, M.A., Sites, T.J., Newton, P.O., Orlando, C.A., and Akeson, W.H. (1987b). The biomechanical and morphological changes in the medial collateral ligament of the rabbit after immobilization and remobilization. *Journal of Bone and Joint Surgery* **69A**:1200–1211.

Woo, S.L.-Y., and Buckwalter, J.A. (1988). *Injury and Repair of the Musculoskeletal Soft Tissues.* American Academy of Orthopaedic Surgeons, Park Ridge, IL.

Woo, S.L.Y., Peterson, R.H., Ohland, K.J., Sites, T.J., and Danto, M.I. (1990). The effects of strain rate on the properties of the medial collateral ligament in skeletally immature and mature rabbits: a biomechanical and histological study. *Journal of Orthopaedic Research* **8**:712–721.

Woo, S.L.-Y., An, K.-N., Arnoczky, S.P., Wayne, J.S., Fithian, D.C., and Myers, B.S. (1994). Anatomy, biology, and biomechanics of tendon, ligaments, and meniscus. In *Orthopaedic Basic Science* (edited by S.R. Simon), 45–87. American Academy of Orthopaedic Surgeons, Park Ridge, IL.

Wright, V., and Radin, E.L. (1993). *Mechanics of Human Joints.* Dekker, New York.

Wu, K., Schubeck, K.E., Frost, H.M., and Villanueva, A.R. (1970). Haversian bone formation rates determined by a new method in a mastodon, and in human diabetes mellitus and osteoporosis. *Calcified Tissue International* **22**:204–219.

Yamada, H. (1973). *Strength of Biological Materials.* Krieger, Huntington.

Yoon, H.S., and Katz, J.L. (1976a). Ultrasonic wave propagation in human cortical bone-I. Theoretical considerations for hexagonal symmetry. *Journal of Biomechanics* **9**:407–412.

Yoon, H.S., and Katz, J.L. (1976b). Ultrasonic wave propagation in human cortical bone. II. Measurements of elastic properties and microhardness. *Journal of Biomechanics* **9**:459–464.

Yoshikawa, T., Mori, S., Santiesteban, A.J., Sun, T.C., Hofstad, E., Chen, J., and Burr, D.B. (1994). The effects of muscle fatigue on bone strain. *Journal of Experimental Biology* **188**:217–233.

Zajac, F.E. (1989). Muscle and tendon: properties, models, scaling, and application to biomechanics and motor control. *Critical Reviews in Biomedical Engineering* **17**:359–411.

Zihlman, A.L., and Hunter, W.S. (1972). A biomechanical interpretation of the pelvis of *Australopithecus*. *Folia Primatologica* **18**:1–19.

Index